Formation Tracking Control for Heterogeneous Swarm Systems

Formation Tracking Control for Heterogeneous Swarm Systems

Yongzhao Hua
Xiwang Dong
Zhang Ren

CRC Press
Taylor & Francis Group
Boca Raton London New York

CRC Press is an imprint of the
Taylor & Francis Group, an **informa** business

First edition published 2022
by CRC Press
6000 Broken Sound Parkway NW, Suite 300, Boca Raton, FL 33487-2742

and by CRC Press
4 Park Square, Milton Park, Abingdon, Oxon, OX14 4RN

CRC Press is an imprint of Taylor & Francis Group, LLC

© 2022 Yongzhao Hua, Xiwang Dong and Zhang Ren

ISBN: 978-1-032-20116-0 (hbk)
ISBN: 978-1-032-20400-0 (pbk)
ISBN: 978-1-003-26347-0 (ebk)

DOI: 10.1201/9781003263470

Publisher's note: This book has been prepared from camera-ready copy provided by the authors.

*To our colleagues
and our friends.*

Contents

Preface

In the past two decades, cooperative control of swarm systems has received significant attention from both the scientific and engineering communities. A swarm system is also known as a multi-agent system, and refers to a system composed of multiple subsystems (agents) with certain communication, calculation, decision-making, and action capabilities through local information interaction. Swarm intelligence is referred to as such because agents in a swarm system can overcome the deficiencies of individual capabilities and complete effectively complex tasks that are difficult for individual agents, which makes the entire swarm system emerge as powerful intelligent behaviours. Formation tracking control of swarm systems is an important technical support system and approach for the emergence of swarm intelligence on a motion control level, and has practical potential in numerous fields, such as cooperative surveillance with unmanned aerial vehicles, coordinated attack with missiles, deep space exploration with satellites, and more. In cross-domain collaborative applications, including air-ground coordination and air-sea coordination, the swarm systems are usually composed of several heterogeneous agents, and the swarm intelligence can be enhanced by the complementary functions of different agents. How to achieve time-varying formation tracking for heterogeneous swarm systems is a crucial technical problem for cross-domain coordination, which has important theoretical value and practical significance.

This book focuses on the time-varying formation tracking control problems for high-order heterogeneous swarm systems. According to the number of leaders and the relative relationship between leaders and followers in the state or output space, the formation tracking control problems in this book can be specifically divided into the formation tracking with a single leader, the formation tracking with multiple leaders, and the formation-containment tracking. Distributed controller design, stability analysis of closed-loop systems, and simulation/experiment results on these specific formation tracking problems for heterogeneous swarm systems are provided. In detail, this book starts from formation tracking control for homogeneous swarm systems in Chapter 3, where time-varying formation analysis and design problems with switching directed topologies and time-varying formation tracking problems with multiple leaders are investigated. Then, in Chapter 4, time-varying formation tracking control problems for weak heterogeneous swarm systems with matched/mismatched disturbances are studied respectively. Furthermore, for high-order heterogeneous swarm systems with a non-autonomous leader of unknown input, a hierarchical formation tracking control strategy composed of the distributed observer and the local tracking controller is provided in Chapter 5 based on the output regulation control and the sliding mode control. In Chapter 6, for heterogeneous swarm systems with multiple leaders, time-varying formation tracking control problems with directed switching topologies and multiple leaders' incomplete information are investigated respectively. Considering heterogeneous swarm systems with different intra-layer cooperative control objectives and inter-layer coordination couplings, the definition and the framework of formation-containment tracking control are presented in Chapter 7. Finally, to overcome the barriers between complex cooperative theories and actual physical systems, time-varying formation tracking control approaches are applied to practical cooperative experiment platforms composed of UAVs and UGVs in Chapter 8, and several formation tracking experiments are carried out to further verify the effectiveness of the theoretical results. In addition, the introduction and preliminaries are given in Chapters 1 and 2 respectively, and Chapter 9 presents the conclusions and future prospects.

We are grateful for the help, support, and efforts of our colleagues and students on this book. Specifically, we are indebted to Prof. Guoqiang Hu at Nanyang Technological University and Prof. Yisheng Zhong at Tsinghua University for their fruitful discussions, professional inspirations, and helpful guidances. We also acknowledge the efforts of Dr. Qingdong Li, Dr. Liang Han, Dr. Jianglong Yu, Dr. Yishi Liu, Dr. Xiaoduo Li, Yangfan Li, Chuang Lu, Siquan Zhou, and Peixuan Shu at Beihang University and Yan Zhou at Tsinghua University. Moreover, we acknowledge the IEEE, Elsevier, and John Wiley & Sons for permitting us to reuse materials from our publications copyrighted by these publishers in this book. Finally, we are thankful to the support of our research in this book by Science and Technology Innovation 2030-Key Project of "New Generation Artificial Intelligence" under Grant 2020AAA0108200, National Natural Science Foundation of China under Grants 62103023, 61922008, 61973013, 61873011, 61803014, 62103016, Fundamental Research Funds for the Central Universities under Grant YWF-21-BJ-J-1145, Defense Industrial Technology Development Program under Grant JCKY2019601C106, and Innovation Zone Project under Grant 18-163-00-TS-001-001-34.

Yongzhao Hua
Xiwang Dong
Zhang Ren
Beihang University, Beijing, China

Symbol Description

$\mathbb{N}$ set of natural numbers

$\mathbb{R}$ set of real numbers

$\mathbb{C}$ set of complex numbers

$\mathbb{R}^n$ set of $n \times 1$ dimension real vectors

$\mathbb{R}^{n \times m}$ set of $n \times m$ dimension real matrices

$\mathrm{Re}\,(s)$ real part of a complex number s

$\mathrm{Im}\,(s)$ imaginary part of a complex number s

I_n $n \times n$ dimension identity matrix

$\mathbf{1}_N$ $N \times 1$ vector with all elements being 1

0 zero number, vector, or matrix with matched dimensions

A^T transpose of a real matrix A

A^H conjugate transpose of a complex matrix A

$\mathrm{rank}\,(A)$ rank of a matrix A

$\lambda_i(A)$ the i-th eigenvalue of a matrix A

$\| \cdot \|_1$ 1-norm of a vector or matrix

$\| \cdot \|$ 2-norm of a vector or matrix

$\| \cdot \|_\infty$ ∞-norm of a vector or matrix

diag block diagonal matrix with diagonal blocks being d_i

$Q > 0$ a positive definite matrix Q

$Q \geqslant 0$ a positive semi-definite matrix Q

$Q < 0$ a negative definite matrix Q

$Q \leqslant 0$ a negative semi-definite matrix Q

$\lambda_{\min}(Q)$ minimum eigenvalue of Q

$\lambda_{\max}(Q)$ maximum eigenvalue of Q

$f^{(k)}(t)$ k times derivative of a function $f(t)$

$\mathrm{sgn}(\cdot)$ sign function

$\mathrm{sig}^\beta(x)$ the function is defined as $\mathrm{sgn}(x)|x|^\beta$

Chapter 1

Introduction

1.1 Scientific and Engineering Background

In the past two decades, cooperative control of swarm systems has received significant attention from both the scientific and engineering communities. A swarm system is also known as a multi-agent system, and refers to a system composed of multiple subsystems (agents) with certain communication, calculation, decision-making, and action capabilities through local information interaction. As shown in Fig. 1.1, swarm system cooperation has demonstrated powerful application potentials in many engineering fields including aerospace, such as cooperative reconnaissance using a group of unmanned aerial vehicles (UAVs), multi-missile saturation attack, deep space detection with multiple micro-satellites, and multi-robot cooperative handling, etc.

Through local perception and simple neighbouring information interaction, agents in a swarm system can overcome the deficiencies of individual capabilities and complete effectively complex tasks that are difficult for individual agents, which makes the entire swarm system emerge as powerful intelligent behaviours. That is known as swarm intelligence [1]. In recent years, swarm intelligence has received extensive attention in various fields. Swarm intelligence was listed as the core technology for autonomous efficiency enhancement and the long-term goal of unmanned system autonomy in 'Unmanned Systems Integrated Roadmap 2017–2042' by the US Department of Defense [2]. In 'New Generation Artificial Intelligence Development Plan' issued by the China State Council, at least two of the eight basic theories are directly related to the coordination of swarm systems, including swarm intelligence theory and autonomous cooperative control and optimal decision theory. The research on swarm intelligence is meaningful for us to solve practical engineering problems and understand the complicated swarming phenomenons in human society, such as panic of crowds and propagation of fashion. Therefore, swarm intelligence is one of the important research areas of the new generation of artificial intelligence, and achievements in this field can provide new concepts, principles and theories for the development of numerous technologies.

For practical swarm systems in engineering applications, including multiple UAVs, robots, vehicles, ships, missiles, and so on, formation control is an important guarantee and realization approach for the emergence of swarm

DOI: 10.1201/9781003263470-1

FIGURE 1.1: Practical engineering applications of swarm systems.

intelligence at the level of motion control [3, 4]. By applying formation control, agents in the swarm system can adjust the relative relationship in the state or output space through neighbouring information interaction, and then the swarm system can form the desired formation configuration. Achieving formation can provide favourable space-time conditions for many cooperative tasks such as source seeking [5], target enclosing [6], and surveillance [7], so the complex tasks can be performed efficiently or cost-effectively by swarm systems. Due to important engineering application prospects, formation control has become a research hotspot and technical difficulty in many scientific communities, especially the control and robotics fields.

In many practical applications, such as cooperative surveillance and target enclosing, besides forming a specific formation shape, the whole swarm system is also required to track a reference trajectory or a specific target for movement. For example, when a group of UAVs perform the cooperative surveillance task, it is important for multiple UAVs to achieve the desired relative position against the target for coordinated observation. Under these application scenarios, formation tracking control problem of swarm system arises. The agents in entire swarm can be divided into two categories, i.e., leaders and followers. The leader can provide a reference command signal for the whole formation, and the follower needs to track the leader's movement in a specific form.

According to the number of leaders and the relative relationship between leaders and followers in the state or output space, the formation tracking control problem can be specifically divided into the formation tracking with a single leader, the formation tracking with multiple leaders, and the formation-containment tracking. Specific definitions of these formation tracking problems are given as follows.

(1) Formation tracking with a single leader

The control object is to make the followers accomplish a desired formation and track the trajectory generated by the leader simultaneously in the state or output space. In the applications of formation tracking, the leader can be a virtual reference trajectory that characterizes the macro movement of the entire formation, or it can denote a cooperative tracking entity and even a non-cooperative target. For example, in the target enclosing scenario [8], the target can be treated as the leader of the UAV swarm system, and the time-varying formation tracking control allows multiple UAVs to form a encirclement situation for the target and to improve the coordination efficacy significantly. When the leader denotes a non-cooperative target, its control input is usually unknown to all the followers and may be time-varying in practice.

(2) Formation tracking with multiple leaders

In some practical applications, there may exist more than one leaders for the swarm system to track [9]. For example, during the cooperative flying of multiple manned/unmanned combat aerial vehicles (CAVs), a fleet of unmanned CAVs can keep desired time-varying tactical formation centred by the convex combination of the positions of all the available manned CAVs to enclose them. In this configuration, those dangerous tasks such as reconnaissance and attack can be accomplished by the unmanned CAVs, and the safety of the manned CAVs can be guaranteed since the defense system of the opponent will be triggered and consumed by the unmanned CAVs flying outside. It should be pointed out that the tracking target for the formation is the convex combination of all the available manned CAVs instead of one of the specific manned CAVs. If some of the manned CAVs fail or crash, the tracking target becomes the convex combination of the remaining available manned CAVs. If the tracking target for the formation is one specific manned CAV, the failure or crash of this manned CAV may destroy the whole mission since no tracking target is available for the formation. Therefore, tracking the convex combination of all the available manned CAVs is more robust than tracking one of the specific manned CAVs. In such scenarios, time-varying formation tracking control problems with multiple leaders arise, where the states/outputs of followers form the desired time-varying formation while tracking the convex combination of those of the leaders.

(3) Formation-containment tracking

In order to improve the cost-effectiveness of collaborative operations, there are usually different configurations for agents in swarm systems. For example, in the application of cooperative transportation for a group of mobile robots across the hazardous areas, to enable safety of the entire swarm and reduce costs effectively, a feasible solution is to install detection and navigation sensors on only some robots and designate these robots as leaders. The remaining robots are regarded as followers. Through cooperative control, the multiple leaders can form the desired formation and track the reference trajectory simultaneously, while the follower robots need to move into the formation formed by the multiple leaders. In this way, it can be ensured that the followers are in the safe area formed by the leaders during the movement, and the cooperative crossing is realized at a relatively small cost. In this scenario, the formation-containment tracking control problem arises [10]. The swarm system has different intra-layer cooperative control objectives and inter-layer coordination couplings. For this tracking problem, multiple leaders are required to form a desired formation tracking, while followers need to enter inside the formation formed by the multiple leaders.

The reasonable coordination of different types of agents can make up for the deficiencies of similar agents, and greatly improve the operation efficiency of the swarm system. For example, as shown in Fig. 1.2, when a group of unmanned ground vehicles (UGVs) perform tasks in mountains or urban environments, it is usually unable to effectively detect and communicate with each other due to occlusion and obstacle. By adding multiple UAVs and performing formation control between UAVs and UGVs, it can provide a wide range of environmental information and communication relay for multiple UGVs. So effective resource and capability complementation of UGVs and UAVs can be achieved [11]. Besides, as given in Fig. 1.2, the detection and communication capability of unmanned surface vehicles (USVs) will also be limited due to the influences of sea waves and curvature of the earth. Especially in the coordinated seabed detection scenario, the characteristics of two-dimensional movement of USVs make the obtained information about the detection target not rich enough. The addition of three-dimensional detection information and communication relay using UAVs can greatly improve the communication quality of the entire swarm system and the detection accuracy, dimension and coverage [12]. In air-ground cooperative enclosing with multiple UAVs and UGVs, air-sea cooperative detection with multiple UAVs and USVs, and other cross-domain mission scenarios, there are agents with completely different dynamic characteristics, which means that the swarm system is heterogeneous. In addition, the formation formed by multiple unmanned vehicles needs to be time-varying according to the changes in mission requirements or environmental situations, and the entire formation also needs to track the reference trajectory or the enclosing target. Time-varying formation tracking control for heterogeneous swarm systems is a key technical problem in the cross-domain collaborative applications including air-ground collaboration and air-sea collaboration, and has important theoretical value and practical significance in numerous fields.

(a) Air–ground coordination

(b) Air–sea coordination

FIGURE 1.2: Cross-domain collaborative applications of heterogeneous swarm systems.

1.2 Literature Review on Formation Tracking Control

Time-varying formation tracking control of high-order heterogeneous swarm systems is studied in this book. Considering that the homogeneous swarm system is a special case of heterogeneous system, and formation control is the research basis of formation tracking control, formation control and formation tracking control for homogeneous swarm system are introduced firstly. Then, the relevant literature of heterogeneous swarm system is reviewed.

1.2.1   Formation Tracking of Homogeneous Swarm System

Formation control problem has attracted attention in the robotics research field since the 1990s, and many classic formation control approaches have been proposed, including behaviour [13], virtual structure [14], and leader-follower [15] based ones. In the behaviour-based formation control strategy, each agent in the swarm system has several predetermined behaviour patterns, such as formation maintenance, collision avoidance, and movement to a specific target. These behaviour patterns constitute a behaviour set. Each action can produce corresponding control effects, and the final controller of the agent is obtained by weighted summation of the control actions of these actions. The basic idea of formation control using virtual structure is to treat the desired formation as a rigid virtual structure. Each agent in the swarm system corresponds to a point on the virtual structure. When the formation moves, the agent tracks the corresponding point in the virtual structure. The leader-follower based formation control method is to assign one or more agents in the swarm system as the leader, and the remaining agents are treated as followers. All the followers communicate directly with the leader, which moves according to the specified path. Followers maintain a specific relative position and angle relationship with the leader such that a desired formation can be achieved. However, the above three control strategies have certain shortcomings. It has been pointed out in [16] that it is difficult to conduct systematic theory analysis for behaviour-based approach, the formation control based on virtual structure cannot be implemented in a distributed manner, and the leader-follower based formation control method has weak robustness.

In [17], Prof. J. Baillieul and Prof. P.J. Antsaklis pointed out that two important changes in current control research are the consideration of network factors and the emphasis on distributed control. As a typical networked system, the core subject of swarm system is to design a distributed controller based on local neighbouring information interaction to achieve the emergence of expected intelligence behaviour. The problem of distributed formation control of swarm system has gradually become a research hotspot in the control field in recent years [4]. With the development of consensus control theory, the consensus-based formation control approach has attracted more and more attention from researchers. A swarm system is said to achieve consensus if all the agents reach an agreement on certain variables of interest, where the variables can be named as coordination variables [18–20]. In [21], Prof. W. Ren proposed a consensus-based formation control approach for second-order swarm system. By setting an offset vector relative to the formation reference, the expected formation can be achieved. It was pointed out in [21] that the traditional behaviour, virtual structure, and leader-follower based formation control methods can be unified into the framework of consensus-based approach. Besides, the consensus-based method can overcome the shortcomings of these three traditional approaches to a certain extent.

A series of researches on consensus-based formation control of swarm system have been proposed recently. However, the existing results mainly focus on homogeneous swarm systems, that is, each agent is required to share the same dynamics. The research status of formation control and formation tracking control for homogeneous swarm system is given in the following.

(1) Leaderless formation control

For swarm systems with first-order or second-order dynamics, the consensus-based formation control approaches were proposed in [22–26]. For UAVs, robots, missiles, and other complex controlled objects, the first-order or second-order model is difficult to describe the dynamics accurately. Thus, it is more practical to study the formation control for high-order swarm system. In [27], time-invariant formation control problem for a class of high-order swarm systems with integral dynamics was studied. In [28], Fax and Murray proposed a time-invariant formation control protocol for general high-order linear swarm system. Feasibility condition for high-order linear swarm system to realize a desired time-invariant formation was given in [29]. To reduce the cost of communication resources between agents, an event-triggered time-invariant formation control protocol was presented in [30].

Considering dynamic task demands and external environments, the formation configuration of swarm system is not always fixed. The desired formation needs to adjust dynamically in real time, which means that time-varying formation control is required [31]. For example, considering the application scenario of multiple UAVs flying in a mountain area, in order to ensure flight safety, UAV swarm needs to change the formation configuration in real time to avoid obstacles. Besides, considering that time-invariant formation can be regarded as a special case of time-varying formation, it is more general to study time-varying formation control problems. Dong et al. studied the time-varying formation problems for high-order swarm system with communication delay in [32], and gave the feasible conditions of time-varying formation and the approach to expand the feasible formation set. Using state space decomposition and piecewise Lyapunov function, the influences of directed switching topologies on the time-varying formation feasibility for high-order swarm system were analyzed in [33]. Group formation control problems were considered in [34], where the high-order swarm system is divided into several groups according to the task requirements, and each group can realize the desired sub-formation configuration. For high-order swarm system, it is usually difficult to measure the full state information of agents, and the acquisition of output information is much easier. Besides, in many practical applications, the swarm system is not required to form a full state formation. In [35] and [36], output time-varying formation control of high-order swarm system was studied by using static output feedback and dynamic output feedback respectively.

In [37], Wang et al. constructed a fully distributed adaptive time-varying state formation control protocol by using the measurement output information, and gave the sufficient conditions to achieve the desired formation. Considering the directed communication topology, [38] further investigated the adaptive time-varying output formation control problems based on output feedback. Due to the relatively simple hardware configuration of each agent, actuator failure may occur for swarm system in a complex adversary environment. In [39], a design approach for fault-tolerant time-varying formation control protocol of high-order linear swarm system with actuator failures was proposed.

(2) Formation tracking control

Formation control mainly focuses on forming and keeping the desired formation shape. In many practical applications, besides forming a specific formation, the whole swarm system is also required to track a reference trajectory or a specific target for movement. In this formation tracking problem, the agents in the swarm system can be divided into leaders and followers. The leader provides a motion reference or instruction signal for the swarm system, and the follower needs to track the leader's movement in a specific form. The control object of formation tracking is to make the followers form a desired formation while tracking the leader's trajectory in the state or output space [40].

Based on the consensus strategy, time-invariant formation tracking problems for multi-robot systems with non-holonomic constraints were studied in [41–43], where the position of each follower maintains a specified offset relative to the leader, and the speed and heading angle of followers are consistent with the leader. Hierarchical consensus approach was applied in [44] to solve time-invariant formation tracking problem for high-order linear swarm systems. The expected formation configuration in [41–44] is time-invariant and cannot be applied directly to time-varying formation tracking problem. In [45], Ren studied the time-varying formation tracking problems for multi-robot systems based on consensus control, where a first-order integrator was used to establish the model of each robot, and a distributed controller was designed using the relative information of neighbour. In [46], a decoupling design method was proposed to construct a rotating circular formation tracking protocol for first-order swarm system around a stationary target point. For multi-UGV swarm systems with non-holonomic constraints, rotating circular formation tracking problems with stationary or moving targets were studied in [47] and [48] respectively. Considering the influences of communication delay, a distributed controller was presented in [49] for second-order swarm system to realize time-varying formation tracking. In [8] and [50], Dong et al. studied respectively the time-varying formation tracking problems for second-order swarm systems under undirected/directed switching communication topologies. The proposed control protocols in [8] and [50] were applied to UAV target enclosing scenario, and several formation flight experiments using a group of quadrotor UAVs were given to verify the effectiveness of the controllers.

In [8, 49, 50], it was assumed that the control input of the leader is always zero, which would limit strictly the type and applicability of the leader. In order to generate a more general reference trajectory and achieve real-time adjustment of the trajectory, the leader needs to have time-varying control inputs. Considering the scenario where the leader represents a non-cooperative target, its control input is usually unknown to all followers. In [51], the influences of leader's unknown input and matched disturbances were considered, and a robust adaptive time-varying formation tracking controller was designed for high-order swarm systems. In [52], a finite-time time-varying formation tracking approach for high-order swarm systems with unmatched disturbances and leader's unknown input was proposed.

For high-order swarm system with multiple leaders, necessary and sufficient conditions to achieve the desired time-varying formation tracking and formation feasibility conditions were proposed in [9] , where a distributed formation tracking protocol was given to make the followers form the desired formation configuration and track the convex combination of multiple leaders at the same time. In [53], practical formation tracking control problems for second-order swarm system with multiple leaders were studied based on adaptive neural network. Considering communication delays, sufficient conditions were presented in [54] for the realization of time-varying formation tracking for high-order swarm systems with random sampling and multiple leaders. It should be pointed out that the formation tracking approaches in [9, 53, 54] all rely on the well-informed follower assumption, where a well-informed follower can communicate with all the leaders and an uninformed follower has no leaders as its neighbour. However, this assumption is too restrictive since it forces some followers to receive from all the leaders directly. In practice, it is more possible and realistic to require that a follower only contains a subset of leaders as its neighbour. How to design distributed time-varying formation tracking controller without requiring the well-informed follower assumption still requires further research attention.

(3) Formation-containment control

According to the relative relationship between multiple leaders and followers in the state or output space, the formation tracking problem for swarm system also includes formation-containment control. If the leaders achieve the desired formation shape and the followers can enter the convex hull formed by the multiple leaders, it is said that the swarm system realizes the expected formation-containment. From the definition of formation-containment, it is a more complex cooperative control problem developed on the basis of formation control and containment control. In the traditional containment control, there is no interaction and collaboration among multiple leaders, and only the followers are required to enter into the convex hull formed by multiple leaders [55–58]. Note that the formation-containment control problem cannot be simply decoupled into a formation problem for leaders and a containment problem for followers because the dynamics of the leaders and the time-varying formation have a coupling effect on the follower's movement.

For first-order swarm system with switching graphs, sufficient conditions to achieve formation-containment were given in [59]. Formation-containment control problem of first-order swarm system with constant delay was studied in [60]. For second-order swarm system in [61], Han et al. proposed a formation-containment control protocol under the influence of time delays. Using only position information, formation-containment control protocol for second-order swarm system was proposed in [62] based on the distributed observer theory. In [63], Wang et al. studied the finite-time formation-containment control problem, where the leaders can form the desired rotating formation and the followers can converge into the convex hull formed by the leaders in a finite time. Dong et al. in [64] proposed a theoretical analysis approach for formation-containment control of multi-UAV system, and carried out outdoor flight experiments using a group of quadrotor UAVs. In [65], formation-containment control for multiple Euler-Lagrange systems was studied based on output feedback, and a distributed controller using sliding mode control and high-gain observer was designed. Formation-containment control problem of multiple Euler-Lagrange systems with actuator saturation constraints was investigated in [66].

For high-order linear swarm system, Dong et al. gave the mathematical definition of formation-containment control and sufficient conditions to realize state formation-containment in [67]. In [68], time delays were further considered, and a formation-containment controller was designed based on the Lyapunov-Krasovskii function. In [69], only the output information of each agent was used to design a static output feedback controller such that output formation-containment control can be achieved by high-order swarm system. In [67–69], the proposed control approaches can only make the swarm system achieve the desired formation-containment, but it cannot effectively control the macroscopic trajectory of the entire swarm system. In [10], Hua et al. introduced a tracking-leader with time-varying input to generate the reference trajectory for the swarm system, and further studied the formation-containment tracking control problem for high-order linear swarm system.

1.2.2　Formation Tracking of Heterogeneous Swarm System

Research on time-varying formation tracking control for heterogeneous swarm systems is still in its infancy, and there are limited related results. In view of the fact that consensus is the basic problem in cooperative control of swarm system, the breakthrough of formation tracking control depends on the research progress of consensus theory. In the following, the relevant research progress on consensus control and cooperative output regulation control for heterogeneous swarm system will be introduced firstly. Then, the literature review on formation control for heterogeneous swarm system is given.

(1) Consensus and cooperative output regulation control

For consensus control, a special type of heterogeneous swarm system composed of first-order and second-order agents was studied firstly, where position consensus can be achieved under several constrains, such as switching topologies, communication delays, only available position information, and so on [70–75]. For practical swarm systems, it is difficult to describe the dynamics of each agent accurately with first-order or second-order models, and higher-order linear models are more general. For high-order heterogeneous linear swarm systems, a distributed proportional-integral (PI) controller was proposed in [76], where the integral term of the relative information of neighbour nodes can be used to compensate for the influence of the heterogeneous system matrix, and then the output of swarm system can achieve static consensus. Directed communication topologies were further considered in [77] using distributed PI controller, and the outputs of all followers can converge to a pre-designed fixed value. In [78], adaptive gains were added to design a fully distributed PI protocol, and sufficient conditions for the heterogeneous swarm system to achieve static output consensus were given.

Inspired by the output regulation control theory for a single system, Wieland et al. studied the leaderless consensus control problem of high-order heterogeneous swarm systems in [79], where necessary and sufficient conditions to achieve output consensus were proposed using the internal model principle. In [80], an output consensus control protocol for high-order heterogeneous swarm systems based on event-triggered or self-triggered mechanisms was presented. A hierarchical control strategy for high-order heterogeneous non-linear swarm systems was provided in [81]. First, a homogeneous reference model was designed for each agent, and the output of each agent was regulated to track the state of its own reference model. Then, a distributed controller was constructed to achieve consensus for the states of homogeneous reference models. Thus the output consensus of the high-order heterogeneous swarm system can be realized in a hierarchical way. In [82], an event-triggered consensus control protocol with adaptive mechanism was designed to enable high-order heterogeneous swarm systems to achieve output consensus in a fully distributed form.

In [79–82], consensus control problems without leaders were considered. When there is a leader, cooperative output regulation problem of high-order heterogeneous swarm systems arises, where the multiple follower subsystems can track the reference trajectory generated by the exo-system (which can be viewed as the leader) or suppress the disturbance generated by the exo-system [83–94]. The consensus tracking problem can be regarded as a special case of the cooperative output adjustment problem. For cooperative output regulation of heterogeneous swarm systems, the main approaches can be roughly divided into two categories: the distributed observer based ones [83–88] and the internal model principle based ones [89–94]. In the distributed observer based approach, the neighbouring relative estimation information was applied to construct a observer for each follower such that the leader's state can be estimated in a distributed way, and then a local tracking controller was designed based on the output regulation strategy. The internal model principle based method is to construct a control protocol through the local virtual regulation error between neighbouring nodes, and use the internal model principle to design the gain matrix. Then, combining graph theory, H_∞ robust control, and small gain principle together, the criterion to achieve cooperative output regulation of heterogeneous swarm systems can be obtained.

(2) Formation tracking control

In robotics field, a group of UAVs and UGVs were considered in [95–98], and formation control problems of the heterogeneous multi-robot system were studied. However, the above results are limited to specific models of UAVs and UGVs and rely on centralized control nodes, which is difficult to be extended to distributed control for general high-order heterogeneous swarm systems. In [99] and [100], based on the distributed PI control approach, affine formation control of the heterogeneous swarm system was considered, and the state of the leader would converge to a predefined constant value, which means that the macroscopic motion of the entire formation is ultimately static.

The development of cooperative output regulation theory in heterogeneous swarm systems has provided new ideas for time-varying formation tracking control problems. Based on the cooperative output regulation strategy, formation tracking problem of high-order linear swarm systems with undirected switching topology was studied in [101]. Considering the influences of undirected jointly connected topologies, a hierarchical formation tracking approach was proposed in [102] to decompose the cooperative control problem of heterogeneous high-order swarm systems into the consensus problem of the upper-layer homogeneous virtual system and the lower-layer subsystem tracking control problems. In [103], Li et al. discussed the time-invariant formation tracking problem of discrete-time heterogeneous non-linear swarm systems using output regulation strategy. In [104], Yaghmaie et al. used a linear command generator to describe the desired time-varying formation configuration, and expanded the exo-system and formation generator into a new augmented system. Then, a formation controller was designed based on the internal model principle, and sufficient conditions for heterogeneous swarm systems to achieve time-varying formation tracking were provided using the H_∞ control theory. In [105], it was assumed that the system matrix of leader is only known to the followers which can directly communicate with it, and an adaptive time-varying output formation tracking protocol was designed based on the time-varying L_2 gain approach. In [101–105], the leader is assumed to be an autonomous system without control input, and the eigenvalue information of the Laplacian matrix is needed to determine the control gains. In [106], a distributed adaptive time-varying output formation tracking protocol was proposed for high-order heterogeneous swarm systems with leader's unknown time-varying inputs, and formation feasibility conditions and sufficient conditions to achieve formation tracking were given.

For heterogeneous swarm systems with multiple leaders, based on the well-informed follower assumption, formation tracking protocols were proposed in [107] and [108], where the outputs of heterogeneous followers can form a given formation and track the convex combination of multiple leaders at the same time. It should be pointed out that the formation approaches in [107] and [108] require the same dynamics for multiple leaders and the existence of well-informed followers. In [109], the well-informed follower assumption was removed, and the formation tracking problems with incomplete information of multiple heterogeneous leaders were further discussed. A distributed observer was designed for each follower to estimate the dynamical matrices and the states of multiple leaders, and a time-varying output formation tracking protocol and a design algorithm were proposed.

(3) Formation-containment control

For high-order heterogeneous swarm systems with multiple leaders, the state or output containment control problems have also attracted a great deal of attention in recent years. In [110], output regulation strategy was applied firstly to the containment control problem for heterogeneous swarm systems, and a state containment control protocol was designed for followers using dynamic compensator. In [111], Chu et al. introduced adaptive gains so that the proposed containment control approach does not depend on any global information of communication topology. In [112] and [113], distributed observers were designed for both the dynamics matrices of multiple leaders and the convex combination of their states, thus the assumption that the leader model is completely known in [110] can be relaxed. Based on the internal model principle, an output containment control protocol of heterogeneous swarm systems was designed in [114], and the criteria for achieving the output containment were given. In [115], singular system control theory was applied to study the containment control of heterogeneous singular linear swarm systems. In [116], it was assumed that the leader may have unknown time-varying input, and an optimal containment controller was designed for heterogeneous swarm systems based on the off-line reinforcement learning strategy. In [117], Qin et al. considered the influences of switching topologies on the containment control of high-order heterogeneous swarm systems, and gave the sufficient conditions on the topologies to achieve containment.

In [110–117], the given controllers can only guarantee that the states or outputs of followers converge to the inside of the convex hull formed by multiple leaders, but there is no interaction and collaboration between the leaders. Thus, it is not possible to adjust effectively the relative relationship among multiple leaders. In [118], formation-containment tracking control problem of high-order heterogeneous swarm systems with intermittent communication was further studied, where a tracking-leader was applied to describe the macroscopic motion of the swarm, and a formation-containment controller was constructed based on the output regulation strategy. In [119], output formation-containment tracking problem of heterogeneous swarm systems with discrete communication was investigated by distributed hybrid active control. In [120], a fully distributed formation-containment tracking controller was proposed based on adaptive control theory. In [118–120], the tracking-leader was assumed to be an autonomous system without control input, which will strictly limit the type and applicability of the motion reference trajectory of the whole swarm. In addition, the desired time-varying formation was required to have the same dynamics as the tracking-leader in [118] and [119], and the same autonomous exo-system was applied to generate the expected formation vectors for all the followers in [120]. These requirements make heterogeneous swarm systems have fewer types of feasible time-varying formations, which would seriously affect the applicable scenarios of formation tracking approaches.

From the relevant literature on cooperative control of heterogeneous swarm systems, we can see that there are still many opportunities and challenges in this field. Firstly, most of the results only consider the consensus problem or cooperative output regulation problem, and the research on time-varying formation tracking of high-order heterogeneous swarm systems is limited. Secondly, for the formation tracking problem with multiple leaders, most existing approaches rely on the well-informed follower assumption. How to relax this assumption (i.e., in the case where multiple leaders' information is not complete) and design distributed time-varying formation tracking controller still need further study. In addition, most of the existing research assumes that the leader does not have control input, and also has strict restrictions on the type of feasible time-varying formation. How to fully analyze the influences of the heterogeneous dynamics of the swarm system on time-varying formation feasibility and propose an approach to expand effectively the set of feasible time-varying formations are still open. Thus, it is significant to further investigate the design and analysis approach for distributed time-varying formation tracking and formation-containment tracking of high-order general heterogeneous swarm systems.

1.3   Key Problems and Challenges

Based on the in-depth analysis of the formation tracking control for multiple UAVs, UGVs, and other practical swarm systems, combined with the relevant research status and progress in Section 1.2, the key problems and challenges to be studied in this book are listed as follow.

(1) Formation tracking control for swarm systems with heterogeneous disturbances

Actual swarm systems such as multiple UAVs and UGVs are often affected by disturbances during the formation movement. For example, airflow disturbances including gust and turbulence have significant impacts on UAV formation flight. The disturbance of each agent in the swarm system is generally different, and it is difficult to accurately measure the disturbances, which means that there exist heterogeneous unknown disturbances. Under the influences of heterogeneous disturbances, even if the agents in the swarm system have the same nominal model, their actual dynamics are essentially different. Thus, it is called as a 'weak heterogeneous' swarm system in this book.

Disturbances will make the swarm system unable to form and maintain the desired formation effectively, or even cause the entire formation to collapse. In addition, in the applications of formation tracking control, the leader can be either a reference trajectory that characterizes the macroscopic motion of the entire formation, or it can denote a cooperative tracking entity and even a non-cooperative target. When the leader represents a non-cooperative target, its control input is unknown to all followers and may be time-varying. The leader's time-varying input will cause unknown additional terms into the dynamics of swarm system, which affects seriously the tracking control performance of the closed-loop system. The existing time-varying formation tracking results generally assume that the leader has no control input and does not consider the influences of disturbances. Therefore, under the condition of unknown disturbances and leader's unknown time-varying input, how to realize the time-varying formation tracking control of the weak heterogeneous swarm system is the key problem to be solved firstly in this book.

(2) Formation tracking control for heterogeneous swarm systems with a leader of unknown input

In practical cross-domain mission scenarios, including air-ground cooperative enclosing with multiple UAVs and UGVs, and air-sea cooperative detection with multiple UAVs and USVs, there are agents with completely different dynamic characteristics, which means that the swarm system is heterogeneous. Generally, high-order heterogeneous swarm systems cannot be written into a compact form through the Kronecker product, which makes the existing time-varying formation tracking approaches based on state space decomposition no longer applicable. It is also difficult to construct directly Lyapunov functions for heterogeneous swarm systems. So there exist great challenges in the analysis and design of the time-varying formation tracking problem of heterogeneous swarm systems. Even for homogeneous swarm systems, not all time-varying formations are feasible, and the formation feasibility constraints reveal the requirement that the desired time-varying formations need to match the dynamics of each agent. Due to the heterogeneous dynamics of the agents, the formation feasibility analysis becomes more difficult. Furthermore, considering the influences of the leader's unknown time-varying input, how to analyze the feasible set of the time-varying formation tracking, compensate for the influences of the leader's unknown input effectively, and give distributed formation tracking controller for heterogeneous swarm systems is a key problem to be solved in this book.

(3) Formation tracking control for heterogeneous swarm systems with multiple leaders of incomplete information

In some practical applications, there may exist more than one leaders for the swarm system to track. For example, during the cooperative flying of multiple manned/unmanned combat aerial vehicles, a fleet of UAVs can keep desired time-varying tactical formation centred by the convex combination of the positions of all the available manned vehicles to enclose them. It is significant and challenging to achieve the time-varying output formation tracking of high-order heterogeneous swarm system with multiple leaders. Since there are multiple leaders that need to be tracked in the swarm system, how to specify effectively the relationship between the formation reference and the multiple leaders and achieve an effective description of the macroscopic motion of the entire formation is a difficult problem that needs to be overcome firstly. Furthermore, in practice, each follower generally can only obtain information of some leaders directly, i.e., multiple leaders' information is incomplete for all the followers. How to design a distributed observer for each follower to estimate the states of multiple leaders without relying on well-informed follower assumption, and then give a time-varying formation tracking analysis and design approach for high-order heterogeneous swarm systems, is another key problem to be studied in this book.

(4) Formation-containment tracking control for heterogeneous swarm systems with inter-layer coordination couplings

In the formation tracking problems with multiple leaders and the containment control problems, it is usually assumed that there is no interaction and collaboration among multiple leaders. However, in practical application scenarios, the leaders also need to coordinate to maintain a desired time-varying formation and track the reference trajectory or a specific target such that the task requirements, such as enclosing and surveillance, can be satisfied. In this case, the leader layer and the follower layer have different collaborative goals, and the leader layer also has a coupling effect on the cooperative goals of the follower layer. How to model, analyze, and design the above-mentioned formation tracking control problem with multiple coordination layers is a complex problem that needs to be further solved. The existing formation-containment control approaches cannot effectively control the macroscopic trajectory of the swarm system effectively, so they cannot meet the needs of many practical applications. It is significant and challenging to give the mathematical definition and control framework of formation-containment tracking, describe the overall reference trajectory of the swarm effectively, and further propose a distributed formation-containment tracking controller and an algorithm to determine control parameters. Therefore, formation-containment tracking control for high-order heterogeneous swarm systems with different intra-layer cooperative control objectives and inter-layer coordination couplings is a challenging problem to be investigated.

1.4 Contents and Outline

Time-varying formation tracking control problems for high-order heterogeneous swarm systems are studied in this book, and the specific cooperative control objects can be divided into formation tracking control with a single leader, formation tracking control with multiple leaders, and formation-containment tracking control. Distributed controller design, stability analysis, and simulation/experiment results on these specific formation tracking problems are provided in this book.

This book includes nine chapters, and the contents and outline are given in Fig. 1.3. From the perspective of the dynamics of each agent, homogeneous swarm systems are considered in Chapter 3, and weak heterogeneous swarm systems with different disturbances are studied in Chapter 4. Chapters 5–7 focus on general heterogeneous swarm systems, where each agent can have a completely different dynamics in both model matrix and state dimension. Experiment results on formation tracking for UAV and UGV swarm systems are given in Chapter 8.

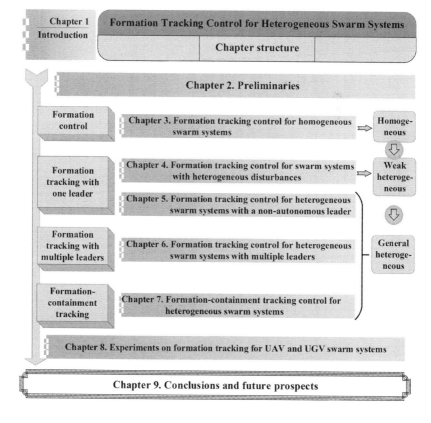

FIGURE 1.3: Chapter structure diagram.

Chapter 1. Introduction. The scientific and engineering background of formation tracking control, literature review, key problems and challenges, and the main contents and outline of the book are given.

Chapter 2. Preliminaries. Basic definitions, concepts, and results on graph theory, matrix theory, inequality theory, linear and non-linear system theory, and finite-time stability theory are introduced. Besides, from the formation control level, the dynamics models of unmanned vehicles including UAVs and UGVs are given respectively.

Chapter 3. Formation tracking control for homogeneous swarm systems. Time-varying formation analysis and design problems for homogeneous swarm systems with general linear dynamics and switching directed interaction topologies are investigated firstly. Necessary and sufficient conditions to achieve time-varying formations are proposed, where a description of the feasible time-varying formation set and approaches to expand the feasible formation set are given. An algorithm to design the formation control protocol for homogeneous swarm systems is presented. Then, time-varying formation tracking problems for homogeneous linear swarm systems with multiple leaders are studied based on the well-informed follower assumption. A formation tracking protocol is constructed using only neighbouring relative information. Necessary and sufficient conditions for swarm systems with multiple leaders to achieve time-varying formation tracking are proposed by utilizing the properties of the Laplacian matrix. An approach to design the formation tracking protocol is presented by solving an algebraic Riccati equation.

Chapter 4. Formation tracking control for swarm systems with heterogeneous disturbances. Time-varying formation tracking control problems for weak heterogeneous swarm systems with matched/mismatched disturbances are studied respectively. For the case with matched disturbances, a robust adaptive time-varying formation tracking protocol and an algorithm to design the parameters in a distributed manner are proposed. Then, formation tracking feasible conditions, an approach to expand the feasible formation set, and sufficient conditions to achieve the desired formation tracking are given. For the case with mismatched disturbances, based on the finite-time disturbance observer, the integral sliding mode control, and the super-twisting algorithm, a continuous time-varying formation tracking protocol using the neighbouring interaction is presented, and the finite-time convergence of the formation tracking errors is proved.

Chapter 5. Formation tracking control for heterogeneous swarm systems with a non-autonomous leader. Time-varying output formation control of high-order heterogeneous swarm systems with layered architecture is proposed firstly. An algorithm to design controller parameters and the time-varying formation feasibility conditions with heterogeneous dynamics are given. Then, considering a non-autonomous leader with unknown input, time-varying formation tracking problem for heterogeneous swarm systems is further studied. Based on the output regulation control and the sliding mode control, a hierarchical formation tracking control strategy composed of the distributed observer and the local tracking controller is provided. Using the neighbouring interaction, a distributed formation tracking protocol with the adaptive compensation capability for the unknown input of the leader is proposed. Considering the features of heterogeneous dynamics, the time-varying formation tracking feasible constraints are provided, and a compensation input is applied to expand the feasible formation set.

Chapter 6. Formation tracking control for heterogeneous swarm systems with multiple leaders. For high-order heterogeneous swarm systems with multiple leaders, time-varying formation tracking control problems with directed switching topologies and multiple leaders' incomplete information are investigated respectively. For the case with directed switching topologies, based on the well-informed follower assumption, a distributed time-varying output formation tracking protocol is designed. Sufficient conditions to achieve formation tracking with multiple leaders are given by using the piecewise Lyapunov stability theory. Furthermore, the well-informed follower assumption is removed, and the formation tracking problems with incomplete information of multiple leaders are discussed. A distributed observer is designed for each follower to estimate the dynamical matrices and the states of multiple leaders, and an adaptive algorithm is proposed to solve the regulator equations in finite time. Then, a fully distributed time-varying output formation tracking protocol and a design algorithm are proposed. It is proved that the desired formation tracking with multiple leaders can be achieved by heterogeneous swarm systems without requiring the well-informed follower assumption.

Chapter 7. Formation-containment tracking control for heterogeneous swarm systems. For high-order heterogeneous swarm systems with different intra-layer cooperative control objectives and inter-layer coordination couplings, the definition and the framework of formation-containment tracking control are presented. A tracking-leader with time-varying input is applied to generate the macroscopic reference trajectory for the whole swarm systems. Considering the influences of switching topologies, based on the robust adaptive estimation approach and the predefined containment control strategy, a distributed formation-containment tracking protocol and a multi-step design algorithm are proposed. With inter-layer coordination couplings, sufficient conditions for heterogeneous swarm systems on switching graphs to achieve formation-containment tracking are given.

Chapter 8. Experiments on formation tracking for UAV and UGV swarm systems. Time-varying formation tracking control approaches proposed in the previous chapters are applied to practical cooperative experiment platforms composed of UAVs and UGVs. How to modify the general formation controllers to meet the characteristics of UAV and UGV swarm system is given, and the formation controller design and stability analysis are provided. Then, the system composition, hardware structure, and software framework of the experimental platform are introduced. Several formation tracking experiments are carried out for UAV and UGV swarm system to further verify the effectiveness of the theoretical results.

Chapter 9. Conclusions and future prospects. The whole work of this book is summarized and some remaining open problems that require further investigation are discussed.

1.5   Conclusions

In this chapter, the backgrounds and motivations of formation tracking control for heterogeneous swarm systems were given firstly. Then, literature reviews on formation control of swarm systems were presented. Key problems and challenges to be investigated were provided. Finally, the main contents and outline of this book were introduced.

Chapter 2

Preliminaries

This chapter introduces the basic concepts and results for graph theory, algebra and matrix theory, linear and non-linear system theory, and finite-time stability theory, which will be used in the following chapters. In addition, from the formation control level, the dynamics models of unmanned vehicles including UAVs and UGVs are given, respectively.

2.1 Notations

As shown in Table 2.1, the following notations will be used in this book.

2.2 Graph Theory

The interaction topology in swarm systems can be denoted by a graph $\mathcal{G} = \{\mathcal{V}, \mathcal{E}, \mathcal{W}\}$, where $\mathcal{V} = \{v_1, v_2, \ldots, v_N\}$ is the set of nodes, $\mathcal{E} \subseteq \{(v_i, v_j) : v_i, v_j \in V; i \neq j\}$ is the set of edges, and $\mathcal{W} = [w_{ij}] \in \mathbb{R}^{N \times N}$ denotes the adjacency matrix with non-negative weights w_{ij}. Let $\varepsilon_{ij} = (v_i, v_j)$ represent an edge from v_i to v_j in the graph $\mathcal{G}$. A directed path from v_1 to v_k is a sequence of ordered edges $(v_1, v_2), (v_2, v_3), \ldots, (v_{k-1}, v_k)$. For two graphs $\mathcal{G}_s$ and $\mathcal{G}$, if $\mathcal{V}_s \subseteq \mathcal{V}$ and $\mathcal{E}_s \subseteq \mathcal{E}$, then $\mathcal{G}_s$ is said to be a subgraph of $\mathcal{G}$.

The weight $w_{ij} > 0$ if and only if $\varepsilon_{ji} \in \mathcal{E}$, and $w_{ij} = 0$ otherwise. Let $\mathcal{N}_i = \{v_j \in \mathcal{V} : (v_j, v_i) \in \mathcal{E}\}$ denote the set of neighbours of node v_i. The in-degree matrix of $\mathcal{G}$ is defined as $\mathcal{D} = \text{diag}\{\deg_{in}(v_1), \ldots, \deg_{in}(v_N)\}$, where $\deg_{in}(v_i) = \sum_{j=1}^{N} w_{ij} (i = 1, 2, \ldots, N)$ is the in-degree of node v_i. The Laplacian matrix L is defined as $L = \mathcal{D} - \mathcal{W}$.

A graph $\mathcal{G}$ is said to be undirected if $\varepsilon_{ij} \in \mathcal{E}$ implies $\varepsilon_{ji} \in \mathcal{E}$ and $w_{ij} = w_{ji}$. An undirected graph is connected if there is an undirected path between every pair of distinct nodes. A directed graph is said to have a spanning tree if there is a root node which has at least one directed path to every other node. A directed graph is strongly connected if there is a directed path from every

DOI: 10.1201/9781003263470-2

TABLE 2.1: Notations used in this book.

$\mathbb{N}$	set of natural numbers
$\mathbb{R}$	set of real numbers
$\mathbb{C}$	set of complex numbers
$\mathbb{R}^n$	set of $n \times 1$ dimension real vectors
$\mathbb{R}^{n \times m}$	set of $n \times m$ dimension real matrices
$\text{Re}(s)$	real part of a complex number s
$\text{Im}(s)$	imaginary part of a complex number s
I_n	$n \times n$ dimension identity matrix
$\mathbf{1}_N$	$N \times 1$ vector with all elements being 1
0	zero number, vector, or matrix with matched dimensions
A^T	transpose of a real matrix A
A^H	conjugate transpose of a complex matrix A
$\text{rank}(A)$	rank of a matrix A
$\lambda_i(A)$	the i-th eigenvalue of a matrix A
$\| \cdot \|_1$	1-norm of a vector or matrix
$\| \cdot \|$	2-norm of a vector or matrix
$\| \cdot \|_\infty$	∞-norm of a vector or matrix
$\text{diag}\{d_1, \ldots, d_n\}$	block diagonal matrix with diagonal blocks being d_i
$Q > 0$	a positive definite matrix Q
$Q \geqslant 0$	a positive semi-definite matrix Q
$Q < 0$	a negative definite matrix Q
$Q \leqslant 0$	a negative semi-definite matrix Q
$\lambda_{\min}(Q)$	minimum eigenvalue of Q
$\lambda_{\max}(Q)$	maximum eigenvalue of Q
$f^{(k)}(t)$	k times derivative of a function $f(t)$
$\text{sgn}(\cdot)$	sign function
$\text{sig}^\beta(x)$	$\text{sgn}(x)|x|^\beta$

node to every other node. Two graph examples are given in Fig. 2.1, where the first one is a connected undirected graph and the second one is a directed graph with spanning tree.

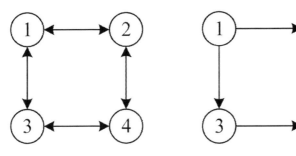

(a) A connected undirected graph

(b) A directed graph with spanning tree

FIGURE 2.1: Examples for undirected and directed graphs.

The properties of Laplacian matrix L on undirected graph and directed graph are given in the following lemmas.

Lemma 2.1 ([18]). *For an undirected graph $\mathcal{G}$ with N nodes, it follows that*
1) L has at least one 0 eigenvalue with $\mathbf{1}_N$ being the associated eigenvector, i.e., $L\mathbf{1}_N = 0$.
2) If $\mathcal{G}$ is connected, then 0 is a simple eigenvalue of L, and all the other $N-1$ eigenvalues are positive.

Lemma 2.2 ([19]). *For a directed graph $\mathcal{G}$ with N nodes, it follows that*
1) L has at least one 0 eigenvalue with $\mathbf{1}_N$ being the associated eigenvector, i.e., $L\mathbf{1}_N = 0$.
2) If $\mathcal{G}$ has a spanning tree, then 0 is a simple eigenvalue of L, and all the other $N-1$ eigenvalues have positive real parts.

2.3 Algebra and Matrix Theory

For matrices $A = [a_{ij}] \in \mathbb{R}^{p \times q}$ and $B \in \mathbb{R}^{m \times n}$, the Kronecker product is defined as

$$A \otimes B = \begin{bmatrix} a_{11}B & a_{12}B & \cdots & a_{1q}B \\ a_{21}B & a_{22}B & \cdots & a_{2q}B \\ \vdots & \vdots & \ddots & \vdots \\ a_{p1}B & a_{p2}B & \cdots & a_{pq}B \end{bmatrix} \in \mathbb{R}^{pm \times qn}.$$

For matrices A, B, C, and D with appropriate dimensions, the Kronecker product has the following properties [121]:

1) $A \otimes (B + C) = A \otimes B + A \otimes C$.
2) $(A \otimes B)^T = A^T \otimes B^T$.
3) $(A \otimes B)(C \otimes D) = (AC) \otimes (BD)$.
4) $(A \otimes B)^{-1} = A^{-1} \otimes B^{-1}$.
5) $\|A \otimes B\| = \|A\| \|B\|$.

Lemma 2.3 ([122]). *Let x and y denote two non-negative real numbers, and positive constants p and q satisfy $\frac{1}{p} + \frac{1}{q} = 1$. Then, it can be verified that*

$$xy \leqslant \frac{x^p}{p} + \frac{y^q}{q}.$$

Lemma 2.4 ([122]). *For any $x_i \in \mathbb{R}$ ($i = 1, 2, \ldots, n$) and $0 < \alpha \leqslant 1$, it holds that*

$$\left(\sum_{i=1}^{n} |x_i| \right)^{\alpha} \leqslant \sum_{i=1}^{n} |x_i|^{\alpha} \leqslant n^{1-\alpha} \left(\sum_{i=1}^{n} |x_i| \right)^{\alpha}.$$

Lemma 2.5 ([123]). *For $\chi \in \mathbb{R}$ and $\zeta \in \mathbb{R}$, if $\eta_1 > 0$ and $\eta_2 > 0$, then*

$$|\chi|^{\eta_1} |\zeta|^{\eta_2} \leqslant \frac{\eta_1 |\chi|^{\eta_1 + \eta_2}}{\eta_1 + \eta_2} + \frac{\eta_2 |\zeta|^{\eta_1 + \eta_2}}{\eta_1 + \eta_2}.$$

Lemma 2.6 (Schur complement [124]). *For a given symmetric matrix $Q = \begin{bmatrix} Q_{11} & Q_{12} \\ Q_{12}^T & Q_{22} \end{bmatrix} \in \mathbb{R}^{n \times n}$, where $Q_{11} \in \mathbb{R}^{r \times r}$, the following three conditions are equivalent:*

1) $Q < 0$.
2) $Q_{11} < 0$, $Q_{22} - Q_{12}^T Q_{11}^{-1} Q_{12} < 0$.
3) $Q_{22} < 0$, $Q_{11} - Q_{12} Q_{22}^{-1} Q_{12}^T < 0$.

2.4   Linear and Non-linear System Theory

The linear time-invariant system is described by

$$\begin{cases} \dot{x}(t) = Ax(t) + Bu(t), \\ y(t) = Cx(t), \end{cases} \tag{2.1}$$

where $x(t) \in \mathbb{R}^n$, $u(t) \in \mathbb{R}^m$, and $y(t) \in \mathbb{R}^q$ represent the state, control input, and output, respectively. Let $A \in \mathbb{R}^{n \times n}$, $B \in \mathbb{R}^{n \times m}$, and $C \in \mathbb{R}^{q \times n}$ denote the system matrix, input matrix, and output matrix, respectively.

Definition 2.1. *For any initial state $x(0)$ in the space $\mathbb{R}^n$, if there exists control input $u(t)$ such that the state $x(t)$ of system (2.1) can converge to the origin in a finite time, then system (2.1) is controllable or (A, B) is controllable.*

Lemma 2.7 ([125]). *If $\mathrm{rank}[B, AB, ..., A^{n-1}B] = n$, then (A, B) is controllable.*

Lemma 2.8 ([125]). *If $\mathrm{rank}[sI - A, B] = n$ ($\forall s \in \mathbb{C}$), then (A, B) is controllable.*

Definition 2.2. *If any initial state $x(0)$ of system (2.1) can be determined by the control input $u(t)$ and output $y(t)$ uniquely in a finite time, then system (2.1) is observable or (C, A) is observable.*

Lemma 2.9 ([125]). *If $\mathrm{rank}[C^T, A^T C^T, ..., (A^{n-1})^T C^T]^T = n$, then (C, A) is observable.*

Lemma 2.10 ([125]). *If $\mathrm{rank}[sI - A^T, C^T]^T = n$ ($\forall s \in \mathbb{C}$), then (C, A) is observable.*

Definition 2.3. *For a matrix $A \in \mathbb{C}^{n \times n}$, if all the eigenvalues of A have negative real parts, then A is called a Hurwitz matrix.*

Definition 2.4. *If there is a matrix $K \in \mathbb{R}^{m \times n}$ such that $A + BK$ is Hurwitz, then system (2.1) is stabilizable or (A, B) is stabilizable.*

Lemma 2.11 ([126]). *System (2.1) is stabilizable if and only if*

$$\mathrm{rank}[sI - A, B] = n \ (\forall s \in \mathbb{C}_+),$$

where $\mathbb{C}_+ = \{ s | s \in \mathbb{C}, \mathrm{Re}(s) \geqslant 0 \}$.

Definition 2.5. *If there is a matrix $F \in \mathbb{R}^{n \times q}$ such that $A + FC$ is Hurwitz, then system (2.1) is detectable or (C, A) is detectable.*

Lemma 2.12 ([126]). *System (2.1) is detectable if and only if*

$$\mathrm{rank}[sI - A^T, C^T]^T = n \ (\forall s \in \mathbb{C}_+).$$

In the following, the stability of autonomous non-linear systems will be given.

Definition 2.6 ([127]). *Consider the autonomous system*

$$\dot{x} = f(x), \tag{2.2}$$

where $f : D \to \mathbb{R}^n$ is a locally Lipschitz map from a domain $D \subset \mathbb{R}^n$ into $\mathbb{R}^n$. The equilibrium point $x = 0$ of (2.2) is stable if, for each $\varepsilon > 0$, there exists $\delta = \delta(\varepsilon) > 0$ such that

$$\|x(0)\| < \delta \Rightarrow \|x(t)\| < \varepsilon, \ \forall t \geqslant 0.$$

The equilibrium point $x = 0$ of (2.2) is asymptotically stable if it is stable and δ can be chosen such that

$$\|x(0)\| < \delta \Rightarrow \lim_{t \to \infty} x(t) = 0.$$

The equilibrium point $x = 0$ of (2.2) is globally asymptotically stable if it is asymptotically stable for all initial states.

Lemma 2.13 ([127]). *Let $x = 0$ be an equilibrium point for (2.2) and $D \subset \mathbb{R}^n$ be a domain containing $x = 0$. Let $V : D \to \mathbb{R}$ be a continuously differentiable function such that*

$$V(0) = 0, V(x) > 0 \text{ in } D - \{0\},$$

$$\dot{V}(x) \leqslant 0 \text{ in } D,$$

then $x = 0$ is stable. Furthermore, if

$$\dot{V}(x) < 0 \text{ in } D - \{0\},$$

then $x = 0$ is asymptotically stable.

Lemma 2.14 ([127]). *Let $x = 0$ be an equilibrium point for (2.2). Let $V : \mathbb{R}^n \to \mathbb{R}$ be a continuously differentiable function such that*

$$V(0) = 0 \text{ and } V(x) > 0, \ \forall x \neq 0,$$

$$\|x\| \to \infty \Rightarrow V(x) \to \infty,$$

$$\dot{V}(x) < 0, \ \forall x \neq 0,$$

then $x = 0$ is globally asymptotically stable.

Lemma 2.15 ([126]). *For the autonomous linear system $\dot{x}(t) = Ax(t)$, the following four conditions are equivalent:*

1) The system is asymptotically stable.

2) The matrix A is Hurwitz.

3) For any given positive definite matrix Q, the Lyapunov equation $A^T P + PA = -Q$ has positive definite solution P.

4) There exists a positive definite matrix P such that $A^T P + PA < 0$.

Lemma 2.16 (LaSalle's invariance principle, [127]). *Define $\Omega \subset D$ as a positively invariant compact set with respect to (2.2). Let $V : D \to \mathbb{R}$ denote a continuously differentiable function which satisfies $\dot{V}(x) \leqslant 0$ in Ω. The set of all points in Ω where $\dot{V}(x) = 0$ is represented by E. Let M be the largest invariant set in E. Then, each solution starting in Ω will approach M when $t \to \infty$.*

Lemma 2.17 ([127]). *Let $x = 0$ be an equilibrium point for (2.2). Define $V : D \to \mathbb{R}$ as a continuously differentiable positive definite function in D including the origin $x = 0$, where $\dot{V}(x) \leqslant 0$. Let $S = \{x \in D \,\big|\, \dot{V}(x) = 0\}$ and suppose that no solution can stay identically in S, other than the trivial solution $x(t) \equiv 0$. Then, the origin is asymptotically stable.*

Lemma 2.18 ([127]). *Let $x = 0$ be an equilibrium point for (2.2). Define $V : \mathbb{R}^n \to \mathbb{R}$ as a continuously differentiable, radially unbounded, positive definite function, where $\dot{V}(x) \leqslant 0$ for all $x \in \mathbb{R}^n$. Let $S = \{x \in \mathbb{R}^n \,\big|\, \dot{V}(x) = 0\}$ and suppose that no solution can stay identically in S, other than the trivial solution $x(t) \equiv 0$. Then, the origin is globally asymptotically stable.*

Lemma 2.19 ([127]). *Let $x = 0$ be an equilibrium point for the non-linear system $\dot{x} = f(x)$, where $f : D \to \mathbb{R}^n$ is continuously differentiable and D is a neighbourhood of the origin. Let*

$$A = \left. \frac{\partial f}{\partial x}(x) \right|_{x=0}.$$

Then, the origin is asymptotically stable if $\operatorname{Re} \lambda_i < 0$ for all eigenvalues of A.

The stability of non-autonomous systems will be further introduced.

Definition 2.7 ([127]). *A continuous function $\alpha : [0, a) \to [0, \infty)$ is said to belong to class $\mathcal{K}$ if it is strictly increasing and $\alpha(0) = 0$. It is said to belong to class $\mathcal{K}_\infty$ if $a = \infty$ and $\alpha(r) \to \infty$ as $r \to \infty$. A continuous function $\beta : [0, a) \times [0, \infty) \to [0, \infty)$ is said to belong to class $\mathcal{KL}$ if, for each fixed s, the mapping $\beta(r, s)$ belongs to class $\mathcal{K}$ with respect to r and, for each fixed r, the mapping $\beta(r, s)$ is decreasing with respect to s and $\beta(r, s) \to 0$ as $s \to \infty$.*

Definition 2.8 ([127]). *Consider the non-autonomous system*

$$\dot{x} = f(t, x), \tag{2.3}$$

where $f : [0, \infty) \times D \to \mathbb{R}^n$ is piecewise continuous in t and locally Lipschitz in x on $[0, \infty) \times D$, and $D \subset \mathbb{R}^n$ is a domain that contains the origin $x = 0$. The equilibrium point $x = 0$ of (2.3) is
1) stable if, for each $\varepsilon > 0$, there exists $\delta = \delta(\varepsilon, t_0) > 0$ such that

$$\|x(t_0)\| < \delta \Rightarrow \|x(t)\| < \varepsilon, \ \forall t \geqslant t_0 \geqslant 0.$$

2) asymptotically stable if it is stable and there exists a positive constant $c = c(t_0)$ such that $x(t) \to 0$ when $t \to \infty$ for all $\|x(t_0)\| < c$.
3) uniformly stable if and only if there exist a class $\mathcal{K}$ function α and a positive constant c, independent of t_0, such that

$$\|x(t)\| \leqslant \alpha(\|x(t_0)\|), \ \forall t \geqslant t_0 \geqslant 0, \ \forall \|x(t_0)\| < c. \tag{2.4}$$

4) uniformly asymptotically stable if and only if there exist a class $\mathcal{KL}$ function β and a positive constant c, independent of t_0, such that

$$\|x(t)\| \leqslant \beta\left(\|x(t_0)\|, t - t_0\right), \; \forall t \geqslant t_0 \geqslant 0, \; \forall \|x(t_0)\| < c. \tag{2.5}$$

5) globally uniformly asymptotically stable if and only if the inequality (2.5) is satisfied for any initial state $x(t_0)$.

6) exponentially stable if there exist positive constants c, k, and λ such that

$$\|x(t)\| \leqslant k\,\|x(t_0)\|\, e^{-\lambda(t-t_0)}, \; \forall \|x(t_0)\| < c. \tag{2.6}$$

7) globally exponentially stable if (2.6) is satisfied for any initial state $x(t_0)$.

Lemma 2.20 ([127]). *Let $x = 0$ be an equilibrium point for (2.3). Define $V : [0, \infty) \times D \to \mathbb{R}$ as a continuously differentiable function such that $\forall t \geqslant 0$ and $\forall x \in D$*

$$W_1(x) \leqslant V(t, x) \leqslant W_2(x), \tag{2.7}$$

$$\frac{\partial V}{\partial t} + \frac{\partial V}{\partial x} f(t, x) \leqslant 0, \tag{2.8}$$

where $W_1(x)$ and $W_2(x)$ are continuous positive definite functions on D. Then, $x = 0$ is uniformly stable. Furthermore, the inequality (2.8) is strengthened to

$$\frac{\partial V}{\partial t} + \frac{\partial V}{\partial x} f(t, x) \leqslant -W_3(x), \tag{2.9}$$

where $W_3(x)$ is a continuous positive definite function on D. Then, $x = 0$ is uniformly asymptotically stable. Moreover, if r and c are chosen such that $B_r = \{\|x\| \leqslant r\} \subset D$ and $c < \min_{\|x\|=r} W_1(x)$, then each trajectory starting in $\{x \in B_r \,|\, W_2(x) \leqslant c\}$ satisfies

$$\|x(t)\| \leqslant \beta\left(\|x(t_0)\|, t - t_0\right), \; \forall t \geqslant t_0 \geqslant 0,$$

for some class $\mathcal{KL}$ function β. Finally, if $D = \mathbb{R}^n$ and $W_1(x)$ is radially unbounded, then $x = 0$ is globally uniformly asymptotically stable.

Lemma 2.21 ([127]). *Let $x = 0$ be an equilibrium point for (2.3). Define $V : [0, \infty) \times D \to \mathbb{R}$ as a continuously differentiable function such that $\forall t \geqslant 0$ and $\forall x \in D$*

$$k_1\|x\|^a \leqslant V(t, x) \leqslant k_2\|x\|^a,$$

$$\frac{\partial V}{\partial t} + \frac{\partial V}{\partial x} f(t, x) \leqslant -k_3\|x\|^a,$$

where k_1, k_2, k_3, and a are positive constants. Then, $x = 0$ is exponentially stable. If the assumptions hold globally, then $x = 0$ is globally exponentially stable.

Definition 2.9 ([127]). *The solutions of (2.3) are*

1) uniformly bounded if there exists a positive constant c, independent of $t_0 \geqslant 0$, and for every $a \in (0, c)$, there exists $\beta = \beta(a) > 0$, independent of t_0, such that

$$\|x(t_0)\| \leqslant a \Rightarrow \|x(t)\| \leqslant \beta, \, \forall t \geqslant t_0. \tag{2.10}$$

2) globally uniformly bounded if (2.10) holds for arbitrarily large a.

3) uniformly ultimately bounded with ultimate bound b if there exist positive constants b and c, independent of $t_0 \geqslant 0$, and for every $a \in (0, c)$, there exists $T = T(a, b) \geqslant 0$, independent of t_0, such that

$$\|x(t_0)\| \leqslant a \Rightarrow \|x(t)\| \leqslant b, \, \forall t \geqslant t_0 + T. \tag{2.11}$$

4) globally uniformly ultimately bounded if (2.11) holds for any arbitrarily large a.

Lemma 2.22 ([127]). *Let $D \subset \mathbb{R}^n$ be a domain containing the origin and $V : [0, \infty) \times D \to \mathbb{R}$ be a continuously differentiable function such that $\forall t \geqslant 0$ and $\forall x \in D$*

$$\alpha_1(\|x\|) \leqslant V(t, x) \leqslant \alpha_2(\|x\|), \tag{2.12}$$

$$\frac{\partial V}{\partial t} + \frac{\partial V}{\partial x} f(t, x) \leqslant -W_3(x), \, \forall \|x\| \geqslant \mu > 0, \tag{2.13}$$

where α_1 and α_2 are class $\mathcal{K}$ functions and $W_3(x)$ is a continuous positive definite function. Choose $r > 0$ such that $B_r \subset D$ and suppose that $\mu < \alpha_2^{-1}(\alpha_1(r))$. Then, there exists a class $\mathcal{KL}$ function β and for each initial state $x(t_0)$ which satisfies $\|x(t_0)\| \leqslant \alpha_2^{-1}(\alpha_1(r))$, there exists $T = T(x(t_0), \mu) \geqslant 0$ such that the solution of (2.3) satisfies

$$\|x(t)\| \leqslant \beta(\|x(t_0)\|, \, t - t_0), \, \forall t_0 \leqslant t \leqslant t_0 + T, \tag{2.14}$$

$$\|x(t)\| \leqslant \alpha_1^{-1}(\alpha_2(\mu)), \, \forall t \geqslant t_0 + T. \tag{2.15}$$

Moreover, if $D = \mathbb{R}^n$ and α_1 belongs to class $\mathcal{K}_\infty$, then (2.14) and (2.15) hold for any initial state $x(t_0)$.

The input-to-state stability is to be presented in the following.

Definition 2.10 ([127]). *Consider the system*

$$\dot{x} = f(t, x, u), \tag{2.16}$$

where $f : [0, \infty) \times \mathbb{R}^n \times \mathbb{R}^m \to \mathbb{R}^n$ is piecewise continuous in t and locally Lipschitz in x and u. The input $u(t)$ is a piecewise continuous bounded function of t for all $t \geqslant 0$. The system (2.16) is said to be input-to-state stable if there exist a class $\mathcal{KL}$ function β and a class $\mathcal{K}$ function γ such that for any initial state $x(t_0)$ and any bounded input $u(t)$, the solution $x(t)$ exists for all $t \geqslant t_0$ and satisfies

$$\|x(t)\| \leqslant \beta(\|x(t_0)\|, t - t_0) + \gamma\left(\sup_{t_0 \leqslant \tau \leqslant t} \|u(\tau)\|\right). \tag{2.17}$$

Lemma 2.23 ([127]). *Let $V : [0, \infty) \times \mathbb{R}^n \to \mathbb{R}$ be a continuously differentiable function such that $\forall (t, x, u) \in [0, \infty) \times \mathbb{R}^n \times \mathbb{R}^m$*

$$\alpha_1 (\|x\|) \leqslant V(t, x) \leqslant \alpha_2 (\|x\|), \tag{2.18}$$

$$\frac{\partial V}{\partial t} + \frac{\partial V}{\partial x} f (t, x, u) \leqslant -W_3 (x), \; \forall \|x\| \geqslant \rho (\|u\|) > 0, \tag{2.19}$$

where α_1 and α_2 are class $\mathcal{K}_\infty$ functions, ρ is a class $\mathcal{K}$ function, and $W_3 (x)$ is a continuous positive definite function on $\mathbb{R}^n$. Then, the system (2.16) is input-to-state stable with $\gamma = \alpha_1^{-1} \circ \alpha_2 \circ \rho$.

Lemma 2.24 ([127]). *If A is Hurwitz, then system (2.1) is input-to-state stable. Moreover, for an input-to-state stable system (2.1), if $\lim_{t \to \infty} u(t) = 0$, then it holds that $\lim_{t \to \infty} x(t) = 0$.*

Lemma 2.25 (Comparison Lemma, [127]). *Consider the scalar differential equation*

$$\dot{u} = f (t, u), \; u (t_0) = u_0,$$

where $f(t, u)$ is continuous in t and locally Lipschitz in u, for all $t \geqslant 0$ and all $u \in J \subset \mathbb{R}$. Let $[t_0, T)$ (T could be infinity) be the maximal interval of existence of the solution $u (t)$, and suppose $u (t) \in J$ for all $t \in [t_0, T)$. Let $v (t)$ be a continuous function whose upper right-hand derivative $D^+v (t)$ satisfies the differential inequality

$$D^+v (t) \leqslant f (t, v(t)), \; v (t_0) \leqslant u_0$$

with $v (t) \in J$ for all $t \in [t_0, T)$. Then, $v (t) \leqslant u (t)$ for all $t \in [t_0, T)$.

Lemma 2.26 (Barbalat lemma, [127]). *For a differentiable function $g(t)$, if $g(t)$ exists a finite limit when $t \to \infty$ and $\dot{g} (t)$ is uniformly continuous, then $\lim_{t \to \infty} \dot{g} (t) = 0$.*

2.5   Finite-time Stability Theory

Consider the following autonomous system:

$$\dot{x}(t) = f (x(t)), \; x (0) = x_0, \tag{2.20}$$

where $x(t) \in \mathbb{R}^n$ is the state, and $f : D \to \mathbb{R}^n$ is continuous on an open neighbourhood $D \subseteq \mathbb{R}^n$ of the origin with $f(0) = 0$.

Definition 2.11. *If the origin of system (2.20) is Lyapunov stable, and there exist an open neighbourhood $U \subseteq D$ and a positive convergence-time function $T(x_0) : U \to \mathbb{R}$ for all $x(0) \in U \backslash \{0\}$ such that*

$$\lim_{t \to T(x_0)} x(t) = 0,$$
$$x(t) = 0, \ \forall t > T(x_0),$$

then the origin of system (2.20) is a finite-time stable equilibrium. Moreover, if $U = D = \mathbb{R}^n$, the origin is a global finite-time stable equilibrium.

Lemma 2.27 ([128]). *For system (2.20), suppose that there is a continuous differentiable positive definite function $V(x) \in \mathbb{R}$. If there exist positive constants c and $0 < \alpha < 1$ such that*

$$\dot{V}(x) + cV^{\alpha}(x) \leqslant 0, \ x \in U \backslash \{0\},$$

then the origin of system (2.20) is finite-time stable. In addition, the upper bound of convergence-time function $T(x_0)$ can be estimated by

$$T(x_0) \leqslant \frac{V^{1-\alpha}(x_0)}{c(1-\alpha)}.$$

2.6 Dynamics Models of Unmanned Vehicles

In practical applications, a group of UAVs and UGVs are the typical representatives for cross-domain heterogeneous swarm systems. Quadrotor UAVs and Mecanum wheel UGVs will be chosen to verify the proposed formation tracking approaches in this book. Before the controller design and analysis, how to build the simplified dynamics models of UAVs and UGVs in the formation control level is given in this section.

Firstly, let us consider the quadrotor UAV as shown in Fig. 2.2. Let $O_n - X_n Y_n Z_n$ and $O_b - X_b Y_b Z_b$ denote the inertial frame and the body frame, respectively. A quadrotor UAV has four actuators, and each actuator is composed of motor and propeller. The thrust and the torque generated by the spinning of propellers are the main control inputs for a quadrotor. For the i-th rotor ($i = 1, 2, 3, 4$), the thrust T_i and the torque M_i can be described by

$$T_i = c_T w_i^2, \tag{2.21}$$

$$M_i = c_M w_i^2, \tag{2.22}$$

where w_i is the rotation speed of the i-th rotor, and c_T and c_M are the lumped thrust coefficient and torque coefficient, respectively.

Based on the thrust and the torque of each rotor, we will build a control allocation model which can give the control force and the control moments

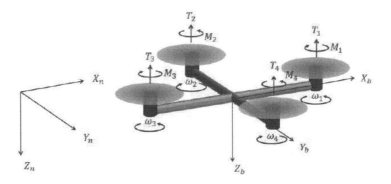

FIGURE 2.2: Structure diagram of a quadrotor UAV.

for a quadrotor. As shown in Fig. 2.2, the quadrotor UAV considered in this section has a '+' configuration. Assume that the distance between the body centre and each rotor is denoted by L. Then, the total control thrust and moments for the quadrotor can be described by

$$
\begin{bmatrix} u_1 \\ u_2 \\ u_3 \\ u_4 \end{bmatrix} = \begin{bmatrix} T_1 + T_2 + T_3 + T_4 \\ L\left(T_2 - T_4\right) \\ L\left(T_1 - T_3\right) \\ M_1 - M_2 + M_3 - M_4 \end{bmatrix} = \begin{bmatrix} c_T & c_T & c_T & c_T \\ 0 & Lc_T & 0 & -Lc_T \\ Lc_T & 0 & -Lc_T & 0 \\ c_M & -c_M & c_M & -c_M \end{bmatrix} \begin{bmatrix} w_1^2 \\ w_2^2 \\ w_3^2 \\ w_4^2 \end{bmatrix},
\tag{2.23}
$$

where u_1 is the total control thrust, and u_2, u_3, and u_4 are the control moments with respect to $O_b X_b$, $O_b Y_b$, and $O_b Z_b$ axes, respectively.

Define x, y, and z as the positions of the centre of mass for the UAV in the inertial frame. Let ϕ, θ, and ψ represent the roll, pitch, and yaw Euler angles, respectively. The position and attitude dynamics of the quadrotor UAV can be described by

$$
\begin{cases}
\ddot{x} = \left(-\sin\phi\sin\psi - \cos\phi\sin\theta\cos\psi\right)\dfrac{u_1}{m} \\[2mm]
\ddot{y} = \left(-\cos\phi\sin\theta\sin\psi + \sin\phi\cos\psi\right)\dfrac{u_1}{m} \\[2mm]
\ddot{z} = -\cos\phi\cos\theta\dfrac{u_1}{m} + g \\[2mm]
\ddot{\phi} = \dfrac{u_2}{I_{xx}} + \dot{\theta}\dot{\psi}\dfrac{I_{yy} - I_{zz}}{I_{xx}} \\[2mm]
\ddot{\theta} = \dfrac{u_3}{I_{yy}} + \dot{\phi}\dot{\psi}\dfrac{I_{zz} - I_{xx}}{I_{yy}} \\[2mm]
\ddot{\psi} = \dfrac{u_4}{I_{zz}} + \dot{\phi}\dot{\theta}\dfrac{I_{xx} - I_{yy}}{I_{zz}}
\end{cases}
\tag{2.24}
$$

where m is the mass, g is the gravitational acceleration, and I_{xx}, I_{yy}, and I_{zz} are the moments of inertia with respect to $O_b X_b$, $O_b Y_b$, and $O_b Z_b$ axes in the body frame. Moreover, for another multi-rotor UAV with different configuration, such as a quadrotor with 'x' configuration or a hexarotor, only the control allocation model (2.23) needs to be adjusted, and the dynamics model (2.24) still has the same form, which makes the proposed models for multi-rotor UAV general and practical.

In the following, the simplified model of a quadrotor UAV around the hovering state will be introduced, where $\phi \approx 0$, $\dot{\phi} \approx 0$, $\theta \approx 0$, $\dot{\theta} \approx 0$, $\dot{\psi} \approx 0$, and $u_1 \approx mg$. Let $\Delta u_1 = u_1 - mg$. Then, the dynamics model (2.24) can be simplified as

$$\begin{cases} \ddot{x} = (-\phi \sin \psi - \theta \cos \psi) \, g \\ \ddot{y} = (-\theta \sin \psi + \phi \cos \psi) \, g \\ \ddot{z} = -\dfrac{u_1}{m} + g = -\dfrac{\Delta u_1}{m} \\ \ddot{\phi} = \dfrac{u_2}{I_{xx}} \\ \ddot{\theta} = \dfrac{u_3}{I_{yy}} \\ \ddot{\psi} = \dfrac{u_4}{I_{zz}} \end{cases} \tag{2.25}$$

From (2.25), we can see that the height, roll, pitch, and yaw channels of a quadrotor UAV can be controlled by u_1, u_2, u_3, and u_4 separately and they all have a double integrator model. Besides, for a fixed yaw angle ψ, the positions x and y in the XY plane can be determined by the roll angle ϕ and the pitch angle θ.

Based on the above observation, the control of a quadrotor UAV can be implemented with an inner-loop and outer-loop structure [129, 130], where the outer-loop drives the UAV towards the desired position while the inner-loop tracks the attitude. In the formation control level, we usually focus on the relative position and velocity relationship between different quadrotor UAVs. Due to the fact that the trajectory dynamics have much larger time constants than the attitude dynamics, the formation tracking controller can be designed in the outer-loop for simplification. Fig. 2.3 shows the control scheme for the two-loop formation tracking structure in the XY plane.

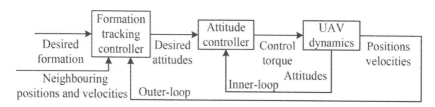

FIGURE 2.3: Two-loop formation tracking control architecture for the UAV swarm system in the XY plane.

Assume that a UAV swarm system is composed of N quadrotors. For the i-th quadrotor $(i = 1, 2, \ldots, N)$, let $p_i = [p_{Xi}, p_{Yi}]^T \in \mathbb{R}^2$ and $v_i = [v_{Xi}, v_{Yi}]^T \in \mathbb{R}^2$ denote the position and velocity vectors in the XY plane, respectively. If the dynamic process of the attitude inner-loop can be ignored and define $u_i = [u_{Xi}, u_{Yi}]^T$, $u_{Xi} = (-\phi_i \sin \psi_i - \theta_i \cos \psi_i) g$, and $u_{Yi} = (-\theta_i \sin \psi_i + \phi_i \cos \psi_i) g$, then from (2.25), we can obtain the approximate UAV dynamics model in the formation control level as follows

$$\begin{cases} \dot{p}_i = v_i, \\ \dot{v}_i = u_i. \end{cases} \tag{2.26}$$

Moreover, if the influence of damping is considered, we can revise the model (2.26) as

$$\begin{cases} \dot{p}_i = v_i, \\ \dot{v}_i = \alpha_{pi} p_i + \alpha_{vi} v_i + u_i, \end{cases} \tag{2.27}$$

where α_{pi} and α_{vi} are two damping constants. By assigning different α_{pi} and α_{vi}, the model (2.27) can reflect approximately the dynamic response process of a quadrotor UAV.

In the following, the dynamics model of a Mecanum wheel UGV will be given. One prominent advantage of Mecanum wheel UGV is the omnidirectional movement ability.

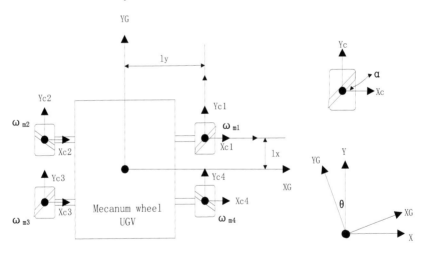

FIGURE 2.4: Structure diagram of a Mecanum wheel UGV.

Fig. 2.4 shows the structure diagram of a Mecanum wheel UGV. Let $O - XY$ and $O_G - X_G Y_G$ denote the inertial frame and the body frame, respectively. The control inputs of the UGV are the rotation speeds of four wheels, i.e., ω_{m1}, ω_{m2}, ω_{m3}, and ω_{m4}. The angle between the roll shaft and the axle of Mecanum wheel is denoted by α. Let v_x and v_y denote the velocities of mass centre along the $O_G X_G$-axis and $O_G Y_G$-axis in the body frame, respectively. The yaw angle is denotes by θ. Define the angular velocity of UGV as ω. Then, we can use v_x, v_y, and ω to describe the three degrees of freedom movement for the UGV in the XY plane.

To guarantee that v_x, v_y, and ω can be controlled separately, the angle α_i of four wheels are usually chosen as $\alpha_1 = 45°$, $\alpha_2 = -45°$, $\alpha_3 = 45°$, and $\alpha_4 = -45°$ in practical applications. Then, the kinematics relationship between the Mecanum wheels and the UGV body can be described by

$$\begin{bmatrix} v_x \\ v_y \\ \omega \end{bmatrix} = \begin{bmatrix} \frac{R}{4}\tan\alpha & -\frac{R}{4}\tan\alpha & -\frac{R}{4}\tan\alpha & \frac{R}{4}\tan\alpha \\ \frac{R}{4} & \frac{R}{4} & \frac{R}{4} & \frac{R}{4} \\ -\frac{R\tan\alpha}{4(l_x+l_y)} & \frac{R\tan\alpha}{4(l_x+l_y)} & -\frac{R\tan\alpha}{4(l_x+l_y)} & \frac{R\tan\alpha}{4(l_x+l_y)} \end{bmatrix} \begin{bmatrix} \omega_{m1} \\ \omega_{m2} \\ \omega_{m3} \\ \omega_{m4} \end{bmatrix}, \quad (2.28)$$

where R is the radius of Mecanum wheel, and l_x and l_y represent the distances between the centre of a Mecanum wheel and the mass centre of UGV along $O_G X_G$-axis and $O_G Y_G$-axis, respectively.

Let (x, y) denote the position of the UGV in the inertial frame. Then, the kinematics model can be denoted by

$$\begin{cases} \dot{x} = v_x \cos\theta - v_y \sin\theta, \\ \dot{y} = v_x \sin\theta + v_y \cos\theta, \\ \dot{\theta} = \omega. \end{cases} \quad (2.29)$$

As shown in (2.29), we can use ω to control θ, and $\begin{bmatrix} \cos\theta & -\sin\theta \\ \sin\theta & \cos\theta \end{bmatrix}$ is invertible for a given θ.

Assume that there exists a UGV swarm system with N Mecanum wheel UGVs. When only relative position relationship is considered in the formation control, the yaw angle θ_i of UAVs can be set to some constants. Let $p_i = [p_{Xi}, p_{Yi}]^T \in \mathbb{R}^2$ denote the position of the i-th UGV in the XY plane. Define $u_i = [u_{Xi}, u_{Yi}]^T$, where $u_{Xi} = v_{xi}\cos\theta_i - v_{yi}\sin\theta_i$ and $u_{Yi} = v_{xi}\sin\theta_i + v_{yi}\cos\theta_i$. Then, in the formation control level, the dynamics model of the i-th Mecanum wheel UGV can be described by

$$\dot{p}_i = u_i. \quad (2.30)$$

Finally, when a heterogeneous UAV and UGV swarm system is considered, we can unify the dynamics models (2.26), (2.27), and (2.30) into the following heterogeneous linear system:

$$\begin{cases} \dot{x}_i = A_i x_i + B_i u_i, \\ y_i = C_i x_i, \end{cases} \quad (2.31)$$

where $x_i \in \mathbb{R}^{n_i}$, $u_i \in \mathbb{R}^{m_i}$, and $y_i \in \mathbb{R}^p$ denote the state, control input, and output vectors, respectively. Since the swarm system is expected to achieve output formation, all the robots are assumed to have the same output dimension. Specifically, for the i-th quadrotor UAV in the XY plane, it can

be verified that $x_i = [p_{Xi}, v_{Xi}, p_{Yi}, v_{Yi}]^T$, $u_i = [u_{Xi}, u_{Yi}]^T$, $y_i = [p_{Xi}, p_{Yi}]^T$,
$A_i = I_2 \otimes \begin{bmatrix} 0 & 1 \\ \alpha_{pi} & \alpha_{vi} \end{bmatrix}$, $B_i = I_2 \otimes \begin{bmatrix} 0 \\ 1 \end{bmatrix}$, and $C_i = I_2 \otimes \begin{bmatrix} 1 & 0 \end{bmatrix}$. For the i-th
Mecanum wheel UGV, we can obtain that $x_i = [p_{Xi}, p_{Yi}]^T$, $u_i = [u_{Xi}, u_{Yi}]^T$,
$y_i = [p_{Xi}, p_{Yi}]^T$, $A_i = I_2 \otimes [0]$, $B_i = I_2$, and $C_i = I_2$.

In this book, we will focus on the high-order heterogeneous swarm system
(2.31) to design several formation tracking controllers and give the related
theoretical analysis. Then, based on UAV and UGV cooperative experiment
platform, a series of simulation and experiment results will be given to verify
the theoretical results.

2.7   Conclusions

Basic concepts and results for graph theory, algebra and matrix theory,
linear and non-linear system theory, and finite-time stability theory were introduced in this chapter, which can be viewed as the related foundation for
the following chapters. Moreover, the dynamics models of quadrotor UAV and
Mecanum wheel UGV were built from the formation control level.

Chapter 3

Formation Tracking Control for Homogeneous Swarm Systems

3.1 Introduction

Since homogeneous swarm system can be viewed as a special case of heterogeneous swarm system, this chapter will study the time-varying formation control and formation tracking control problems for homogeneous swarm systems. The main object of this chapter is to introduce basic definitions and control protocols for time-varying formation tracking problem, which can be viewed as the research basis for subsequent chapters. Leaderless formation control and leader-follower formation tracking problems will be discussed respectively in this chapter, and the main contents are given as follows.

Time-varying formation control problems for homogeneous swarm systems with switching directed interaction topologies are investigated firstly. Necessary and sufficient conditions for general linear swarm systems on switching graphs to achieve time-varying formations are proposed, where a description of the feasible time-varying formation set and approaches to expand the feasible formation set are given. An explicit expression of the time-varying formation reference function is derived to describe the macroscopic movement of the whole formation. An approach to assign the motion modes of the formation reference is provided, and an algorithm to design the formation protocol is presented. In the case where the given time-varying formation belongs to the feasible formation set, it is proven that by designing the formation protocol using the proposed algorithm, time-varying formation can be achieved by swarm systems with general linear dynamics and switching directed topologies if the dwell time is larger than a positive threshold.

Furthermore, time-varying formation tracking problems for homogeneous linear swarm systems with multiple leaders are studied, where the states of followers form a predefined time-varying formation while tracking the convex combination of the states of multiple leaders. Followers are classified into well-informed ones and uninformed ones, where the neighbour set of the former contains all the leaders while the latter contains no leaders. A formation tracking protocol is constructed using only neighbouring relative information. Necessary and sufficient conditions for swarm systems with multiple leaders

DOI: 10.1201/9781003263470-3

to achieve time-varying formation tracking are proposed by utilizing the properties of the Laplacian matrix, where the formation tracking feasibility constraints are also given. An approach to design the formation tracking protocol is presented by solving an algebraic Riccati equation.

3.2   Formation Control with Switching Directed Topologies

In this section, time-varying formation analysis and design problems for general linear swarm systems on switching directed topologies are investigated. Firstly, necessary and sufficient conditions to achieve time-varying formations under the influences of switching directed topologies are proposed. Then, an explicit expression of the time-varying formation reference function is derived to describe the macroscopic movement of the whole formation. An approach to assign the motion modes of the formation reference is provided. Moreover, an algorithm consisting of four steps to design the formation protocol is presented. It is proven that time-varying formation can be achieved by swarm systems with general linear dynamics and switching directed topologies by using the proposed controller if the dwell time is larger than a positive threshold. Finally, numerical simulations are presented to demonstrate the effectiveness of the theoretical results.

3.2.1   Problem Description

Consider a group of N homogeneous agents. Suppose that each agent has the general linear dynamics described by

$$\dot{x}_i(t) = Ax_i(t) + Bu_i(t), \ i \in \{1, 2, \ldots, N\}, \qquad (3.1)$$

where $A \in \mathbb{R}^{n \times n}$, $B \in \mathbb{R}^{n \times m}$, $x_i(t) \in \mathbb{R}^n$ and $u_i(t) \in \mathbb{R}^m$ are the state and the control input of the i-th agent, respectively. The matrix B is of full column rank, i.e., $\text{rank}(B) = m$, which means that the columns of B are independent with each other and there exist no redundant control input components.

The directed interaction topology of the swarm system is assumed to be switching and there exists an infinite sequence of uniformly bounded non-overlapping time intervals $[t_k, t_{k+1})$ ($k \in \mathbb{N}$), with $t_1 = 0$, $0 < \tau_0 \leqslant t_{k+1} - t_k \leqslant \tau_1$, and $\mathbb{N}$ being the set of natural numbers. The time sequence t_k ($k \in \mathbb{N}$) is called the switching sequence, at which the interaction topology changes. τ_0 is named as the dwell time, during which the interaction topology keeps fixed. Let $\sigma(t) : [0, +\infty) \to \{1, 2, \ldots, p\}$ be a switching signal whose value at time t is the index of the topology. Define $G_{\sigma(t)}$ and $L_{\sigma(t)}$ as the corresponding interaction topology and Laplacian matrix at $\sigma(t)$. Let $N^i_{\sigma(t)}$ be the neighbour set of the ith agent at $\sigma(t)$. Consider the following assumption for the switching graphs.

Assumption 3.1. *Each possible topology $G_{\sigma(t)}$ contains a spanning tree.*

The desired time-varying formation is specified by vector $h(t) = [h_1^T(t), h_2^T(t), \ldots, h_N^T(t)]^T \in \mathbb{R}^{nN}$ with $h_i(t)$ $(i = 1, 2, \ldots, N)$ piecewise continuously differentiable. It should be pointed out that $h(t)$ is only used to characterize the desired time-varying formation rather than providing reference trajectory for each agent to follow. Assume that $h_i(t)$ and $\dot{h}_i(t)$ are uniformly continuous.

Definition 3.1. *Swarm system (3.1) is said to achieve time-varying formation $h(t)$ if for any given bounded initial states, there exists a vector-valued function $r(t) \in \mathbb{R}^n$ such that*

$$\lim_{t \to \infty} (x_i(t) - h_i(t) - r(t)) = 0 \ (i = 1, 2, \ldots, N),$$

where $r(t)$ is called the formation reference function.

Definition 3.1 shows that $h_i(t)$ $(i = 1, 2, \ldots, N)$ can be used to specify the desired formation configuration and $r(t)$ is applied to describe the macroscopic movement of the entire time-varying formation. An illustration example is given to clarify the meanings of $h_i(t)$ and $r(t)$ more clearly.

Illustrative example 3.1. *Consider a swarm system with four agents. These agents move in the XY plane and their positions are required to accomplish a constant diamond formation with side length equal to $\sqrt{5}l$. To specify the desired formation shape, the vector $h = [h_1, h_2, h_3, h_4]^T$ can be chosen as $h_1 = [-l, 0]^T$, $h_2 = [0, 2l]^T$, $h_3 = [l, 0]^T$ and $h_4 = [0, -2l]^T$. The geometric relationships of $x_i(t)$, h_i $(i = 1, 2, 3, 4)$ and $r(t)$ are shown in Fig. 3.1. If Definition 3.1 is satisfied, it follows from (3.3) that $\lim_{t \to \infty}((x_i(t) - x_j(t)) - (h_i - h_j)) = 0$ $(i, j = 1, 2, 3, 4)$, which implies that the two diamonds specified by $x_i(t)$ and h_i are congruent. Thus, the expected diamond formation is accomplished by the swarm system. Furthermore, when the formation is achieved, one can obtain from Fig. 3.1 that h_i stands for the offset of $x_i(t)$ relative to $r(t)$ (i.e., $h_i = x_i(t) - r(t)$) and the formation reference function $r(t)$ can be applied to describe the macroscopic movement of the entire formation.*

Consider the following time-varying formation control protocol with switching directed interaction topologies:

$$\begin{aligned} u_i(t) = &K_1 x_i(t) + K_2(x_i(t) - h_i(t)) \\ &+ \alpha K_3 \sum_{j \in N_{\sigma(t)}^i} w_{ij}((x_j(t) - h_j(t)) - (x_i(t) - h_i(t))) + v_i(t), \end{aligned} \quad (3.2)$$

where $i = 1, 2, \ldots, N$, K_1, K_2, $K_3 \in \mathbb{R}^{m \times n}$ are constant gain matrices, α is the positive coupling strength, and $v_i(t) \in \mathbb{R}^m$ represents the formation compensation signal dependent on $h_i(t)$.

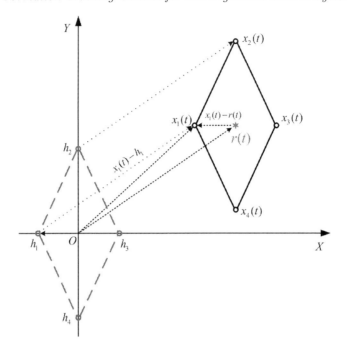

FIGURE 3.1: Illustration example for a constant diamond formation.

Remark 3.1. *In protocol (3.2), the gain matrix K_1 and compensation signal $v_i(t)$ $(i = 1, 2, \ldots, N)$ will be used to expand the feasible time-varying formation set. The gain matrix K_2 will be used to specify the motion modes of the time-varying formation reference $r(t)$. The gain matrix K_3 and the constant α can be used to drive the states of swarm system (3.1) to achieve the desired time-varying formation under switching directed topologies. It should be pointed out that K_1, K_2, and $v_i(t)$ $(i = 1, 2, \ldots, N)$ are dispensable for swarm system (3.1) to achieve some time-varying formations.*

Let $x(t) = [x_1^T(t), x_2^T(t), \ldots, x_N^T(t)]^T$ and $v(t) = [v_1^T(t), v_2^T(t), \ldots, v_N^T(t)]^T$. Under protocol (3.2) with switching directed topologies, swarm system (3.1) can be written in a compact form as follows

$$
\begin{aligned}
\dot{x}(t) = {}&\left(I_N \otimes (A + BK_1 + BK_2) - \alpha L_{\sigma(t)} \otimes BK_3\right) x(t) \\
&+ \left(\alpha L_{\sigma(t)} \otimes BK_3 - I_N \otimes BK_2\right) h(t) + (I_N \otimes B) v(t).
\end{aligned}
\tag{3.3}
$$

This section mainly focuses on the following two problems for swarm system (3.3) with switching directed interaction topologies: (i) under what conditions the time-varying formation specified by $h(t)$ can be achieved, and (ii) how to design the formation control protocol (3.2).

3.2.2   Time-varying Formation Analysis

In this subsection, firstly, necessary and sufficient conditions for swarm system (3.3) with switching directed interaction topologies to achieve time-varying formation specified by $h(t)$ are presented. Then, an explicit expression of the formation reference function is given to describe the macroscopic movement of the whole formation.

Let $\phi_i(t) = x_i(t) - h_i(t)$ and $\phi(t) = [\phi_1^T(t), \phi_2^T(t), \ldots, \phi_N^T(t)]^T$. Then, swarm system (3.3) with switching directed interaction topologies can be rewritten as

$$
\begin{aligned}
\dot{\phi}(t) =& \left(I_N \otimes (A+BK_1+BK_2) - \alpha L_{\sigma(t)} \otimes BK_3\right)\phi(t) \\
&+ (I_N \otimes (A+BK_1)) h(t) - (I_N \otimes I_n)\dot{h}(t) + (I_N \otimes B)v(t).
\end{aligned} \tag{3.4}
$$

Let $U = [\tilde{u}_1, \tilde{U}] \in \mathbb{R}^{N \times N}$ be a non-singular matrix with $\tilde{u}_1 = \mathbf{1}_N$ and $\tilde{U} = [\tilde{u}_2, \tilde{u}_3, \ldots, \tilde{u}_N]$. Let $U^{-1} = [\bar{u}_1^T, \bar{U}^T]^T$ with $\bar{U} = [\bar{u}_2^T, \bar{u}_3^T, \ldots, \bar{u}_N^T]^T$ and $\bar{u}_i \in \mathbb{R}^{1 \times N}$. Then one has $U^{-1}L_{\sigma(t)}U = \begin{bmatrix} \bar{u}_1 \\ \bar{U} \end{bmatrix} L_{\sigma(t)} \begin{bmatrix} \tilde{u}_1 & \tilde{U} \end{bmatrix} = \begin{bmatrix} 0 & \bar{u}_1 L_{\sigma(t)}\tilde{U} \\ 0 & \bar{U} L_{\sigma(t)}\tilde{U} \end{bmatrix}$. If Assumption 3.1 is satisfied, it follows from Lemma 2.2 that all the eigenvalues of $\bar{U} L_{\sigma(t)}\tilde{U}$ have positive real parts, which means that $\bar{U} L_{\sigma(t)}\tilde{U}$ is nonsingular.

Let $\theta(t) = (U^{-1} \otimes I_n)\phi(t) = [\theta_1^T, \theta_2^T, \ldots, \theta_N^T]^T$ and $\vartheta(t) = [\theta_2^T, \theta_3^T, \ldots, \theta_N^T]^T$. Then, swarm system (3.4) can be transformed into

$$
\begin{aligned}
\dot{\theta}_1(t) =& (A+BK_1+BK_2)\,\theta_1(t) - \alpha(\bar{u}_1 L_{\sigma(t)}\tilde{U}) \otimes BK_3 \vartheta(t) \\
&+ (\bar{u}_1 \otimes (A+BK_1)) h(t) - (\bar{u}_1 \otimes I_n)\dot{h}(t) + (\bar{u}_1 \otimes B)v(t),
\end{aligned} \tag{3.5}
$$

$$
\begin{aligned}
\dot{\vartheta}(t) =& \left(I_{N-1} \otimes (A+BK_1+BK_2) - \alpha(\bar{U} L_{\sigma(t)}\tilde{U}) \otimes BK_3\right)\vartheta(t) \\
&+ (\bar{U} \otimes (A+BK_1)) h(t) - (\bar{U} \otimes I_n)\dot{h}(t) + (\bar{U} \otimes B)v(t).
\end{aligned} \tag{3.6}
$$

Since $\mathrm{rank}(B) = m$, there exists a non-singular matrix $T = [\tilde{B}^T, \bar{B}^T]^T$ with $\tilde{B} \in \mathbb{R}^{m \times n}$ and $\bar{B} \in \mathbb{R}^{(n-m) \times n}$ such that $\tilde{B}B = I_m$ and $\bar{B}B = 0$. Let $h_{ij}(t) = h_i(t) - h_j(t)$ and $v_{ij}(t) = v_i(t) - v_j(t)$ $(i, j \in \{1, 2, \ldots, N\})$. The following theorem presents a necessary and sufficient condition for swarm system (3.3) to achieve time-varying formation specified by $h(t)$.

Theorem 3.1. *Swarm system (3.3) with switching directed interaction topologies achieves time-varying formation specified by $h(t)$ if and only if $\forall i \in \{1, 2, \ldots, N\}$, the following formation feasibility condition holds*

$$
\lim_{t \to \infty} \left(\bar{B}Ah_{ij}(t) - \bar{B}\dot{h}_{ij}(t)\right) = 0, \; j \in N_{\sigma(t)}^i; \tag{3.7}
$$

and the switched linear system described by

$$
\dot{\bar{\vartheta}}(t) = \left(I_{N-1} \otimes (A+BK_1+BK_2) - \alpha(\bar{U} L_{\sigma(t)}\tilde{U}) \otimes BK_3\right)\bar{\vartheta}(t), \tag{3.8}
$$

is asymptotically stable.

Proof. Define auxiliary variables $\phi_C(t)$ and $\phi_{\bar{C}}(t)$ as

$$\phi_C(t) = (U \otimes I_n)[\theta_1^T(t), 0]^T, \tag{3.9}$$

$$\phi_{\bar{C}}(t) = (U \otimes I_n)[0, \vartheta^T(t)]^T. \tag{3.10}$$

It can be shown that $[\theta_1^T(t), 0]^T = e_1 \otimes \theta_1(t)$, where $e_1 \in \mathbb{R}^N$ has 1 as its first component and 0 elsewhere. Therefore,

$$\phi_C(t) = (U \otimes I_n)(e_1 \otimes \theta_1(t)) = Ue_1 \otimes \theta_1(t) = \mathbf{1}_N \otimes \theta_1(t). \tag{3.11}$$

Note that $\theta(t) = [\theta_1^T(t), \vartheta^T(t)]^T$ and $\phi(t) = (U \otimes I_n)\theta(t)$. From (3.9) and (3.10), one has

$$\phi(t) = \phi_C(t) + \phi_{\bar{C}}(t). \tag{3.12}$$

Since $U \otimes I_n$ is nonsingular, it follows from (3.9) and (3.10) that $\phi_C(t)$ and $\phi_{\bar{C}}(t)$ are linearly independent. From (3.11) and (3.12), one gets

$$\phi_{\bar{C}}(t) = \phi(t) - \mathbf{1}_N \otimes \theta_1(t). \tag{3.13}$$

From (3.10), (3.13) and the fact that $U \otimes I_n$ is nonsingular, one gets that $\lim_{t\to\infty}(\phi(t) - \mathbf{1}_N \otimes \theta_1(t)) = 0$ if and only if $\lim_{t\to\infty}\vartheta(t) = 0$. Note that $\phi(t) - \mathbf{1}_N \otimes \theta_1(t)$ can be rewritten as $x_i(t) - h_i(t) - \theta_1(t)$ $(i = 1, 2, \ldots, N)$. Therefore, swarm system (3.3) achieves time-varying formation if and only if

$$\lim_{t\to\infty} \vartheta(t) = 0, \tag{3.14}$$

which means that $\vartheta(t)$ describes the time-varying formation error. Note that $h_i(t)$ and $\dot{h}_i(t)$ are uniformly continuous. From (3.6), one gets that for any given bounded initial states, (3.14) holds if and only if

$$\lim_{t\to\infty} \left((\bar{U} \otimes B)v(t) + (\bar{U} \otimes (A+BK_1)) h(t) - (\bar{U} \otimes I_n)\dot{h}(t) \right) = 0, \tag{3.15}$$

and the switched linear system described by (3.8) is asymptotically stable.

In the following it will be proven that condition (3.15) is equivalent to condition (3.7).

Necessity: If condition (3.7) holds, one has that $\forall i \in \{1, 2, \ldots, N\}$ and $j \in N_{\sigma(t)}^i$

$$\lim_{t\to\infty} \left(\bar{B}(A + BK_1)h_{ij}(t) - \bar{B}\dot{h}_{ij}(t) + \bar{B}Bv_{ij}(t) \right) = 0. \tag{3.16}$$

For any $i \in \{1, 2, \ldots, N\}$ and $j \in N_{\sigma(t)}^i$, one can find $v_i(t)$ and $v_j(t)$ satisfying

$$\lim_{t\to\infty} \left(\tilde{B}(A + BK_1)h_{ij}(t) - \tilde{B}\dot{h}_{ij}(t) + v_{ij}(t) \right) = 0. \tag{3.17}$$

It follows from (3.16) and (3.17) that

$$\lim_{t\to\infty} \left(T(A + BK_1)h_{ij}(t) - T\dot{h}_{ij}(t) + TBv_{ij}(t) \right) = 0. \tag{3.18}$$

Pre-multiplying the both sides of (3.18) by T^{-1}, one gets that $\forall i \in \{1, 2, \ldots, N\}$ and $j \in N^i_{\sigma(t)}$

$$\lim_{t\to\infty} \left((A + BK_1)h_{ij}(t) - \dot{h}_{ij}(t) + Bv_{ij}(t) \right) = 0. \tag{3.19}$$

From (3.19), one can obtain

$$\lim_{t\to\infty} \left(\left(L_{\sigma(t)} \otimes (A+BK_1) \right) h(t) - \left(L_{\sigma(t)} \otimes I_n \right) \dot{h}(t) + \left(L_{\sigma(t)} \otimes B \right) v(t) \right) = 0. \tag{3.20}$$

Substituting $L_{\sigma(t)} = U \begin{bmatrix} 0 & \bar{u}_1 L_{\sigma(t)} \tilde{U} \\ 0 & \bar{U} L_{\sigma(t)} \tilde{U} \end{bmatrix} U^{-1}$ into (3.20) and pre-multiplying the both sides of (3.20) by $U^{-1} \otimes I_n$ lead to

$$\lim_{t\to\infty} \left(\left(\bar{U} L_{\sigma(t)} \tilde{U} \bar{U} \otimes (A+BK_1) \right) h(t) - \left(\bar{U} L_{\sigma(t)} \tilde{U} \bar{U} \otimes I_n \right) \dot{h}(t) \right.$$
$$\left. + \left(\bar{U} L_{\sigma(t)} \tilde{U} \bar{U} \otimes B \right) v(t) \right) = 0. \tag{3.21}$$

Since $\bar{U} L_{\sigma(t)} \tilde{U}$ is invertible, pre-multiplying the both sides of (3.21) by $(\bar{U} L_{\sigma(t)} \tilde{U})^{-1} \otimes I_n$ yields

$$\lim_{t\to\infty} \left(\left(\bar{U} \otimes (A + BK_1) \right) h(t) - \left(\bar{U} \otimes I_n \right) \dot{h}(t) + \left(\bar{U} \otimes B \right) v(t) \right) = 0,$$

that is, condition (3.15) is required.
Sufficiency: Recall that rank($\bar{U}$) $= N - 1$. Without loss of generality, let $\bar{U} = [\hat{u}, \hat{U}]$, where $\hat{u} \in \mathbb{R}^{(N-1) \times 1}$ and $\hat{U} \in \mathbb{R}^{(N-1) \times (N-1)}$ is of full rank. If condition (3.15) holds, one has

$$\lim_{t\to\infty} ([\hat{u}, \hat{U}] \otimes B)v(t) + \left([\hat{u}, \hat{U}] \otimes (A+BK_1) \right) h(t) - ([\hat{u}, \hat{U}] \otimes I_n)\dot{h}(t) = 0. \tag{3.22}$$

Due to $\bar{U}\mathbf{1} = 0$, one gets $\hat{u} = -\hat{U}\mathbf{1}$. Let $\hat{h}(t) = [h_2^T(t), h_3^T(t), \ldots, h_N^T(t)]^T$ and $\hat{v}(t) = [v_2^T(t), v_3^T(t), \ldots, v_N^T(t)]^T$. Then it follows from (3.22) that

$$\lim_{t\to\infty} \left(\hat{U} \otimes I_n \right) \left(\bar{\Upsilon} - \tilde{\Upsilon} \right) = 0, \tag{3.23}$$

where $\bar{\Upsilon} = (I_{N-1} \otimes (A + BK_1))\hat{h}(t) - (I_{N-1} \otimes I_n)\dot{\hat{h}}(t) + (I_{N-1} \otimes B)\hat{v}(t)$ and $\tilde{\Upsilon} = (\mathbf{1} \otimes (A + BK_1))h_1(t) - (\mathbf{1} \otimes I_n)\dot{h}_1(t) + (\mathbf{1} \otimes B)v_1(t)$. Note that $\hat{U}$ is invertible. Pre-multiplying the both sides of (3.23) by $\hat{U}^{-1} \otimes I_n$ yields

$$\lim_{t\to\infty} \left((A+BK_1)h_{i1}(t) - \dot{h}_{i1}(t) + Bv_{i1}(t) \right) = 0 \ (i = 2, 3, \ldots, N). \tag{3.24}$$

From (3.24), it can be obtained that for any $i \in \{1, 2, \ldots, N\}$ and $j \in N^i_{\sigma(t)}$

$$\lim_{t \to \infty} \left((A + BK_1)h_{ij}(t) - \dot{h}_{ij}(t) + Bv_{ij}(t) \right) = 0. \qquad (3.25)$$

Pre-multiplying the both sides of (3.25) by T gives $\lim\limits_{t \to \infty} \left(\bar{B}Ah_{ij}(t) - \bar{B}\dot{h}_{ij}(t) \right) = 0$ $(i = 1, 2, \ldots, N; j \in N^i_{\sigma(t)})$. Therefore, condition (3.15) is equivalent to condition (3.7). Based on the above analysis, the conclusion of Theorem 3.1 can be obtained. □

Remark 3.2. *Condition (3.7) reveals that for any $i \in \{1, 2, \ldots, N\}$ and $j \in N^i_{\sigma(t)}$, $Ah_{ij}(t) - \dot{h}_{ij}(t)$ must belong to the right null space or the kernel of $\bar{B}$, which means that not all the time-varying formation can be achieved by general high-order swarm systems with switching directed topologies. In other words, constraint (3.7) describes the feasible time-varying formation set which is determined by the dynamics of each agent and the switching directed topologies, and only the ones belonging to the feasible formation set can be achieved. From (3.7) and (3.19), one sees that the application of $v(t)$ expands the feasible formation set. Condition (3.8) is an asymptotic stability constraint for a switched linear system. For the switched linear system like (3.8), no testable necessary and sufficient criteria for the asymptotic stability of the system have been obtained in the literature, and the best stability result attained so far is that if for any $\sigma(t)$, $I_{N-1} \otimes (A + BK_1 + BK_2) - \alpha(\bar{U}L_{\sigma(t)}\tilde{U}) \otimes BK_3$ is Hurwitz and the dwell time is large enough, then system (3.8) is asymptotically stable [131]. Based on the results of Theorem 3.1, testable sufficient conditions for swarm system (3.1) under protocol (3.2) to achieve time-varying formation will be further presented in Section 3.2.3.*

In the case where $v(t) \equiv 0$, the following corollary can be obtained directly from Theorem 3.1.

Corollary 3.1. *In the case where $v(t) \equiv 0$, swarm system (3.3) with switching directed interaction topologies achieves time-varying formation specified by $h(t)$ if and only if $\forall i \in \{1, 2, \ldots, N\}$*

$$\lim_{t \to \infty} \left((A + BK_1)h_{ij}(t) - \dot{h}_{ij}(t) \right) = 0, \quad j \in N^i_{\sigma(t)}, \qquad (3.26)$$

and the switched linear system

$$\dot{\bar{\vartheta}}(t) = \left(I_{N-1} \otimes (A + BK_1 + BK_2) - \alpha(\bar{U}L_{\sigma(t)}\tilde{U}) \otimes BK_3 \right) \bar{\vartheta}(t)$$

is asymptotically stable.

Remark 3.3. *From constraint (3.26) in Corollary 3.1, one sees that in the case where $v(t) \equiv 0$, K_1 can be applied to expand the feasible time-varying formation set. Formation feasibility problems for general linear swarm systems* *to achieve time-invariant formations with fixed topologies were discussed in [29]. By choosing $v(t) \equiv 0$, $\dot{h}(t) \equiv 0$, $K_1 = 0$, $K_2 = 0$, $\alpha = 1$, fixed topologies and appropriate U, Theorem 1 in [29] can be treated as a special case of Corollary 3.1.*

The formation reference represents the macroscopic movement of the whole formation. The following theorem reveals the effects of switching directed interaction topologies, dynamics of each agent, initial states of all the agents and time-varying formation on the evolution of the formation reference.

Theorem 3.2. *If swarm system (3.3) with switching directed interaction topologies achieves time-varying formation specified by $h(t)$, then the formation reference function $r(t)$ satisfies*

$$\lim_{t \to \infty} \left(r(t) - (r_0(t) + r_{\vartheta}(t) + r_v(t) + r_h(t)) \right) = 0,$$

where

$$r_0(t) = e^{(A+BK_1+BK_2)t}(\bar{u}_1 \otimes I_n)x(0),$$

$$r_{\vartheta}(t) = -\int_0^t e^{(A+BK_1+BK_2)(t-\tau)}\alpha \left(\bar{u}_1 L_{\sigma(t)} \tilde{U} \right) \otimes (BK_3)\, \vartheta(\tau)d\tau,$$

$$r_v(t) = \int_0^t \left(e^{(A+BK_1+BK_2)(t-\tau)}(\bar{u}_1 \otimes B)v(\tau) \right)d\tau,$$

$$r_h(t) = -(\bar{u}_1 \otimes I_n)h(t) - \int_0^t e^{(A+BK_1+BK_2)(t-\tau)}(\bar{u}_1 \otimes BK_2)h(\tau)d\tau.$$

Proof. If swarm system (3.3) achieves time-varying formation specified by $h(t)$, it follows from Theorem 3.1 that the formation error converges to zero at $t \to \infty$; that is, $\lim_{t\to\infty}\vartheta(t) = 0$. From (3.10) and (3.13), one gets

$$\lim_{t \to \infty} (\phi_i(t) - \theta_1(t)) = 0 \ (i = 1, 2, \ldots, N). \tag{3.27}$$

It holds that

$$\theta_1(0) = (\bar{u}_1 \otimes I_n)(x(0) - h(0)). \tag{3.28}$$

It can be obtained that

$$
\begin{aligned}
&\int_0^t e^{(A+BK_1+BK_2)(t-\tau)}(\bar{u}_1 \otimes I_n)\dot{h}(\tau)d\tau \\
&= e^{(A+BK_1+BK_2)(t-\tau)}(\bar{u}_1 \otimes I_n)h(\tau)\big|_{\tau=0}^{\tau=t} \\
&\quad -\int_0^t \frac{d}{d\tau}\left(e^{(A+BK_1+BK_2)(t-\tau)} \right)(\bar{u}_1 \otimes I_n)h(\tau)d\tau \\
&= (\bar{u}_1 \otimes I_n)h(t) - e^{(A+BK_1+BK_2)t}(\bar{u}_1 \otimes I_n)h(0) \\
&\quad -\int_0^t e^{(A+BK_1+BK_2)(t-\tau)}\left(-(A+BK_1+BK_2) \right)(\bar{u}_1 \otimes I_n)h(\tau)d\tau,
\end{aligned}
\tag{3.29}
$$

and

$$(A+BK_1+BK_2)(\bar{u}_1 \otimes I_n)h(\tau) = (\bar{u}_1 \otimes (A+BK_1+BK_2))\, h(\tau). \tag{3.30}$$

In virtue of (3.5) and (3.27)-(3.30), one can derive the conclusions of Theorem 3.2. □

Remark 3.4. *Theorem 3.2 shows an explicit expression of the formation reference function $r(t)$ which describes the macroscopic movement of the whole time-varying formation. From Theorem 3.2, one sees that $r(t)$ is jointly determined by $r_0(t)$, $r_\vartheta(t)$, $r_v(t)$, and $r_h(t)$, where $r_0(t)$ is the nominal component determined by the dynamics of each agent and initial states, $r_\vartheta(t)$ describes the effect of switching directed topologies and the time-varying formation error, $r_v(t)$ and $r_h(t)$ represent the contributions of $v(t)$ and $h(t)$ to $r(t)$, respectively. It should be pointed out that although we can obtain the explicit expression of the formation reference, the trajectory of the formation reference cannot be specified arbitrarily in advance. However, from Theorem 3.2, K_2 can be used to specify the motion modes of the formation reference by assigning the eigenvalues of $A+BK_1+BK_2$ at the desired places in the complex plane. Moreover, if $h(t) \equiv 0$, $r(t)$ becomes the explicit expression of the consensus function for general linear swarm systems with switching directed interaction topologies, which has not been obtained before.*

3.2.3   Time-varying Formation Protocol Design

In this subsection, firstly an algorithm to design the time-varying formation protocol (3.2) is proposed. Then it is proven that using the algorithm, time-varying formation can be achieved by swarm system (3.3) with switching directed topologies if the formation feasibility condition is satisfied and the dwell time is larger than a positive threshold.

Since the interaction topology $G_{\sigma(t)}$ has a spanning tree, from Lemma 2.2 and the structure of U, one knows that the real parts of all the eigenvalues of $\bar{U}L_{\sigma(t)}\tilde{U}$ are positive. Let $\hat{\mu}_{\sigma(t)} = \min\{\mathrm{Re}(\lambda_i(\bar{U}L_{\sigma(t)}\tilde{U})), i = 1, 2, \ldots, N - 1\}$, where $\lambda_i(\bar{U}L_{\sigma(t)}\tilde{U})$ represents the ith eigenvalue of $\bar{U}L_{\sigma(t)}\tilde{U}$. Then from Lemma 3 in [132], it can be obtained that for any $0 < \mu_{\sigma(t)} < \hat{\mu}_{\sigma(t)}$, there exists a symmetric positive definite matrix $\Xi_{\sigma(t)} \in \mathbb{R}^{(N-1)\times(N-1)}$ such that

$$\left(\bar{U}L_{\sigma(t)}\tilde{U}\right)^T \Xi_{\sigma(t)} + \Xi_{\sigma(t)}\left(\bar{U}L_{\sigma(t)}\tilde{U}\right) > 2\mu_{\sigma(t)}\Xi_{\sigma(t)}. \tag{3.31}$$

Lemma 3.1 ([133]). *For any positive definite matrix $M_1 \in \mathbb{R}^{n \times n}$ and symmetric matrix $M_2 \in \mathbb{R}^{n \times n}$, it holds that $x^T(t)M_2x(t) \leqslant \lambda_{\max}(M_1^{-1}M_2)x^T(t)M_1x(t)$.*

In the following, a design procedure with four steps is presented to determine the control parameters in time-varying formation control protocol (3.2).

Algorithm 3.1. *The time-varying formation control protocol (3.2) with switching directed topologies can be designed in the following procedure:*
***Step 1**: Check the time-varying formation feasibility condition (3.7). If it is satisfied, $v_i(t)$ $(i = 1, 2, \ldots, N)$ can be determined by solving the equation (3.17) and K_1 can be any constant matrix with appropriate dimension (e.g., $K_1 = 0$). From (3.17), $v_i(t)$ $(i = 1, 2, \ldots, N)$ are not unique. One can firstly*

specify a $v_k(t)$ ($k \in \{1, 2, \ldots, N\}$), and then determine the other $v_j(t)$ ($j \in \{1, 2, \ldots, N\}, j \neq k$) by equation (3.17). If the feasibility condition (3.7) is not satisfied, then the time-varying formation specified by $h(t)$ is not feasible and the algorithm stops.

If it is required that $v(t) \equiv 0$, solve the time-varying formation feasibility condition (3.26) for K_1. If there exists a K_1 satisfying (3.26), then continue, otherwise the time-varying formation specified by $h(t)$ is not feasible and the algorithm stops.

Step 2: *Choose K_2 to specify the motion modes of the formation reference $r(t)$ by placing the eigenvalues of $A + BK_1 + BK_2$ at the desired places in the complex plane. If (A, B) is controllable, the existence of K_2 can be guaranteed.*

Step 3: *For a given $\beta > 0$, solve the following linear matrix inequality for a symmetric positive definite matrix P:*

$$(A + BK_1 + BK_2)P + P(A + BK_1 + BK_2)^T - BB^T + \beta P < 0. \qquad (3.32)$$

Then, K_3 can be given by $K_3 = B^T P^{-1}$. It can be verified that if (A, B) is controllable, then inequality (3.32) is feasible for any given $\beta > 0$.

Step 4: *Choose a coupling strength α satisfying that $\alpha > 1/(2\bar{\mu})$ where $\bar{\mu} = \min\{\mu_{\sigma(t)}, \sigma(t) \in \{1, 2, \ldots, p\}\}$.*

Based on Algorithm 3.1, the following theorem can be obtained.

Theorem 3.3. *In the case where the time-varying formation feasibility condition (3.7) in Theorem 3.1 is satisfied, swarm system (3.3) with switching directed interaction topologies achieves time-varying formation specified by $h(t)$ if the formation control protocol (3.2) is designed by Algorithm 3.1 and the dwell time of the switching directed topologies satisfies*

$$\tau_0 > \frac{\ln \gamma}{\beta}, \qquad (3.33)$$

where $\gamma = \max\{\lambda_{\max}(\Xi_i^{-1}\Xi_j), i, j \in \{1, 2, \ldots, p\}, i \neq j\}$ with $\lambda_{\max}(\Xi_i^{-1}\Xi_j)$ being the largest eigenvalue of $\Xi_i^{-1}\Xi_j$.

Proof. Consider the stability of the switched linear system (3.8). Choose the following piecewise Lyapunov functional candidate

$$V(t) = \bar{\vartheta}^T(t) \left(\Xi_{\sigma(t)} \otimes P^{-1} \right) \bar{\vartheta}(t), \ \Xi_{\sigma(t)} \in \{\Xi_1, \Xi_2, \ldots, \Xi_p\}, \qquad (3.34)$$

where $\Xi_{\sigma(t)}$ and P are defined in (3.31) and (3.32). Note that the interaction topology $G_{\sigma(t)}$ is fixed for $t \in [t_1, t_2)$. Taking the time derivative of $V(t)$ along the trajectories of switched linear system (3.8), one has that $\forall t \in [t_1, t_2)$,

$$\dot{V}(t) = \bar{\vartheta}^T(t) \left(\Xi_{\sigma(t)} \otimes \Psi - \alpha \Phi_{\sigma(t)} \right) \bar{\vartheta}(t), \qquad (3.35)$$

where $\Psi = (A + BK_1 + BK_2)^T P^{-1} + P^{-1}(A + BK_1 + BK_2)$ and $\Phi_{\sigma(t)} = (\bar{U}L_{\sigma(t)}\tilde{U})^T \Xi_{\sigma(t)} \otimes (BK_3)^T P^{-1} + \Xi_{\sigma(t)}(\bar{U}L_{\sigma(t)}\tilde{U}) \otimes P^{-1}BK_3$. Substituting $K_3 = B^T P^{-1}$ into (3.35) gives

$$\dot{V}(t) = \bar{\vartheta}^T(t) \left(\Xi_{\sigma(t)} \otimes \Psi - \alpha \bar{\Phi}_{\sigma(t)} \right) \bar{\vartheta}(t), \tag{3.36}$$

where $\bar{\Phi}_{\sigma(t)} = \left((\bar{U}L_{\sigma(t)}\tilde{U})^T \Xi_{\sigma(t)} + \Xi_{\sigma(t)}(\bar{U}L_{\sigma(t)}\tilde{U}) \right) \otimes P^{-1}BB^T P^{-1}$. Let $\tilde{\vartheta}(t) = (I_{N-1} \otimes P)\bar{\vartheta}(t)$. It follows from (3.36) that

$$\dot{V}(t) = \tilde{\vartheta}^T(t) \left(\Xi_{\sigma(t)} \otimes \tilde{\Psi} - \alpha \tilde{\Phi}_{\sigma(t)} \right) \tilde{\vartheta}(t), \tag{3.37}$$

where $\tilde{\Psi} = P(A + BK_1 + BK_2)^T + (A + BK_1 + BK_2)P$ and $\tilde{\Phi}_{\sigma(t)} = \left((\bar{U}L_{\sigma(t)}\tilde{U})^T \Xi_{\sigma(t)} + \Xi_{\sigma(t)}(\bar{U}L_{\sigma(t)}\tilde{U}) \right) \otimes BB^T$. From (3.31), (3.32), and (3.37), one gets

$$\dot{V}(t) \leqslant \tilde{\vartheta}^T(t) \left(\Xi_{\sigma(t)} \otimes (BB^T - \beta P) - \alpha \left(2\mu_{\sigma(t)}\Xi_{\sigma(t)} \right) \otimes BB^T \right) \tilde{\vartheta}(t). \tag{3.38}$$

Substituting $\alpha > 1/(2\bar{\mu})$ into (3.38) yields $\dot{V}(t) \leqslant -\beta\tilde{\vartheta}^T(t) \left(\Xi_{\sigma(t)} \otimes P \right) \tilde{\vartheta}(t)$. Note that $\tilde{\vartheta}(t) = (I_{N-1} \otimes P^{-1})\bar{\vartheta}(t)$. One has that $\forall t \in [t_1, t_2)$

$$\dot{V}(t) \leq -\beta\bar{\vartheta}^H(t) \left(\Xi_{\sigma(t)} \otimes P^{-1} \right) \bar{\vartheta}(t) = -\beta V(t). \tag{3.39}$$

Since swarm system (3.2) switches at $t = t_2$, it follows from (3.39) that

$$V(t_2^-) < e^{-\beta(t_2 - t_1)}V(t_1) < e^{-\beta\tau_0}V(t_1). \tag{3.40}$$

Because $\bar{\vartheta}(t)$ is continuous, from (3.34) and Lemma 3.1, it can be obtained that

$$V(t_2) \leqslant \gamma V(t_2^-). \tag{3.41}$$

From (3.40) and (3.41), one gets $V(t_2) < \gamma e^{-\beta\tau_0}V(t_1) = e^{(\ln\gamma - \beta\tau_0)}V(0)$. Let $\nu = \beta - (\ln\gamma)/\tau_0$. If inequality (3.33) holds, then $\nu > 0$ and $V(t_2) < e^{-\nu\tau_0}V(0)$. For an arbitrarily given $t > t_2$, there exists a positive integer b satisfying $b \geqslant 2$. When $t \in (t_b, t_{b+1})$, using recursion approach, one has

$$V(t) < e^{-(\beta(t - t_b) + (b-1)\nu\tau_0)}V(0) < e^{-(b-1)\nu\tau_0}V(0). \tag{3.42}$$

Note that $t \leqslant b\tau_1$ and $b \geqslant 2$. It follows from (3.42) that $\forall t \in (t_b, t_{b+1})$

$$V(t) < e^{-\frac{(b-1)\tau_0\nu}{b\tau_1}t}V(0) < e^{-\frac{\tau_0\nu}{2\tau_1}t}V(0). \tag{3.43}$$

If $t = t_{b+1}$, it can be obtained that

$$V(t) < e^{-\frac{\tau_0\nu}{\tau_1}t}V(0). \tag{3.44}$$

From (3.43) and (3.44), one gets $\lim_{t\to\infty} \bar{\vartheta}(t) = 0$. Since the formation feasibility condition (3.7) is satisfied and the switched linear system (3.8) is asymptotically stable, it follows from Theorem 3.1 that swarm system (3.3) with switching directed interaction topologies achieves time-varying formation specified by $h(t)$. This completes the proof of Theorem 3.3. $\square$

Remark 3.5. *In the case where $h(t) \equiv 0$, the problems discussed in this section become consensus problems. Necessary and sufficient conditions for general linear swarm systems with switching directed topologies to achieve consensus and the consensus protocol design procedure can be obtained from Theorem 3.1 and Algorithm 3.1 directly. An explicit expression of the consensus function and a positive threshold for the dwell time can be derived from Theorems 3.2 and 3.3 respectively. Furthermore, if $h(t) \equiv 0$ and all the possible topologies have the same root as the leader, then all the results in this section can be applied to deal with the consensus tracking problems for general linear swarm systems with switching directed topologies.*

Remark 3.6. *From Theorems 3.1, 3.2, and 3.3, one sees that the non-singular transformation matrix U is a useful tool to deal with the time-varying formation control problems in this section. The construction of U utilizes the common property for all the possible switching topologies $G_{\sigma(t)}$ $(\sigma(t) = 1, 2, \ldots, p)$ containing a spanning tree that 0 is the simple eigenvalue of the corresponding Laplacian matrices $L_{\sigma(t)}$ with the associated right eigenvector $\mathbf{1}_N$. For theoretical analysis, U can be any non-singular matrix with $\mathbf{1}_N$ as one of its columns. Without loss of generality, we choose the first column $\tilde{u}_1 = \mathbf{1}_N$ in this section. Note that the choice of U is not unique. For simplicity, one can choose $U = \begin{bmatrix} 1 & 0 \\ \mathbf{1}_{N-1} & I_{N-1} \end{bmatrix}$ as used in the following simulation example.*

3.2.4 Numerical Simulations

In this subsection, a numerical example is given to illustrate the effectiveness of theoretical results obtained in the previous sections.

Consider a third-order swarm system with six agents, where the dynamics of each agent is described by (3.1) with $x_i(t) = [x_{i1}(t), x_{i2}(t), x_{i3}(t)]^T$ ($i = 1, 2, \ldots, 6$) and $A = \begin{bmatrix} 0 & -4 & 1 \\ 2 & 2 & -1 \\ 3 & 5 & 7 \end{bmatrix}$, $B = \begin{bmatrix} 0 \\ 0 \\ 1 \end{bmatrix}$. Suppose that there are three different 0-1 weighted directed topologies, namely, G_1, G_2, and G_3 as shown in Fig. 3.2. These six agents are required to keep a periodic time-varying parallel hexagon formation and at the same time keep rotation around the

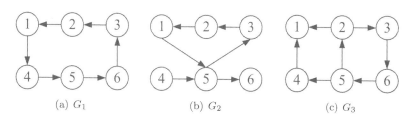

(a) G_1 (b) G_2 (c) G_3

FIGURE 3.2: Switching directed interaction topologies.

time-varying formation reference $r(t) = [r_1(t), r_2(t), r_3(t)]^T$. The time-varying formation is specified by

$$h_i(t) = \begin{bmatrix} 15\cos\left(2t + \frac{(i-1)\pi}{3}\right) \\ 15\sin\left(2t + \frac{(i-1)\pi}{3}\right) \\ 30\cos\left(2t + \frac{(i-1)\pi}{3}\right) \end{bmatrix} \quad (i = 1, 2, \cdots, 6).$$

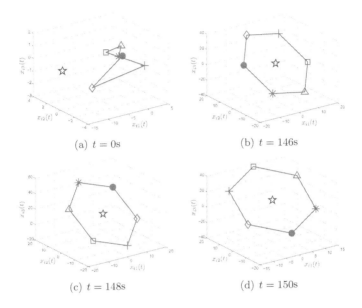

(a) $t = 0s$ (b) $t = 146s$

(c) $t = 148s$ (d) $t = 150s$

FIGURE 3.3: State snapshots of the six agents and the formation reference.

As $B \in \mathbb{R}^{3\times1}$ and rank$(B) = 1$, one gets that B is of full column rank. Choose $\tilde{B} = [0, 0, 1]$, $\bar{B} = \begin{bmatrix} 1 & 0 & 0 \\ 0 & 1 & 0 \end{bmatrix}$, and $U = \begin{bmatrix} 1 & 0 \\ 1_{N-1} & I_{N-1} \end{bmatrix}$. It can be verified that the formation feasibility constraint (3.7) in Theorem 3.1 is satisfied. According to Algorithm 3.1, gain matrix K_1 can be chosen as $K_1 = [0, 0, 0]$ and $v_i(t)$ can be solved from (3.17) as $v_i(t) = -285\sin\left(2t + \frac{\pi}{3}(i-1)\right) + 15\cos\left(2t + \frac{\pi}{3}(i-1)\right)$, where $i = 1, 2, \cdots, 6$. In the case where $K_2 = 0$, one can obtain the eigenvalues of $A + BK_1 + BK_2$ are 7.0439, $0.9780 + 3.0438j$ and $0.9780 - 3.0438j$ with $j^2 = -1$, which means that the motion modes of the formation reference are unstable and the whole formation will diverge exponentially. To keep the whole time-varying formation moving in a visual range, one can assign the motion modes of the formation reference to be oscillated using the approach in Step 2 of Algorithm 3.1. To this end, choose $K_2 = [3.8125, 0.0625, -10]$ to assign the eigenvalues of $A + BK_1 + BK_2$ at -1, $0.5j$ and $-0.5j$. Choose $\beta = 0.2$. Solving the inequality (3.32), one gets $K_3 = [-1.0241, -9.974, 2.9112]$. It can be obtained that

$\alpha > 2.7651$ and $\tau_0 > 12.6762$s. Therefore, choose $\alpha = 3$ and the dwell time to be 15s.

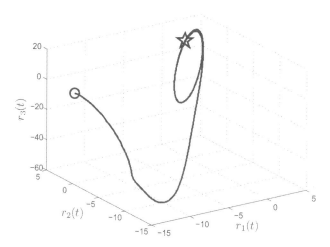

FIGURE 3.4: Trajectory of $r(t)$.

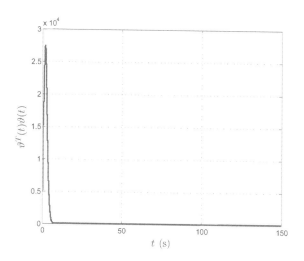

FIGURE 3.5: Curve of the formation error.

Let the initial states of the six agents generated by $x_{ij}(0) = i(\Theta - 0.5)$ ($i = 1, 2, \ldots, 6; j = 1, 2, 3$) with Θ being a random value between 0 and 1. Fig. 3.3 displays the snapshots of the six agents and the formation reference at $t = 0$s, $t = 146$s, $t = 148$s, and $t = 150$s, where the states of the six agents and the formation reference are denoted by the triangle, asterisk, dot, plus, square, diamond and pentagram respectively. Fig. 3.4 shows the trajectory of the formation reference, where the initial state is denoted by the circle. Fig. 3.5

depicts the curve of the formation error $\vartheta(t)$. From Figs. 3.3-3.5, the following phenomena can be observed: (i) the states of the six agents keep a parallel hexagon formation, (ii) the edge of the parallel hexagon is time-varying, (iii) the formation reference moves along a circle and lies in the centre of the time-varying formation, and (iv) the parallel hexagon is keeping rotation around the formation reference. Therefore, the desired time-varying formation is achieved by swarm system (3.3) under switching directed interaction topologies.

3.3 Formation Tracking Control with Multiple Leaders

In this section, more than one leader are introduced to specify the macroscopic movement trajectory of the swarm systems, and time-varying formation tracking problems for linear swarm systems with multiple leaders are studied. The states of followers are required to form a predefined time-varying formation and track the convex combination of the states of multiple leaders at the same time. Firstly, a formation tracking protocol is constructed using only neighbouring relative information. Then, by utilizing the properties of the Laplacian matrix, necessary and sufficient conditions for swarm systems with multiple leaders to achieve time-varying formation tracking are proposed, and the formation tracking feasibility constraints are also given. An approach to design the formation tracking protocol is presented by solving an algebraic Riccati equation. Finally, numerical simulations are provided to demonstrate the effectiveness of the theoretical results.

3.3.1 Problem Description

Consider a swarm system with N agents. The interaction topology of the swarm system can be described by a weighted directed graph $G = \{\mathcal{V}, \mathcal{E}, \mathcal{W}\}$. The agent i in the swarm system can be represented as the node v_i in G. For $i, j \in \{1, 2, \ldots, N\}$, the interaction channel from agent i to agent j is denoted by the edge ε_{ij}, and the corresponding interaction strength is denoted by w_{ji}. Agents in the swarm system are classified into leaders and followers.

Definition 3.2. *An agent is called a leader if it has no neighbour and a follower if it has at least one neighbour. A follower is called a well-informed one if its neighbour set contains all the leaders and an uninformed one if its neighbour set contains no leaders.*

Suppose that there are M $(M < N)$ followers and $N - M$ leaders. Let $F = \{1, 2, \ldots, M\}$ and $E = \{M + 1, M + 2, \ldots, N\}$ be the follower and leader subscript sets, respectively. For any $i, j \in \{1, 2, \ldots, N\}$, w_{ij} is defined as

$$w_{ij} = \begin{cases} 0, & i = j \text{ or } e_{ji} \notin \mathscr{E}, \\ b_j > 0, & j \in E \text{ and } e_{ji} \in \mathscr{E}, \\ a_{ij} > 0, & j \notin E \text{ and } e_{ji} \in \mathscr{E}, \end{cases} \qquad (3.45)$$

where b_j and a_{ij} are known positive constants. From (3.45), the interaction strengths from the same leader to different well-informed followers are identical. This technical assumption is mild. For example, in the special case where $b_j = 1$ and $a_{ij} = 1$, the interaction topology becomes the 0-1 weighted ones used widely in the existing results.

The dynamics of the leaders and followers are described by

$$\begin{cases} \dot{x}_k(t) = A x_k(t), & k \in E, \\ \dot{x}_i(t) = A x_i(t) + B u_i(t), & i \in F, \end{cases} \qquad (3.46)$$

where $x_k(t) \in \mathbb{R}^n$ and $x_i(t) \in \mathbb{R}^n$ are the states of leader k and follower i, respectively, $A \in \mathbb{R}^{n \times n}$, $B \in \mathbb{R}^{n \times m}$ with rank$(B) = m$, and $u_i(t) \in \mathbb{R}^m$ is the control input of the follower i. From Definition 3.2, one gets that the Laplacian matrix L has the form of $L = \begin{bmatrix} L_1 & L_2 \\ 0 & 0 \end{bmatrix}$ with $L_1 \in \mathbb{R}^{M \times M}$ and $L_2 \in \mathbb{R}^{M \times (N-M)}$.

The time-varying formation for the followers is specified by a vector $h_F(t) = [h_1^T(t), h_2^T(t), \ldots, h_M^T(t)]^T \in \mathbb{R}^{Mn}$, where $h_i(t) \in \mathbb{R}^n$ ($i \in F$) is the piecewise continuously differentiable offset vector with respect to the formation reference. It should be pointed out that $h_i(t) \in \mathbb{R}^n$ ($i \in F$) has the same dimension as the state $x_i(t)$ and includes the offsets corresponding to all the components of $x_i(t)$. Besides, it is required that $h_i(t)$ and $\dot{h}_i(t)$ are uniformly continuous.

Definition 3.3. *Swarm system (3.46) with multiple leaders is said to achieve time-varying formation tracking if for any given bounded initial states, there exist positive constants α_k ($k \in E$) satisfying $\sum_{k=M+1}^{N} \alpha_k = 1$ such that*

$$\lim_{t \to \infty} \left(x_i(t) - h_i(t) - \sum_{k=M+1}^{N} \alpha_k x_k(t) \right) = 0 \ (i \in F). \qquad (3.47)$$

Definition 3.3 reveals that when the time-varying formation tracking is realized, the states of the M followers reach an agreement on the formation reference, namely, the convex combination of the states of the $N - M$ leaders and keep the time-varying offset $h_F(t)$ with respect to it. In the case where there exists only one leader, i.e., $M = N-1$, Definition 3.3 becomes the definition for time-varying formation tracking with one leader, and equation (3.47) becomes $\lim_{t \to \infty} (x_i(t) - h_i(t) - x_k(t)) = 0$ ($i \in F, k \in E$). In the case where $\lim_{t \to \infty} \sum_{i=1}^{M} h_i(t) = 0$, it follows from (3.47) that $\lim_{t \to \infty} (\sum_{i=1}^{M} x_i(t)/M - \sum_{k=M+1}^{N} \alpha_k x_k(t)) = 0$, which means that $\sum_{k=M+1}^{N} \alpha_k x_k(t)$ lies in the centre of the time-varying formation (i.e., $\sum_{i=1}^{M} x_i(t)/M$) specified by $h_F(t)$. Therefore, by choosing appropriate $h_F(t)$, Definition 3.3 becomes the definitions for target enclosing or target pursuing with one target (see, e.g., [6]) and multiple targets, respectively.

Fig. 3.6 shows a formation tracking example for four followers and one leader in the XY plane. When the equation (3.47) is satisfied, it follows that the two squares specified by $x_i(t)$ and h_i are congruent. Thus, these four followers can achieve the desired square formation specified by $h_F = [h_1^T, h_2^T, h_3^T, h_4^T]^T$ and track the trajectory of the leader $x_0(t)$ at the same time.

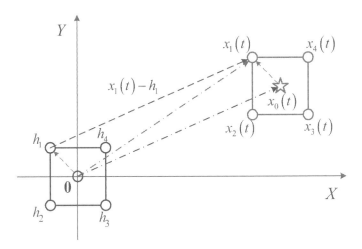

FIGURE 3.6: Illustration example for a square formation tracking.

Note that rank$(B) = m$, there exists a non-singular matrix $T = [\tilde{B}^T, \bar{B}^T]^T$ with $\tilde{B} \in \mathbb{R}^{m \times n}$ and $\bar{B} \in \mathbb{R}^{(n-m) \times n}$ such that $\tilde{B}B = I_m$ and $\bar{B}B = 0$. Consider the following time-varying formation tracking protocol:

$$
\begin{aligned}
u_i(t) = K \sum_{j=1}^{M} w_{ij} \left((x_i(t) - h_i(t)) - (x_j(t) - h_j(t)) \right) \\
+ K \sum_{k=M+1}^{N} w_{ik} \left((x_i(t) - h_i(t)) - x_k(t) \right) + v_i(t),
\end{aligned}
\tag{3.48}
$$

where $i \in F$, K is a constant gain matrix with appropriate dimension and $v_i(t)$ is the time-varying formation tracking compensational signal given by $v_i(t) = -\bar{B}(Ah_i(t) - \dot{h}_i(t))$.

Remark 3.7. *Protocol (3.48) is constructed by the neighbouring relative formation error term, the available neighbouring tracking error term, and the time-varying formation tracking compensational signal $v_i(t)$ determined by the dynamics of the agent and the time-varying formation vector $h_i(t)$. The role* *of $v_i(t)$ is to expand the time-varying formation feasibility condition by compensating the time-varying formation vector $h_i(t)$.*

Let $x_F(t) = [x_1^T(t), x_2^T(t), \dots, x_M^T(t)]^T \in \mathbb{R}^{Mn}$, $x_E(t) = [x_{M+1}^T(t), x_{M+2}^T(t), \dots, x_N^T(t)]^T \in \mathbb{R}^{(N-M)n}$ and $v_F(t) = [v_1^T(t), v_2^T(t), \dots, v_M^T(t)]^T \in \mathbb{R}^{Mm}$. Under protocol (3.48), swarm system (3.46) can be written as

$$
\begin{cases}
\dot{x}_F(t) = (I_M \otimes A + L_1 \otimes BK)x_F(t) + (I_M \otimes B)v_F(t) \\
\qquad\quad + (L_2 \otimes BK)x_E(t) - (L_1 \otimes BK)h_F(t), \\
\dot{x}_E(t) = (I_{N-M} \otimes A)x_E(t).
\end{cases}
\tag{3.49}
$$

This section mainly focuses on the following two problems for swarm system (3.46) under protocol (3.48): (i) under what conditions the time-varying formation tracking can be achieved; and (ii) how to design protocol (3.48) to achieve time-varying formation tracking.

3.3.2   Time-varying Formation Tracking Analysis and Design

In this subsection, firstly, necessary and sufficient conditions for swarm system (3.46) with multiple leaders to achieve time-varying formation tracking under protocol (3.48) are derived. Then, an approach to design the formation tracking protocol (3.48) is presented.

Assumption 3.2. *For any given follower, it is a well-informed one or an uninformed one. For each uninformed follower, there exists at least one well-informed follower that has a directed path to it.*

Remark 3.8. *To achieve the desired time-varying formation tracking with multiple leaders, all the followers should reach an agreement on the formation reference and then keep the offset $h_i(t)$ ($i \in F$) with respect to the formation reference. It is well-known in the research of containment control that if not all the informed followers are the well-informed ones, then the states of the followers will converge to different convex combinations of the states of all the leaders [3] (see the subsequent Illustrative example 3.2). In these cases, it is impossible for the followers to keep the desired time-varying formation as there exist multiple different formation references. Assumption 3.2 is required for all the followers to reach an agreement on the formation reference. Considering the practical applications, the well-informed followers are those with powerful sensors and communication devices while the uninformed followers are those with poorer sensors and communication devices, it is reasonable that the well-informed followers can obtain the information of all the leaders while the uninformed followers can only obtain the information of the neighbouring followers.*

To show that Assumption 3.2 is necessary for all the followers to reach an agreement on the formation reference, a comparative illustrative example is provided as follows.

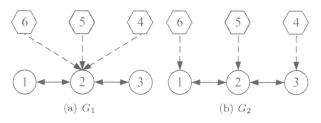

(a) G_1 (b) G_2

FIGURE 3.7: Interaction topologies in Illustrative example 3.2.

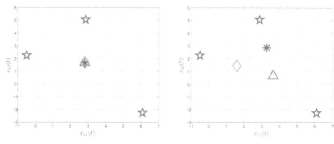

(a) Result for the case under G_1 (b) Result for the case under G_2

FIGURE 3.8: Snapshots of the six agents with $h_i(t) \equiv 0$ and $v_i(t) \equiv 0$ $(i \in F)$.

Illustrative example 3.2. *Consider a second-order swarm system consisting of three leaders and three followers, where the dynamics of each agent are described by (3.46) with $x_i(t) = [x_{i1}(t), x_{i2}(t)]^T$, $A = \begin{bmatrix} 0 & 1 \\ -1 & 0 \end{bmatrix}$, and $B = \begin{bmatrix} 0 \\ 1 \end{bmatrix}$. Let $h_i(t) \equiv 0$ and $v_i(t) \equiv 0$ $(i = 1, 2, 3)$. Then protocol (3.48) contains no formation information and becomes the containment protocol. Consider the following two interaction topologies G_1 and G_2 shown in Fig. 3.7 with 0-1 weights, respectively, where G_1 satisfies Assumption 3.2 while G_2 does not. For the two simulation cases under G_1 and G_2, let the gain matrix K in protocol (3.48) be $K = [-0.7524, -2.4563]$ and $K = [-0.4142, -1.3522]$, respectively. Fig. 3.8 shows the snapshots of the six agents at $t = 20s$, where the states of the leaders are marked by pentagrams, and those of the followers are marked by "$*$", "$\diamond$", and "$\triangle$". From Figs. 3.7 and 3.8, one sees that in the case where the topology satisfies Assumption 3.2, the states of the three followers reach an agreement which is named as the formation reference; otherwise, the states of the three followers converge to different convex combinations of those of the three leaders. Therefore, Assumption 3.2 is required for all the followers to reach an agreement on the formation reference.*

Lemma 3.2 ([134]). *If for each follower, there exists at least one leader that has a directed path to it, then*

(i) All the eigenvalues of L_1 have positive real parts;

(ii) Each entry of $-L_1^{-1}L_2$ is nonnegative, and each row of $-L_1^{-1}L_2$ has a sum equal to one.

Based on Lemma 3.2, the following results can be obtained.

Lemma 3.3. *If the directed interaction topology G satisfies Assumption 3.2, then all the rows of $-L_1^{-1}L_2$ are identical and equal to $[b_{M+1}, b_{M+2}, \ldots, b_N]/\sum_{k=M+1}^{N} b_k$.*

Proof. Since Assumption 3.2 is satisfied, the conclusions of Lemma 3.2 hold directly. From Lemma 3.2, one gets $-L_1^{-1}L_2 \mathbf{1}_{N-M} = \mathbf{1}_M$, which means that

$$L_1\mathbf{1}_M + L_2\mathbf{1}_{N-M} = 0. \tag{3.50}$$

Firstly, analyze the influence of each single leader on the followers individually. Assume that there exists only one leader, e.g., leader j ($j \in E$). To avoid confusion, denote by $\bar{w}_{ij}$ the interaction strength from the leader j to follower i ($i \in F$), and reassign $\bar{w}_{ij} = \sum_{k=M+1}^{N} b_k$ ($i \in F, e_{ji} \in \mathcal{E}$) and $\bar{w}_{ij} = 0$ otherwise. Then similar to (3.50), for the single leader case, it holds from Lemma 3.2 that

$$L_1\mathbf{1}_M + \begin{bmatrix} -\bar{w}_{1j} & -\bar{w}_{2j} & \cdots & -\bar{w}_{Mj} \end{bmatrix}^T = 0. \tag{3.51}$$

It follows from (3.51) that

$$-L_1^{-1}\begin{bmatrix} -\bar{w}_{1j} & -\bar{w}_{2j} & \cdots & -\bar{w}_{Mj} \end{bmatrix}^T = \mathbf{1}_M. \tag{3.52}$$

Let $\bar{e}_j \in \mathbb{R}^{N-M}$ ($j \in E$) with 1 as its $(j-M)$th component and 0 elsewhere. Post-multiplying the both sides of (3.52) by $\bar{e}_j^T(b_j/\sum_{k=M+1}^{N} b_k)$ gives

$$-L_1^{-1}[-\bar{w}_{1j}, -\bar{w}_{2j}, \ldots, -\bar{w}_{Mj}]^T \bar{e}_j^T(b_j/\sum_{k=M+1}^{N} b_k) \\ = \mathbf{1}_M \bar{e}_j^T(b_j/\sum_{k=M+1}^{N} b_k). \tag{3.53}$$

It can be obtained that

$$L_2 = \sum_{j=M+1}^{N}([-\bar{w}_{1j}, -\bar{w}_{2j}, \ldots, -\bar{w}_{Mj}]^T \bar{e}_j^T(b_j/\sum_{k=M+1}^{N} b_k)), \tag{3.54}$$

and

$$\sum_{j=M+1}^{N}(\mathbf{1}_M \bar{e}_j^T(b_j/\sum_{k=M+1}^{N} b_k)) \\ = (1/\sum_{k=M+1}^{N} b_k)(\mathbf{1}_M \otimes [b_{M+1}, b_{M+2}, \ldots, b_N]). \tag{3.55}$$

For all $j \in E$, adding the both sides of (3.53) and submitting (3.54) and (3.55) into the summation, one has

$$-L_1^{-1}L_2 = (1/\sum_{k=M+1}^{N} b_k)(\mathbf{1}_M \otimes [b_{M+1}, b_{M+2}, \ldots, b_N]). \tag{3.56}$$

Therefore, the conclusion of Lemma 3.3 can be obtained. □

Remark 3.9. *From (3.56), each entry of* $-L_1^{-1}L_2$ *is nonnegative and each row of* $-L_1^{-1}L_2$ *has a sum equal to one; that is, the conclusion (ii) in Lemma 3.2 can be obtained directly from (3.56). Therefore, Lemma 3.3 reveals further properties of the Laplacian matrix* L *and can be treated as an extension to the Lemma 3.2 under Assumption 3.2.*

Define $U_F \in \mathbb{C}^{M \times M}$ to be a non-singular matrix such that $U_F^{-1}L_1U_F = J_F$, where J_F is the Jordan canonical form of L_1 with diagonal entries λ_i $(i = 1, 2, \dots, M)$ satisfying $\mathrm{Re}(\lambda_1) \leq \mathrm{Re}(\lambda_2) \leq \cdots \leq \mathrm{Re}(\lambda_M)$.

Theorem 3.4. *Suppose that Assumption 3.2 holds. Swarm system (3.46) with multiple leaders achieves time-varying formation tracking asymptotically under protocol (3.48) if and only if for any* $i \in F$, $A + \lambda_i BK$ *is Hurwitz and the following formation tracking feasibility condition is satisfied*

$$\lim_{t \to \infty} (\bar{B}Ah_i(t) - \bar{B}\dot{h}_i(t)) = 0. \tag{3.57}$$

Proof. Define $\theta_i(t) = x_i(t) - h_i(t)$ $(i \in F)$ and $\theta_F(t) = [\theta_1^T(t), \theta_2^T(t), \dots, \theta_M^T(t)]^T$. Then, swarm system (3.49) can be transformed into

$$\begin{aligned}\dot{\theta}_F(t) = & (I_M \otimes A + L_1 \otimes BK)\theta_F(t) + (L_2 \otimes BK)x_E(t) \\ & + (I_M \otimes A)h_F(t) - (I_M \otimes I_n)\dot{h}_F(t) \\ & + (I_M \otimes B)v_F(t),\end{aligned} \tag{3.58}$$

$$\dot{x}_E(t) = (I_{N-M} \otimes A)x_E(t). \tag{3.59}$$

Sufficiency: Let

$$\phi_i(t) = \sum_{j=1}^{M} w_{ij}\left(\theta_i(t) - \theta_j(t)\right) + \sum_{k=M+1}^{N} w_{ik}\left(\theta_i(t) - x_k(t)\right) \ (i \in F),$$

and $\phi_F(t) = [\phi_1^T(t), \phi_2^T(t), \dots, \phi_M^T(t)]^T$. Then one gets

$$\phi_F(t) = (L_1 \otimes I_n)\theta_F(t) + (L_2 \otimes I_n)x_E(t). \tag{3.60}$$

It follows from (3.60) that

$$\theta_F(t) = \left(L_1^{-1} \otimes I_n\right)\phi_F(t) - \left(L_1^{-1}L_2 \otimes I_n\right)x_E(t). \tag{3.61}$$

Taking the time derivative of (3.60) and then submitting (3.58), (3.59) and (3.61) into the derivative of (3.60) yields

$$\begin{aligned}\dot{\phi}_F(t) = & (I_M \otimes A + L_1 \otimes BK)\phi_F(t) + (L_1 \otimes A)h_F(t) \\ & - (L_1 \otimes I_n)\dot{h}_F(t) + (L_1 \otimes B)v_F(t).\end{aligned} \tag{3.62}$$

Let $\bar{\phi}_F(t) = (U_F^{-1} \otimes I_n)\phi_F(t) = [\bar{\phi}_1^H(t), \bar{\phi}_2^H(t), \dots, \bar{\phi}_M^H(t)]^H$. Then system (3.62) can be converted into

$$\begin{aligned}\dot{\bar{\phi}}_F(t) = & (I_M \otimes A + J_F \otimes BK)\bar{\phi}_F(t) + (U_F^{-1}L_1 \otimes A)h_F(t) \\ & - (U_F^{-1}L_1 \otimes I_n)\dot{h}_F(t) + (U_F^{-1}L_1 \otimes B)v_F(t).\end{aligned} \tag{3.63}$$

If condition (3.57) holds, then for any $i \in F$

$$\lim_{t \to \infty} (\bar{B}Ah_i(t) - \bar{B}\dot{h}_i(t) + \bar{B}Bv_i(t)) = 0. \tag{3.64}$$

Let $v_i(t) = -(\tilde{B}Ah_i(t) - \tilde{B}\dot{h}_i(t))$. It holds that

$$\tilde{B}Ah_i(t) - \tilde{B}\dot{h}_i(t) + \tilde{B}Bv_i(t) = 0. \tag{3.65}$$

From (3.64), (3.65) and the fact that $T = [\tilde{B}^T, \bar{B}^T]^T$ is nonsingular, one gets

$$\lim_{t \to \infty} (Ah_i(t) - \dot{h}_i(t) + Bv_i(t)) = 0 \ (i \in F), \tag{3.66}$$

which means that

$$\lim_{t \to \infty} ((I_M \otimes A)h_F(t) - (I_M \otimes I_n)\dot{h}_F(t) + (I_M \otimes B)v_F(t)) = 0. \tag{3.67}$$

Pre-multiplying the both sides of (3.67) by $U_F^{-1}L_1 \otimes I_n$ yields

$$\lim_{t \to \infty} \left((U_F^{-1}L_1 \otimes A)h_F(t) - (U_F^{-1}L_1 \otimes I_n)\dot{h}_F(t) \right. \\ \left. + (U_F^{-1}L_1 \otimes B)v_F(t) \right) = 0. \tag{3.68}$$

From the structure of J_F, one gets that if $A + \lambda_i BK$ $(i \in F)$ is Hurwitz, then $I_M \otimes A + J_F \otimes BK$ is Hurwitz. Recall that (3.68) holds. Therefore, system (3.63) is asymptotically stable. Due to that U_F is nonsingular, one gets

$$\lim_{t \to \infty} \phi_F(t) = 0. \tag{3.69}$$

From (3.60) and (3.69), one has

$$\lim_{t \to \infty} \left(x_F(t) - h_F(t) - \left(-L_1^{-1}L_2 \otimes I_n\right) x_E(t) \right) = 0. \tag{3.70}$$

Since Assumption 3.2 holds, it follows from Lemma 3.3 and (3.70) that

$$\lim_{t \to \infty} \left(x_i(t) - h_i(t) - \sum_{j=M+1}^{N} \left(\frac{b_j}{\sum_{k=M+1}^{N} b_k} x_j(t) \right) \right) = 0, \tag{3.71}$$

which means that swarm system (3.46) with multiple leaders achieves time-varying formation tracking under protocol (3.48).

Necessity: The necessity is proven by contradiction. Suppose that swarm system (3.46) with multiple leaders achieves time-varying formation tracking under protocol (3.48). Then there exists positive constant α_k $(k \in E)$ satisfying $\sum_{k=M+1}^{N} \alpha_k = 1$ such that (3.47) holds. Let $x_E(0) = 0$. From (3.59), $\lim_{t \to \infty} x_E(t) = 0$, which means that the convex combination of $x_k(t)$ $(k \in E)$ converges to zero as $t \to \infty$. It follows from Definition 3.3 that

$\lim_{t\to\infty} \theta_F(t) = 0$. If the condition that for any $i \in F$, $A + \lambda_i BK$ is Hurwitz is not satisfied, $I_M \otimes A + L_1 \otimes BK$ is not Hurwitz. From (3.58), for any given initial state $\theta_F(0)$ and any feasible time-varying formation specified by $h_F(t)$, the limit of $\theta_F(t)$ as $t \to \infty$ is nonzero or does not exist. A contradiction is obtained. Thus, the condition that for any $i \in F$, $A + \lambda_i BK$ is Hurwitz is necessary. If condition (3.57) is not satisfied, it can be obtained that $\lim_{t\to\infty}((I_M \otimes A)h_F(t) - (I_M \otimes I_n)\dot{h}_F(t) + (I_M \otimes B)v_F(t))$ is nonzero. It follows from (3.58) that for any given initial state $\theta_F(0)$, the limit of $\theta_F(t)$ as $t \to \infty$ is nonzero or does not exist. A contradiction is obtained. Therefore, the necessity is proven. The proof for Theorem 3.4 is completed. □

In the case where there exists only one leader, i.e., $M = N - 1$, Theorem 3.4 presents the criteria for achieving time-varying formation tracking with one leader, and the following corollary can be obtained.

Corollary 3.2. *Suppose that Assumption 3.2 holds. Swarm system (3.46) with one leader achieves time-varying formation tracking asymptotically under protocol (3.48) if and only if for any $i \in F$, $A + \lambda_i BK$ is Hurwitz and the formation tracking feasibility condition (3.57) is satisfied.*

Remark 3.10. *As shown in [32] and [135] that even for formation control and tracking control, the feasible formation and feasible trajectory must be compatible with the dynamics of the agent. Similar conclusions also apply to the time-varying formation tracking problems discussed in this section. From (3.57), the formation tracking feasibility condition is dependent on the desired time-varying formation and the dynamics of each agent. In the case where $v_F(t) \equiv 0$, condition (3.57) becomes $\lim_{t\to\infty}(Ah_i(t) - \dot{h}_i(t)) = 0$, which indicates that $h_i(t)$ can have its explicit dynamics described by $\dot{h}_i(t) = Ah_i(t) + \varepsilon(t)$ with $\varepsilon(t) \to 0$ as $t \to 0$. Although formation tracking problems for linear swarm systems with one leader and undirected topologies were discussed in [44], the formation is time-invariant and formation tracking feasibility was not considered. In the case where $\lim_{t\to\infty} \sum_{i=1}^{M} h_i(t) = 0$ or $h_F(t) \equiv 0$, necessary and sufficient conditions for linear swarm systems to achieve target enclosing or consensus tracking with multiple leaders or one leader can be obtained from Theorem 3.4 and Corollary 3.2, respectively. In [136], the dynamics of each agent is first-order and the criteria are only sufficient.*

To interpret intuitively the reasonability and physical meaning behind the formation tracking feasibility condition, consider the following illustrative example.

Illustrative example 3.3. *Consider a multi-vehicle system with M follower vehicles and $N - M$ leader vehicles. Suppose that the dynamics of each vehicle are described by double integrators; that is,*

$$A = I_{\hat{n}} \otimes \begin{bmatrix} 0 & 1 \\ 0 & 0 \end{bmatrix}, \quad B = I_{\hat{n}} \otimes \begin{bmatrix} 0 \\ 1 \end{bmatrix},$$

with $\hat{n} \geq 1$ being the dimension of space. For the convenience of description, it is assumed that the multi-vehicle system moves in one-dimensional space ($\hat{n} = 1$) as the results can be extended to higher-dimensional space directly using the Kronecker product. For simplicity, assume that $M = N-1$, $v_i(t) \equiv 0$ ($i = 1, 2, \ldots, N-1$) and the formation for the followers to be time-invariant; that is, there exists only one leader and $h_i(t)$ can be rewritten as h_i. Let $h_i = [h_{ix}, h_{iv}]^T$ ($i \in F$). One gets that the formation tracking feasibility condition (3.57) under the above specified parameters becomes $Ah_i = 0$ ($i \in F$); that is,

$$\begin{bmatrix} h_{iv} \\ 0 \end{bmatrix} = 0 \ (i \in F). \tag{3.72}$$

Therefore, for the desired time-invariant formation tracking with one leader, there is no constraint on the component h_{ix} for the position, but the component h_{iv} for the velocity must be zero. It follows from Definition 3.3 that when the time-invariant formation tracking is achieved under the constraint (3.72), then

$$\lim_{t \to \infty} \begin{bmatrix} x_{ix}(t) - h_{ix} - x_{Nx}(t) \\ x_{iv}(t) - x_{Nv}(t) \end{bmatrix} = 0 \ (i \in F),$$

which means that the positions of the $N-1$ follower vehicles can form any time-invariant formation specified by h_{ix} ($i \in F$) with respect to the position of the leader vehicle and the velocities of the $N-1$ follower vehicles reach an agreement with the velocity of the leader vehicle. Therefore, for the multi-vehicle system to achieve time-invariant formation tracking with one leader, the formation tracking feasibility condition (3.57) is reasonable.

Based on Theorem 3.4, an approach to design the protocol (3.48) is proposed in the following theorem.

Theorem 3.5. *Suppose that Assumption 3.2 holds. If condition (3.57) in Theorem 3.4 holds and (A, B) is stabilizable, swarm system (3.46) with multiple leaders achieves time-varying formation tracking asymptotically by protocol (3.48) with $v_i(t) = -(\tilde{B}Ah_i(t) - \tilde{B}\dot{h}_i(t))$ ($i \in F$) and $K = -\delta[\mathrm{Re}(\lambda_1)]^{-1}R^{-1}B^T P$, where $\delta > 0.5$ is a given constant and P is the positive solution to the following algebraic Riccati equation*

$$PA + A^T P - PBR^{-1}B^T P + Q = 0, \tag{3.73}$$

with $R = R^T > 0$ and $Q = Q^T > 0$ given constant matrices.

Proof. Consider the stability of the following subsystem:

$$\dot{\varphi}_i(t) = (A + \lambda_i BK)\varphi_i(t) \ (i \in F). \tag{3.74}$$

Construct the following Lyapunov candidate function

$$V_i(t) = \varphi_i^H(t)P\varphi_i(t) \ (i \in F).$$

Taking the derivative of $V_i(t)$ along the trajectory of subsystem (3.74) gives

$$\dot{V}_i(t) = \varphi_i^H(t)(PA + A^T P + \lambda_i^H (BK)^T P + \lambda_i PBK)\varphi_i(t). \tag{3.75}$$

Substituting $K = -\delta[\text{Re}(\lambda_1)]^{-1}R^{-1}B^T P$ and $PA + A^T P = PBR^{-1}B^T P - Q$ into (3.75) one has

$$
\begin{aligned}
\dot{V}_i(t) = {} & (1 - 2\delta[\text{Re}(\lambda_1)]^{-1}\text{Re}(\lambda_i))\varphi_i^H(t)(PBR^{-1}B^T P)\varphi_i(t) \\
& - \varphi_i^H(t)Q\varphi_i(t).
\end{aligned} \tag{3.76}
$$

From the definition of λ_i ($i \in F$), one has that for any $i \in F$, $\text{Re}(\lambda_1) \le \text{Re}(\lambda_i)$, which means that $1 - 2\delta[\text{Re}(\lambda_1)]^{-1}\text{Re}(\lambda_i) < 0$. Since $P = P^T > 0, R = R^T > 0$ and $Q = Q^T > 0$, it can be verified that $PBR^{-1}B^T P \ge 0$ and

$$(1 - 2\delta[\text{Re}(\lambda_1)]^{-1}\text{Re}(\lambda_i))(PBR^{-1}B_2 P) - Q < 0. \tag{3.77}$$

From (3.76) and (3.77), one gets that $\dot{V}_i(t) < 0$, which means that $\lim_{t \to \infty} \varphi_i(t) = 0$ and $A + \lambda_i BK$ ($i \in F$) is Hurwitz. Since condition (3.57) is satisfied, it holds from Theorem 3.4 that swarm system (3.46) with multiple leaders achieves time-varying formation tracking by the designed protocol (3.48). This completes the proof of Theorem 3.5. □

Remark 3.11. *Theorem 3.4 shows that the design of the gain matrix K can be transformed into stabilizing the subsystem described by $\dot{\varphi}_i(t) = (A + \lambda_i BK)\varphi_i(t)$ ($i \in F$) using K. In Theorem 3.5, K can be given by $K = \delta[\text{Re}(\lambda_1)]^{-1}R^{-1}BTP$, where the matrix P represents the feature of the subsystem dynamics (A, B) and $\text{Re}(\lambda_1)$ represents the contribution of the complex eigenvalue λ_i ($i \in F$), which requires the knowledge of the network topology. In practical implementation, for the Riccati equation described by (3.73), the desired P can be obtained using the are() function in Matlab, i.e., are(A,B,Q). The approach for determining the gain matrix K is not unique and the approach in Theorem 3.5 is only one of them. In the case with only one leader, i.e., $M = N - 1$, Theorem 3.5 can be applied to determine the gain matrix of protocol (3.48) for swarm system with one leader to achieve the formation tracking.*

3.3.3   Numerical Simulation

In this subsection, a numerical example is given to illustrate the effectiveness of theoretical results.

Consider a third-order swarm system with three leaders and ten followers. The dynamics of each agent are described by (3.46) with

$$
A = \begin{bmatrix} 0 & 1 & 1 \\ 1 & 2 & 1 \\ -2 & -6 & -3 \end{bmatrix}, \quad B = \begin{bmatrix} 0 & 1 \\ -1 & 0 \\ 0 & 0 \end{bmatrix}.
$$

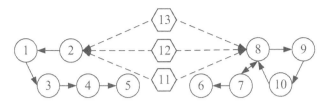

FIGURE 3.9: Directed interaction topology G_3.

The directed interaction topology G_3 with 0-1 weights is shown in Fig. 3.9. The states of the ten followers are required to keep the time-varying formation described by the following $h_i(t)$ ($i \in F$) while tracking the convex combination of the states of the three leaders:

$$h_i(t) = \begin{bmatrix} 15\sin\left(t + \frac{(i-1)\pi}{5}\right) \\ -15\cos\left(t + \frac{(i-1)\pi}{5}\right) \\ 30\cos\left(t + \frac{(i-1)\pi}{5}\right) \end{bmatrix} \quad (i = 1, 2, \ldots, 10).$$

From the above $h_i(t)$, if the desired formation tracking is achieved, then the states of the ten followers will form a parallel decagon and keep rotating around the convex combination of the states of the three leaders. Choose $\tilde{B} = \begin{bmatrix} 0 & -1 & 0 \\ 1 & 0 & 0 \end{bmatrix}$ and $\bar{B} = \begin{bmatrix} 0 & 0 & 1 \end{bmatrix}$. It can be verified that the formation tracking feasibility condition (3.57) in Theorem 3.4 is satisfied. Choose $\delta = 0.55$, $R = I$ and $Q = I$. Using the approach in Theorem 3.5, one gets $v_i(t) = 0$ and the gain matrix $K = \begin{bmatrix} 1.0053 & 3.9625 & 0.5575 \\ -1.2320 & -1.0053 & -0.3637 \end{bmatrix}$.

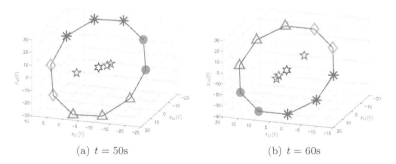

(a) $t = 50$s   (b) $t = 60$s

FIGURE 3.10: State snapshots of all the agents and the convex combination of the leaders.

Fig. 3.10 shows the state snapshots of the leaders, followers and the convex combination of the leaders at $t = 50$s and $t = 60$s, where the states of the followers are marked by "$*$", "$\bullet$", "$\diamond$" and "$\triangle$", respectively, and the states of the leaders and their convex combination are denoted by pentagrams and hexagram, respectively. From Fig. 3.10, one sees that (i) the states of the ten followers form a parallel decagon while keeping rotating around the states of the three leaders, and (ii) the states of the three leaders are time-varying and their convex combination lies in the centre of the parallel decagon. Therefore, the desired time-varying formation tracking with multiple leaders is achieved.

3.4   Conclusions

Leaderless time-varying formation control and leader-follower time-varying formation tracking problems for homogeneous swarm systems were studied respectively in this chapter. Firstly, necessary and sufficient conditions for general linear swarm systems with switching directed topologies to achieve time-varying formations were presented. A description of the feasible time-varying formation set and an explicit expression of the time-varying formation reference function were proposed. Approaches to expand the feasible formation set and assign the motion modes of the formation reference were given. An algorithm to design the formation protocol was presented and the stability analysis of switched systems was provided using piecewise Lyapunov theory. Furthermore, time-varying formation tracking problems for linear swarm systems with multiple leaders were investigated. A formation tracking protocol was constructed using only neighbouring relative information. Necessary and sufficient conditions for linear swarm systems with multiple leaders to achieve time-varying formation tracking were derived by utilizing the properties of the Laplacian matrix. Approaches to design the formation tracking protocol were proposed by solving an algebraic Riccati equation. The results in this chapter are mainly based on [33] and [9].

Chapter 4

Formation Tracking Control for Swarm Systems with Heterogeneous Disturbances

4.1   Introduction

In Chapter 3, formation control problems for high-order swarm systems with known and identical dynamics are considered, where it is assumed that there is not any disturbance. However, actual swarm systems such as multiple UAVs and UGVs are often affected by disturbances during the formation movement. For example, airflow disturbances including gust and turbulence have significant impacts on UAV formation flight. The disturbance of each agent in the swarm system is generally different, and it is difficult to accurately measure the disturbances, which means that there exist heterogeneous unknown disturbances. Under the influences of heterogeneous disturbances, even if the agents in the swarm system have the same nominal model, their actual dynamics are essentially different. Thus, it is called as a 'weak heterogeneous' swarm system in this book.

Disturbances will make the swarm system unable to form and maintain the desired formation effectively, or even cause the entire formation to collapse. In addition, in the applications of formation tracking control, the leader can be either a reference trajectory that characterizes the macroscopic motion of the entire formation, or it can denote a cooperative tracking entity and even a non-cooperative target. When the leader represents a non-cooperative target, its control input is unknown to all followers and may be time-varying. The leader's time-varying input will cause unknown additional terms into the dynamics of swarm system, which affects seriously the tracking control performance of the closed-loop system. The existing time-varying formation tracking results in [8,9,49,50] generally assume that the leader has no control input and does not consider the influences of disturbances. Therefore, under the condition of unknown disturbances and leader's unknown time-varying input, how to realize the time-varying formation tracking control of the weak heterogeneous swarm systems is a key problem to be solved.

In this chapter, high-order swarm systems with unknown heterogeneous disturbances and leader's time-varying input are considered. The following

DOI: 10.1201/9781003263470-4

two time-varying formation tracking problems are studied respectively: i) robust adaptive formation tracking with matched disturbances; ii) finite-time formation tracking with mismatched disturbances.

The main contents of this chapter are summarized as follows. For the case with matched disturbances, a robust adaptive time-varying formation tracking protocol and an algorithm to design the parameters in a distributed manner are proposed. Then, formation tracking feasible conditions, an approach to expand the feasible formation set, and sufficient conditions to achieve the desired formation tracking are given. For the case with mismatched disturbances, based on the finite-time disturbance observer, the integral sliding mode control, and the super-twisting algorithm, a continuous time-varying formation tracking protocol using the neighbouring interaction is presented, and the finite-time convergence of the formation tracking errors is proved.

4.2   Robust Adaptive Formation Tracking with Matched Disturbances

This section investigates the time-varying formation robust tracking problems for high-order linear swarm systems with heterogeneous disturbances and a leader of unknown control input. Firstly, a time-varying formation robust tracking protocol with a totally distributed form is proposed utilizing the neighbourhood state information. With the adaptive updating mechanism, neither any global knowledge about the communication topology nor the upper bounds of the parameter uncertainties, external disturbances, actuator faults, and leader's unknown input are required in the proposed controller. Then, in order to determine the control parameters, an algorithm with four steps is presented, where feasible conditions for the followers to accomplish the expected time-varying formation tracking are provided. Furthermore, based on the Lyapunov stability theory, it is proved that the formation tracking error can converge to zero asymptotically. Finally, a continuous formation tracking controller is proposed to avoid the large chattering of control inputs, under which the tracking errors and adaptive parameters are proved to be uniformly ultimately bounded.

4.2.1   Problem Description

Assume that there exists a swarm system with $N + 1$ agents, where the leader is denoted by 0, and the followers are labeled by $1, 2, \ldots, N$.

Consider the following model for the leader:

$$\dot{x}_0(t) = Ax_0(t) + Bu_0(t), \tag{4.1}$$

where $x_0(t) \in \mathbb{R}^n$ is the state and $u_0(t) \in \mathbb{R}^m$ is the control input of the leader. $A \in \mathbb{R}^{n \times n}$ and $B \in \mathbb{R}^{n \times m}$ are constant known matrices, where $\text{rank}(B) = m$. It is assumed that $x_0(t)$ is bounded. Under the influences of unknown parameter uncertainties, external disturbances, and actuator faults, the dynamics of follower i is described by

$$\dot{x}_i(t) = (A + \Delta A_i(t))\, x_i(t) + B\left(\rho_i(t)u_i(t) + f_{bi}(t)\right) + d_i(t), \qquad (4.2)$$

where $x_i(t) \in \mathbb{R}^n$ and $u_i(t) \in \mathbb{R}^m$ are the state and control input of follower i, $i = 1, 2, \ldots, N$. $\Delta A_i(t) \in \mathbb{R}^{n \times n}$ denotes unknown parameter uncertainties, $d_i(t) \in \mathbb{R}^m$ represents unknown external disturbances, $f_{bi}(t) \in \mathbb{R}^m$ is the actuator output bias, and $\rho_i(t) = \text{diag}\{\rho_{i1}(t), \rho_{i2}(t), \ldots, \rho_{im}(t)\}$ with $0 < \rho_{ij}(t) \leqslant 1$ $(j = 1, 2, \ldots, m)$ standing for the effectiveness factor of the j-th actuator channel. By using different values for $\rho_i(t)$ and $f_{bi}(t)$, several common fault modes including loss of effectiveness fault and bias fault can be described by the model (4.2) [39, 137].

The general case where the leader's time-varying control input $u_0(t)$ can be unknown and nonzero is considered in this section, which is more difficult to deal with than $u_0(t) \equiv 0$ in [8, 49, 50]. $u_0(t)$ is required to satisfy the following bounded assumption.

Assumption 4.1. *The leader's control input $u_0(t)$ is bounded, and there exists an unknown positive constant μ such that $\|u_0(t)\| \leqslant \mu$.*

Assume that $\Delta A_i(t)$ and $d_i(t)$ for follower i $(i = 1, 2, \ldots, N)$ satisfy the following matching and bounded condition, which are widely used in the literature [138–140].

Assumption 4.2. *There exist matrix $N_i(t)$ and vector $\bar{d}_i(t)$ such that $\Delta A_i(t) = BN_i(t)$ and $d_i(t) = B\bar{d}_i(t)$, $i = 1, 2, \ldots, N$. Moreover, there are unknown positive constants α_i and γ_i such that $\|N_i(t)\| \leq \alpha_i$ and $\|\bar{d}_i(t)\| \leq \gamma_i$.*

It is required that the actuator faults of follower i $(i = 1, 2, \ldots, N)$ satisfy the following bounded assumption [39, 137].

Assumption 4.3. *The unknown actuator output bias $f_{bi}(t)$ is bounded and there is an unknown positive constant b_i such that $\|f_{bi}(t)\| \leqslant b_i$. The unknown actuator effectiveness factor $\rho_{ij}(t)$ is bounded and there is an unknown positive constant $\underline{\rho}_{ij}$ such that $\underline{\rho}_{ij} \leqslant \rho_{ij}(t) \leqslant 1$, $i = 1, 2, \ldots, N$, $j = 1, 2, \ldots, m$.*

Let $h_i(t) \in \mathbb{R}^n$ $(i = 1, 2, \ldots, N)$ denote piecewise continuously differentiable vectors, then the expected time-varying formation of followers can be specified by a vector $h_F(t) = [h_1^T(t), h_2^T(t), \ldots, h_N^T(t)]^T$.

Definition 4.1. *For any given bounded initial states, if*

$$\lim_{t \to \infty} (x_i(t) - h_i(t) - x_0(t)) = 0, \quad i = 1, 2, \ldots N, \qquad (4.3)$$

then the expected time-varying state formation tracking is accomplished by swarm system (4.1) and (4.2).

4.2.2 Robust Adaptive Formation Tracking Controller Design and Stability Analysis

Let $\bar{\mathcal{G}}$ denote the topology among $N + 1$ agents and $\mathcal{G}_F$ represent the topology among N followers.

Assumption 4.4. *There exists a spanning tree in the graph $\bar{\mathcal{G}}$, where the leader is the root node. In addition, the topology $\mathcal{G}_F$ among N followers is undirected.*

Let $\bar{L}$ represent the Laplacian matrix for the graph $\bar{\mathcal{G}}$. According to the leader-follower topology structure, $\bar{L}$ can be divided into

$$\bar{L} = \begin{bmatrix} 0 & 0_{1 \times N} \\ L_2 & L_1 \end{bmatrix},$$

where $L_2 = [-w_{10}, -w_{20}, \ldots, -w_{N0}]^T$ and $L_1 \in \mathbb{R}^{N \times N}$. According to Lemma 2.1 and Lemma 2.2, it can be verified that L_1 is a positive definitive matrix under Assumption 4.4.

For a desired time-varying formation specified by $h_F(t)$, let $\delta_i(t)$ represent the formation tracking neighbourhood error of the follower i ($i \in \{1, 2, \ldots, N\}$), and $\delta_i(t)$ is defined by

$$\delta_i(t) = w_{i0}(x_i(t) - h_i(t) - x_0(t)) + \sum_{j=1}^{N} w_{ij}((x_i(t) - h_i(t)) - (x_j(t) - h_j(t))),$$

$$(4.4)$$

where $w_{i0} > 0$ if there is an edge ε_{0i} from the leader to the follower i in the graph $\bar{\mathcal{G}}$ and $w_{i0} = 0$ otherwise. Let $\sigma_i(t) \in \mathbb{R}$ be positive bounded and uniform continuous function such that $\lim\limits_{t \to \infty} \int_{t_0}^{t} \sigma_i(\tau)\mathrm{d}\tau \leqslant \bar{\sigma}_i < \infty$, where $\bar{\sigma}_i$ denotes a positive scalar.

For follower i ($i \in \{1, 2, \ldots, N\}$), consider the following time-varying formation tracking controller:

$$u_i(t) = -\hat{c}_i(t) B^T P \delta_i(t) - r_{Fi}(t) - r_{Di}(t), \qquad (4.5)$$

where

$$r_{Fi}(t) = \frac{\hat{\alpha}_i^2(t)(\|x_i(t)\| + \|v_i(t)\|)^2 B^T P \delta_i(t)}{\hat{\alpha}_i(t)(\|x_i(t)\| + \|v_i(t)\|)\|B^T P \delta_i(t)\| + \sigma_i(t)},$$

$$r_{Di}(t) = \frac{\hat{\beta}_i^2(t) B^T P \delta_i(t)}{\hat{\beta}_i(t)\|B^T P \delta_i(t)\| + \sigma_i(t)}.$$

$\hat{c}_i(t)$, $\hat{\alpha}_i(t)$, and $\hat{\beta}_i(t)$ are adaptive parameters updated by the neighbourhood error $\delta_i(t)$, where $\hat{c}_i(t)$ denotes time-varying coupling weight, $\hat{\alpha}_i(t)$ and

$\hat{\beta}_i(t)$ represent the adaptive estimations for the upper bounds $\frac{\bar{\alpha}_i}{\underline{\rho}_i}$ and $\frac{\bar{\beta}_i}{\underline{\rho}_i}$ respectively, $\bar{\alpha}_i = \max\{\alpha_i, 1\}$, $\bar{\beta}_i = \gamma_i + b_i + \mu$, $\underline{\rho}_i = \min\{\underline{\rho}_{i1}, \underline{\rho}_{i2}, \ldots, \underline{\rho}_{im}\}$. $v_i(t) \in \mathbb{R}^m$ represents the formation tracking compensational input determined by $h_F(t)$. $P \in \mathbb{R}^{n \times n}$ denotes a positive definite matrix to be determined.

In the proposed formation tracking controller (4.5), the first term is applied to drive the followers to achieve the desired formation tracking with ideal nominal models. The main role of second term $r_{Fi}(t)$ is to compensate for the influences of parameter uncertainties and time-varying formations. The third term $r_{Di}(t)$ is applied to suppress the unknown input of the leader and the external disturbance and the actuator output bias of each follower. Compared with the existing results in [8, 9, 49, 50], it is more difficult to deal with the time-varying formation and its derivative in this section due to the unknown actuator effectiveness factors in (4.2).

Let $Y = [\tilde{B}^T, \bar{B}^T]^T \in \mathbb{R}^{n \times n}$ denote a non-singular matrix, where $\tilde{B} \in \mathbb{R}^{m \times n}$ and $\bar{B} \in \mathbb{R}^{(n-m) \times n}$ satisfy $\tilde{B}B = I_m$ and $\bar{B}B = 0$. The existence of matrix Y is guaranteed by the condition rank $(B) = m$. An algorithm is given to design the parameters in the proposed formation tracking controller (4.5) as follows.

Algorithm 4.1. *The following four steps are used to design the time-varying formation tracking protocol (4.5).*

*   ***Step 1.*** *For follower i $(i \in \{1, 2, \ldots, N\})$ and a desired time-varying formation $h_F(t)$, check the following time-varying formation tracking feasible condition:*

$$\bar{B}Ah_i(t) - \bar{B}\dot{h}_i(t) = 0. \tag{4.6}$$

If the feasibility condition (4.6) is satisfied, then continue; else $h_F(t)$ is not feasible under the protocol (4.5) and the algorithm stops.

*   ***Step 2.*** *The formation tracking compensational input $v_i(t)$ is calculated by*

$$v_i(t) = \tilde{B}\left(Ah_i(t) - \dot{h}_i(t)\right), \quad i = 1, 2, \ldots, N. \tag{4.7}$$

*   ***Step 3.*** *Solve the following algebraic Riccati equation (ARE) for a positive definite matrix P:*

$$A^T P + PA - PBB^T P + Q = 0, \tag{4.8}$$

where $Q \in \mathbb{R}^{n \times n}$ is a positive definite matrix. There exists a positive definite matrix P to the ARE (4.8) if and only if (A, B) is stabilizable [138].

*   ***Step 4.*** *Adaptive parameters $\hat{c}_i(t)$, $\hat{\alpha}_i(t)$, and $\hat{\beta}_i(t)$ are generated by the following updating laws:*

$$\dot{\hat{c}}_i(t) = -\eta_{1i}\sigma_i(t)\hat{c}_i(t) + \eta_{1i}\left\|B^T P\delta_i(t)\right\|^2, \tag{4.9}$$

$$\dot{\alpha}_i(t) = -\eta_{2i}\sigma_i(t)\,\hat{\alpha}_i(t) + \eta_{2i}\left(\|x_i(t)\| + \|v_i(t)\|\right)\left\|B^T P\delta_i(t)\right\|, \qquad (4.10)$$

$$\dot{\hat{\beta}}_i(t) = -\eta_{3i}\sigma_i(t)\,\hat{\beta}_i(t) + \eta_{3i}\left\|B^T P\delta_i(t)\right\|, \qquad (4.11)$$

where η_{1i}, η_{2i}, and η_{3i} denote any positive constants. The initial values of adaptive parameters are finite with $\hat{c}_i(t_0) > 0$, $\hat{\alpha}_i(t_0) > 0$, and $\hat{\beta}_i(t_0) > 0$.

As shown in [9,33], even for ideal nominal swarm systems without disturbances, not all time-varying formations can be accomplished. The feasibility condition (4.6) reveals the requirement on the desired time-varying formation to be compatible with the agent dynamics. From (4.7), we see that the compensational input $v_i(t)$ can expand the feasible formation set effectively.

The computation of the minimum non-zero eigenvalue of Laplacian matrix is avoided in this section by using the adaptive weights $\hat{c}_i(t)$. Moreover, with the analytical estimates $\hat{\alpha}_i(t)$ and $\hat{\beta}_i(t)$, the bound information of $\Delta A_i(t)$, $d_i(t)$, $\rho_i(t)$, $f_{bi}(t)$, and $u_0(t)$ can be unknown in this section. Therefore, the proposed adaptive formation tracking protocol (4.5) is determined by each agent in a totally distributed form, which requires neither any global information of the graph nor the upper bounds of disturbances, faults, and leader's unknown input.

The following theorem guarantees that time-varying formation tracking can be accomplished by swarm system (4.1) and (4.2).

Theorem 4.1. *Suppose that Assumptions 4.1–4.4 hold. If the desired time-varying formation $h_F(t)$ satisfies the feasibility condition (4.6) and (A, B) is stabilizable, then swarm system (4.1) and (4.2) can achieve the desired time-varying formation tracking under the distributed controller (4.5) determined by Algorithm 4.1.*

Proof. For follower i ($i \in \{1, 2, \ldots, N\}$), substituting the controller (4.5) into (4.2) gives

$$\dot{x}_i(t) = (A + BN_i(t))\,x_i(t) + B\rho_i(t)\left(-\hat{c}_i(t)\,B^T P\delta_i(t) - r_{Fi}(t) - r_{Di}(t)\right)$$
$$+ B\left(f_{bi}(t) + \bar{d}_i(t)\right). \qquad (4.12)$$

Let $x(t) = [x_1^T(t), x_2^T(t), \ldots, x_N^T(t)]^T$, $\delta(t) = [\delta_1^T(t), \delta_2^T(t), \ldots, \delta_N^T(t)]^T$, $r_F(t) = [r_{F1}^T(t), r_{F2}^T(t), \ldots, r_{FN}^T(t)]^T$, $r_D(t) = [r_{D1}^T(t), r_{D2}^T(t), \ldots, r_{DN}^T(t)]^T$, $f_b(t) = [f_{b1}^T(t), f_{b2}^T(t), \ldots, f_{bN}^T(t)]^T$, and $\bar{d}(t) = [\bar{d}_1^T(t), \bar{d}_2^T(t), \ldots, \bar{d}_N^T(t)]^T$. System (4.12) can be rewritten as the following compact form:

$$\dot{x}(t) = (I_N \otimes A)\,x(t) - \left(\hat{C} \otimes B\right)\Xi\left(I_N \otimes B^T P\right)\delta(t)$$
$$- (I_N \otimes B)\,\Xi\,(r_F(t) + r_D(t)) + (I_N \otimes B)\left(\Lambda x(t) + f_b(t) + \bar{d}(t)\right), \qquad (4.13)$$

where $\hat{C} = \text{diag}\{\hat{c}_1(t), \hat{c}_2(t), \ldots, \hat{c}_N(t)\}$, $\Xi = \text{diag}\{\rho_1(t), \rho_2(t), \ldots, \rho_N(t)\}$, and $\Lambda = \text{diag}\{N_1(t), N_2(t), \ldots, N_N(t)\}$.

Let $\varsigma(t) = [\varsigma_1^T(t), \varsigma_2^T(t), \ldots, \varsigma_N^T(t)]^T$, where $\varsigma_i(t) = x_i(t) - h_i(t) - x_0(t)$ $(i = 1, 2, \ldots, N)$ denotes the formation tracking error for follower i. According to (4.1) and (4.13), it holds that

$$
\begin{aligned}
\dot{\varsigma}(t) =& (I_N \otimes A)\varsigma(t) - \left(\hat{C} \otimes B\right) \Xi \left(I_N \otimes B^T P\right) \delta(t) \\
& - (I_N \otimes B) \Xi (r_F(t) + r_D(t)) + (I_N \otimes B)\left(\Lambda x(t) + f_b(t) + \bar{d}(t)\right) \\
& - (\mathbf{1}_N \otimes B) u_0(t) + (I_N \otimes A) h_F(t) - \dot{h}_F(t).
\end{aligned} \tag{4.14}
$$

Consider the following Lyapunov function candidate:

$$
V_1(t) = \varsigma^T(t)(L_1 \otimes P)\varsigma(t) + \sum_{i=1}^{N} \frac{\underline{\rho}_i \tilde{c}_i^2(t)}{\eta_{1i}} + \sum_{i=1}^{N} \frac{\underline{\rho}_i \tilde{\alpha}_i^2(t)}{\eta_{2i}} + \sum_{i=1}^{N} \frac{\underline{\rho}_i \tilde{\beta}_i^2(t)}{\eta_{3i}}, \tag{4.15}
$$

where $\tilde{c}_i(t) = \hat{c}_i(t) - \frac{c}{\underline{\rho}_i}$, $\tilde{\alpha}_i(t) = \hat{\alpha}_i(t) - \frac{\bar{\alpha}_i}{\underline{\rho}_i}$, $\tilde{\beta}_i(t) = \hat{\beta}_i(t) - \frac{\bar{\beta}_i}{\underline{\rho}_i}$, c is a positive constant to be determined, $\underline{\rho}_i = \min\{\underline{\rho}_{i1}, \underline{\rho}_{i2}, \ldots, \underline{\rho}_{im}\}$, $\bar{\alpha}_i = \max\{\alpha_i, 1\}$, $\bar{\beta}_i = \gamma_i + b_i + \mu$.

Take the time derivative of $V_1(t)$ along the trajectory of (4.14), and substitute adaptive updating laws (4.9)-(4.11) into it. Then, we have

$$
\begin{aligned}
\dot{V}_1(t) =& \varsigma^T(t)\left(L_1 \otimes (PA + A^T P)\right)\varsigma(t) - 2\varsigma^T(t)(L_1\hat{C} \otimes PB)\Xi(I_N \otimes B^T P)\delta(t) \\
& - 2\varsigma^T(t)(L_1 \otimes PB)\Xi(r_F(t) + r_D(t)) + 2\varsigma^T(t)(L_1 \otimes PB)(\Lambda x(t) + f_b(t) + \bar{d}(t)) \\
& - 2\varsigma^T(t)(L_1\mathbf{1}_N \otimes PB)u_0(t) + 2\varsigma^T(t)(L_1 \otimes P)((I_N \otimes A)h_F(t) - \dot{h}_F(t)) \\
& + 2\sum_{i=1}^{N} \underline{\rho}_i \tilde{c}_i(t)\left(-\sigma_i(t)\hat{c}_i(t) + \left\|B^T P \delta_i(t)\right\|^2\right) \\
& + 2\sum_{i=1}^{N} \underline{\rho}_i \tilde{\alpha}_i(t)\left(-\sigma_i(t)\hat{\alpha}_i(t) + (\|x_i(t)\| + \|v_i(t)\|)\left\|B^T P \delta_i(t)\right\|\right) \\
& + 2\sum_{i=1}^{N} \underline{\rho}_i \tilde{\beta}_i(t)\left(-\sigma_i(t)\hat{\beta}_i(t) + \left\|B^T P \delta_i(t)\right\|\right). \tag{4.16}
\end{aligned}
$$

From the definitions of $\delta_i(t)$ and $\varsigma_i(t)$ $(i = 1, 2, \ldots, N)$, one gets that $\delta(t) = (L_1 \otimes I_n)\varsigma(t)$. Since Assumption 4.3 holds, it follows that

$$
\begin{aligned}
& -2\varsigma^T(t)\left(L_1\hat{C} \otimes PB\right) \Xi \left(I_N \otimes B^T P\right) \delta(t) + 2\sum_{i=1}^{N} \underline{\rho}_i \tilde{c}_i(t)\left\|B^T P \delta_i(t)\right\|^2 \\
&= -2\sum_{i=1}^{N} \hat{c}_i(t)\delta_i^T(t)PB\rho_i(t)B^T P\delta_i(t) + 2\sum_{i=1}^{N} \underline{\rho}_i \left(\hat{c}_i(t) - \frac{c}{\underline{\rho}_i}\right)\left\|B^T P \delta_i(t)\right\|^2 \\
&\leqslant -2c\sum_{i=1}^{N} \delta_i^T(t)PBB^T P\delta_i(t). \tag{4.17}
\end{aligned}
$$

According to the forms of $r_D(t)$ and $r_F(t)$, one has

$$- 2\varsigma^T(t) (L_1 \otimes PB) \Xi r_D(t) + 2\sum_{i=1}^{N} \underline{\rho}_i \tilde{\beta}_i(t) \left\| B^T P \delta_i(t) \right\|$$

$$\leqslant 2\sum_{i=1}^{N} \underline{\rho}_i \frac{\hat{\beta}_i(t) \left\| B^T P \delta_i(t) \right\| \sigma_i(t)}{\hat{\beta}_i(t) \left\| B^T P \delta_i(t) \right\| + \sigma_i(t)} - 2\sum_{i=1}^{N} \bar{\beta}_i \left\| B^T P \delta_i(t) \right\|$$

$$\leqslant 2\sum_{i=1}^{N} \underline{\rho}_i \sigma_i(t) - 2\sum_{i=1}^{N} \bar{\beta}_i \left\| B^T P \delta_i(t) \right\|, \tag{4.18}$$

$$- 2\varsigma^T(t) (L_1 \otimes PB) \Xi r_F(t) + 2\sum_{i=1}^{N} \underline{\rho}_i \tilde{\alpha}_i(t) \left(\left\| x_i(t) \right\| + \left\| v_i(t) \right\| \right) \left\| B^T P \delta_i(t) \right\|$$

$$\leqslant 2\sum_{i=1}^{N} \underline{\rho}_i \sigma_i(t) - 2\sum_{i=1}^{N} \bar{\alpha}_i \left(\left\| x_i(t) \right\| + \left\| v_i(t) \right\| \right) \left\| B^T P \delta_i(t) \right\|. \tag{4.19}$$

Substituting the inequalities (4.17)-(4.19) into (4.16) yields

$$\dot{V}_1(t) \leqslant \varsigma^T(t) \left(L_1 \otimes \left(PA + A^T P \right) \right) \varsigma(t) - 2c \sum_{i=1}^{N} \delta_i^T(t) PBB^T P \delta_i(t)$$

$$- 2\sum_{i=1}^{N} \bar{\beta}_i \left\| B^T P \delta_i(t) \right\| - 2\sum_{i=1}^{N} \bar{\alpha}_i \left(\left\| x_i(t) \right\| + \left\| v_i(t) \right\| \right) \left\| B^T P \delta_i(t) \right\|$$

$$+ 2\varsigma^T(t) (L_1 \otimes PB) \left(\Lambda x(t) + f_b(t) + \bar{d}(t) \right) - 2\varsigma^T(t) (L_1 \mathbf{1}_N \otimes PB) u_0(t)$$

$$+ 2\varsigma^T(t) (L_1 \otimes P) \left((I_N \otimes A) h_F(t) - \dot{h}_F(t) \right) + 4\sum_{i=1}^{N} \underline{\rho}_i \sigma_i(t)$$

$$- 2\sum_{i=1}^{N} \underline{\rho}_i \sigma_i(t) \left(\tilde{c}_i(t) \hat{c}_i(t) + \tilde{\alpha}_i(t) \hat{\alpha}_i(t) + \tilde{\beta}_i(t) \hat{\beta}_i(t) \right). \tag{4.20}$$

In light of Assumptions 4.1-4.3, it holds that

$$-2\varsigma^T(t) (L_1 \mathbf{1}_N \otimes PB) u_0(t) = -2\sum_{i=1}^{N} \delta_i^T(t) PBu_0(t) \leqslant 2\mu \sum_{i=1}^{N} \left\| B^T P \delta_i(t) \right\|,$$

$$2\varsigma^T(t) (L_1 \otimes PB) \left(f_b(t) + \bar{d}(t) \right) = 2\sum_{i=1}^{N} \delta_i^T(t) PB \left(f_{bi}(t) + \bar{d}_i(t) \right)$$

$$\leqslant 2\left(\gamma_i + b_i \right) \sum_{i=1}^{N} \left\| B^T P \delta_i(t) \right\|.$$

Note that $\bar{\beta}_i = \gamma_i + b_i + \mu$. We can obtain

$$2\varsigma^T(t)(L_1 \otimes PB)\left(f_b(t) + \bar{d}(t) - 1_N \otimes u_0(t)\right) - 2\sum_{i=1}^{N}\bar{\beta}_i\left\|B^T P\delta_i(t)\right\| \leqslant 0. \tag{4.21}$$

Under Assumption 4.2, it can be verified that

$$2\varsigma^T(t)(L_1 \otimes PB)\Lambda x(t) = 2\sum_{i=1}^{N}\delta_i^T(t)PBN_i(t)x_i(t)$$

$$\leqslant 2\sum_{i=1}^{N}\alpha_i\left\|x_i(t)\right\|\left\|B^T P\delta_i(t)\right\|. \tag{4.22}$$

Since the desired formation $h_F(t)$ satisfies the feasibility condition (4.6), we have

$$\bar{B}Ah_i(t) - \bar{B}\dot{h}_i(t) - \bar{B}Bv_i(t) = 0. \tag{4.23}$$

It follows from (4.7) that

$$\tilde{B}Ah_i(t) - \tilde{B}\dot{h}_i(t) - \tilde{B}Bv_i(t) = 0. \tag{4.24}$$

Since $Y = [\tilde{B}^T, \bar{B}^T]^T$ is a non-singular matrix, it holds from (4.23) and (4.24) that

$$Ah_i(t) - \dot{h}_i(t) - Bv_i(t) = 0, \; i = 1, 2, \ldots, N. \tag{4.25}$$

Then, we have

$$2\varsigma^T(t)(L_1 \otimes P)\left((I_N \otimes A)h_F(t) - \dot{h}_F(t)\right) = 2\sum_{i=1}^{N}\delta_i^T(t)PBv_i(t)$$

$$\leqslant 2\sum_{i=1}^{N}\left\|v_i(t)\right\|\left\|B^T P\delta_i(t)\right\|. \tag{4.26}$$

Since $\bar{\alpha}_i = \max\{\alpha_i, 1\}$, it can be verified from (4.22) and (4.26) that

$$2\varsigma^T(t)(L_1 \otimes PB)\Lambda x(t) + 2\varsigma^T(t)(L_1 \otimes P)\left((I_N \otimes A)h_F(t) - \dot{h}_F(t)\right)$$

$$- 2\sum_{i=1}^{N}\bar{\alpha}_i\left(\left\|x_i(t)\right\| + \left\|v_i(t)\right\|\right)\left\|B^T P\delta_i(t)\right\| \leqslant 0. \tag{4.27}$$

According to the well-known Young's inequality, one has $-\tilde{c}_i(t)\,\hat{c}_i(t) = -\tilde{c}_i^2(t) - \tilde{c}_i(t)\frac{c}{\rho_i} \leqslant \frac{c^2}{4\rho_i^2}$, $-\tilde{\alpha}_i(t)\,\hat{\alpha}_i(t) \leqslant \frac{\bar{\alpha}_i^2}{4\rho_i^2}$, $-\tilde{\beta}_i(t)\,\hat{\beta}_i(t) \leqslant \frac{\bar{\beta}_i^2}{4\rho_i^2}$. Substituting (4.21) and (4.27) into (4.20) gives

$$\dot{V}_1(t) \leqslant \varsigma^T(t)\left(L_1 \otimes \left(PA + A^T P\right)\right)\varsigma(t) - 2c\varsigma^T(t)\left(L_1^2 \otimes PBB^T P\right)\varsigma(t)$$
$$+ \sum_{i=1}^{N} k_i\sigma_i(t), \tag{4.28}$$

where $k_i = 4\rho_i + \frac{c^2}{2\rho_i} + \frac{\bar{\alpha}_i^2}{2\rho_i} + \frac{\bar{\beta}_i^2}{2\rho_i}$, $i = 1, 2, \ldots, N$.

Note that $L_1 > 0$ under Assumption 4.4. Let $\lambda_i > 0$ $(i = 1, 2, \ldots, N)$ denote the eigenvalues of L_1. Choose a unitary matrix $U \in \mathbb{R}^{N \times N}$ such that $U^T L_1 U = \mathrm{diag}\{\lambda_1, \lambda_2, \ldots, \lambda_N\} \triangleq J_L$. Let $\psi(t) = \left(U^T \otimes I_n\right)\varsigma(t) = [\psi_1^T(t), \psi_2^T(t), \ldots, \psi_N^T(t)]^T$. Then, it follows from (4.28) that

$$\dot{V}_1(t) \leqslant \psi^T(t)\left(J_L \otimes \left(A^T P + PA\right) - 2cJ_L^2 \otimes PBB^T P\right)\psi(t) + \sum_{i=1}^{N} k_i\sigma_i(t)$$
$$= \sum_{i=1}^{N}\lambda_i\psi_i^T(t)\left(A^T P + PA - 2c\lambda_i PBB^T P\right)\psi_i(t) + \sum_{i=1}^{N} k_i\sigma_i(t). \tag{4.29}$$

Choose sufficiently large c such that $c > \frac{1}{2\min_{i=1,\ldots,N}\{\lambda_i\}}$. It can be verified from (4.8) that $A^T P + PA - 2c\lambda_i PBB^T P \leqslant -Q$, where Q is a positive matrix. It follows from (4.29) that

$$\dot{V}_1(t) \leqslant -\underline{\lambda}\|\psi(t)\|^2 + \sum_{i=1}^{N} k_i\sigma_i(t), \tag{4.30}$$

where $\underline{\lambda} = \min_{i=1,\ldots,N}\{\lambda_i\}\lambda_{\min}(Q)$. Since $\lim_{t\to\infty}\int_{t_0}^{t}\sigma_i(\tau)\mathrm{d}\tau \leqslant \bar{\sigma}_i < \infty$, taking the integral of (4.30) from t_0 to t gives

$$V_1(t) \leqslant V_1(t_0) - \int_{t_0}^{t}\underline{\lambda}\|\psi(\tau)\|^2\mathrm{d}\tau + \int_{t_0}^{t}\sum_{i=1}^{N} k_i\sigma_i(\tau)\mathrm{d}\tau$$
$$\leqslant V_1(t_0) + \sum_{i=1}^{N} k_i\bar{\sigma}_i. \tag{4.31}$$

From (4.31), we can obtain that $V_1(t)$ is uniformly bounded, which implies that $\psi_i(t)$, $\tilde{c}_i(t)$, $\tilde{\alpha}_i(t)$, and $\tilde{\beta}_i(t)$ $(i = 1, 2, \ldots, N)$ are uniformly bounded.

Since $V_1(t) \geqslant 0$, it holds from (4.31) that $\int_{t_0}^{t}\underline{\lambda}\|\psi(\tau)\|^2\mathrm{d}\tau \leqslant V(t_0) + \sum_{i=1}^{N} k_i\bar{\sigma}_i$. From (4.14), one can obtain that $\dot{\psi}(t)$ is bounded, which further shows that $\psi(t)$ is uniformly continuous. According to Barbalat lemma [141], it can be verified that $\lim_{t\to\infty}\psi(t)=0$. Since $\psi(t)=\left(U^T\otimes I_n\right)\varsigma(t)$, we have $\lim_{t\to\infty}\varsigma(t)=0$, i.e., $\lim_{t\to\infty}\left(x_i(t)-h_i(t)-x_0(t)\right)=0$, $i=1,2,\ldots,N$. Based on Definition 4.1, one has that the desired formation tracking is achieved by swarm (4.1) and (4.2) under the distributed controller (4.5). This completes the proof of Theorem 4.1. $\qquad\square$

Theorem 4.1 shows that swarm system can achieve the desired time-varying formation tracking under the influences of several types of disturbances. The proposed controller (4.5) is fully distributed with no need for any global information of the graph and the bounds of the unknown disturbances. Due to the possible loss of effectiveness faults, there are unknown actuator effectiveness factors in the dynamics model of each follower, which makes it more difficult to compensate for the time-varying formation and its derivative in this section. Compared with [8,9,49,50], by using time-varying control input $u_0(t)$, the leader (4.1) can generate more general reference trajectories, and could also represent a non-cooperative target for practical applications.

For any uniform continuous bounded function $\sigma_i(t)$ $(i=1,2,\ldots,N)$ satisfying $\lim_{t\to\infty}\int_{t_0}^t\sigma_i(\tau)\mathrm{d}\tau\leqslant\bar{\sigma}_i<\infty$, one gets that $\lim_{t\to\infty}\sigma_i(t)=0$. Considering the structures of $r_{Fi}(t)$ and $r_{Di}(t)$ in the proposed controller (4.5), the control inputs will have large chattering when $t\to\infty$. To avoid this phenomenon, we will modify the formation tracking controller, and the following theorem shows the uniform ultimate boundedness of $\varsigma_i(t)$, $\hat{c}_i(t)$, $\hat{\alpha}_i(t)$, and $\hat{\beta}_i(t)$, $i=1,2,\ldots,N$.

Theorem 4.2. *The function $\sigma_i(t)$ in the formation controller (4.5) and the adaptive laws (4.9), (4.10), and (4.11) is replaced by a positive constant $\check{\sigma}_i$ $(i=1,2,\ldots,N)$. Suppose that Assumptions 4.1-4.4 hold. If the desired time-varying formation $h_F(t)$ satisfies the feasibility condition (4.6) and (A,B) is stabilizable, then the formation tracking error $\varsigma(t)$ and the adaptive parameters $\hat{c}_i(t)$, $\hat{\alpha}_i(t)$, $\hat{\beta}_i(t)$ $(i=1,2,\ldots,N)$ are uniformly ultimately bounded and converge to the following bounded set exponentially:*

$$D=\left\{\varsigma(t),\tilde{c}_i(t),\tilde{\alpha}_i(t),\tilde{\beta}_i(t):V_1(t)<\frac{1}{\varepsilon}\sum_{i=1}^N\bar{\kappa}_i\check{\sigma}_i\right\},\qquad(4.32)$$

where $V_1(t)$ is defined as (4.15), and positive constants $\bar{\kappa}_i=4\underline{\rho}_i+\frac{c^2}{\underline{\rho}_i}+\frac{\bar{\alpha}_i^2}{\underline{\rho}_i}+\frac{\bar{\beta}_i^2}{\underline{\rho}_i}$, $c>\frac{1}{2\min_{i=1,\ldots,N}\{\lambda_i\}}$, and $\varepsilon\leqslant\min_{i=1,\ldots,N}\left\{\check{\sigma}_i\eta_{1i},\check{\sigma}_i\eta_{2i},\check{\sigma}_i\eta_{3i},\frac{\lambda_{\min}(Q)}{\lambda_{\max}(P)}\right\}$.

Proof. Follow the similar steps as the proof of Theorem 4.1. It holds that

$$\dot{V}_1(t) \leqslant -\varepsilon V_1(t) + \varepsilon V_1(t) + \varsigma^T(t)\left(L_1 \otimes \left(PA + A^T P\right)\right)\varsigma(t)$$

$$- 2c\varsigma^T(t)\left(L_1^2 \otimes PBB^T P\right)\varsigma(t) + 4\sum_{i=1}^{N}\underline{\rho}_i \check{\sigma}_i$$

$$- 2\sum_{i=1}^{N}\underline{\rho}_i \check{\sigma}_i\left(\tilde{c}_i(t)\hat{c}_i(t) + \tilde{\alpha}_i(t)\hat{\alpha}_i(t) + \tilde{\beta}_i(t)\hat{\beta}_i(t)\right). \qquad (4.33)$$

According to Young's inequality, one has that $-\tilde{c}_i(t)\hat{c}_i(t) = -\tilde{c}_i^2(t) - \tilde{c}_i(t)\frac{c}{\underline{\rho}_i} \leqslant -\frac{1}{2}\tilde{c}_i^2(t) + \frac{c^2}{2\underline{\rho}_i^2}$, $-\tilde{\alpha}_i(t)\hat{\alpha}_i(t) \leqslant -\frac{1}{2}\tilde{\alpha}_i^2(t) + \frac{\bar{\alpha}_i^2}{2\underline{\rho}_i^2}$, $-\tilde{\beta}_i(t)\hat{\beta}_i(t) \leqslant -\frac{1}{2}\tilde{\beta}_i^2(t) + \frac{\bar{\beta}_i^2}{2\underline{\rho}_i^2}$. Substituting (4.15) into (4.33) yields

$$\dot{V}_1(t) \leqslant -\varepsilon V_1(t) + \varsigma^T(t)\left(L_1 \otimes (PA + A^T P) - 2cL_1^2 \otimes PBB^T P + \varepsilon L_1 \otimes P\right)\varsigma(t)$$

$$+ \sum_{i=1}^{N}\underline{\rho}_i\left(\frac{\varepsilon}{\eta_{1i}} - \check{\sigma}_i\right)\tilde{c}_i^2(t) + \sum_{i=1}^{N}\underline{\rho}_i\left(\frac{\varepsilon}{\eta_{2i}} - \check{\sigma}_i\right)\tilde{\alpha}_i^2(t)$$

$$+ \sum_{i=1}^{N}\underline{\rho}_i\left(\frac{\varepsilon}{\eta_{3i}} - \check{\sigma}_i\right)\tilde{\beta}_i^2(t) + \sum_{i=1}^{N}\left(4\underline{\rho}_i + \frac{c^2}{\underline{\rho}_i} + \frac{\bar{\alpha}_i^2}{\underline{\rho}_i} + \frac{\bar{\beta}_i^2}{\underline{\rho}_i}\right)\check{\sigma}_i. \quad (4.34)$$

Using the same transformation $\psi(t) = \left(U^T \otimes I_n\right)\varsigma(t)$ as Theorem 4.1, since $c > \frac{1}{2\min_{i=1,\dots,N}\{\lambda_i\}}$ and $\varepsilon \leqslant \frac{\lambda_{\min}(Q)}{\lambda_{\max}(P)}$, it can be verified from (4.8) that

$$\varsigma^T(t)\left(L_1 \otimes \left(PA + A^T P\right) - 2cL_1^2 \otimes PBB^T P + \varepsilon L_1 \otimes P\right)\varsigma(t) \leqslant 0. \quad (4.35)$$

Furthermore, for $\varepsilon \leqslant \min_{i=1,\dots,N}\{\check{\sigma}_i\eta_{1i}, \check{\sigma}_i\eta_{2i}, \check{\sigma}_i\eta_{3i}\}$, one gets

$$\dot{V}_1(t) \leqslant -\varepsilon V_1(t) + \sum_{i=1}^{N}\bar{\kappa}_i\check{\sigma}_i, \qquad (4.36)$$

where $\bar{\kappa}_i = 4\underline{\rho}_i + \frac{c^2}{\underline{\rho}_i} + \frac{\bar{\alpha}_i^2}{\underline{\rho}_i} + \frac{\bar{\beta}_i^2}{\underline{\rho}_i}$. According to Comparison lemma [127], one has

$$V_1(t) \leqslant \left(V_1(t_0) - \frac{1}{\varepsilon}\sum_{i=1}^{N}\bar{\kappa}_i\check{\sigma}_i\right)e^{-\varepsilon(t-t_0)} + \frac{1}{\varepsilon}\sum_{i=1}^{N}\bar{\kappa}_i\check{\sigma}_i. \qquad (4.37)$$

Thus, $V_1(t)$ converges to the bounded set D in (4.32) exponentially. This completes the proof of Theorem 4.2. $\qquad \square$

If $\check{\sigma}_i$ is chosen to be sufficiently small such that

$$\max_{i=1,\dots,N}\{\check{\sigma}_i\eta_{1i}, \check{\sigma}_i\eta_{2i}, \check{\sigma}_i\eta_{3i}\} \leqslant \frac{\lambda_{\min}(Q)}{\lambda_{\max}(P)},$$

then it holds from (4.34) that

$$\dot{V}_1(t) \leqslant -\varepsilon V_1(t) - \left(\frac{\lambda_{\min}(Q)}{\lambda_{\max}(P)} - \varepsilon \right) \min_{i=1,\dots,N} \{\lambda_i\} \lambda_{\min}(P) \|\varsigma(t)\|^2 + \sum_{i=1}^{N} \bar{\kappa}_i \check{\sigma}_i.$$

Let $\varphi_{\min} = \left(\frac{\lambda_{\min}(Q)}{\lambda_{\max}(P)} - \varepsilon \right) \min_{i=1,\dots,N} \{\lambda_i\} \lambda_{\min}(P)$. If $\|\varsigma(t)\|^2 \geqslant \frac{1}{\varphi_{\min}} \sum_{i=1}^{N} \bar{\kappa}_i \check{\sigma}_i$, one gets that $\dot{V}_1(t) \leqslant -\varepsilon V_1(t)$. Thus, $\|\varsigma(t)\|^2$ will converge to the upper bound $\frac{1}{\varphi_{\min}} \sum_{i=1}^{N} \bar{\kappa}_i \check{\sigma}_i$. In practical applications, if the actuator performance is allowed, one can choose $\check{\sigma}_i$ $(i = 1, 2, \dots, N)$ relatively small to make the formation error $\varsigma(t)$ as small as possible. If the formation error is small enough to meet the actual requirements, then the practical time-varying formation tracking is said to be accomplished by the swarm system.

4.2.3 Simulation Example

Consider a swarm system with 9 agents, where the leader is denoted by 0, and the followers are labeled by $1, 2, \dots, 8$. The graph of swarm system is shown in Fig. 4.1, where the weights of edges are assumed to be 0 or 1.

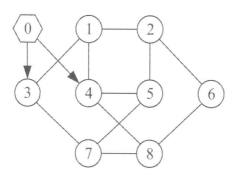

FIGURE 4.1: Interaction topology.

The dynamics models of the leader and the follower are described by (4.1) and (4.2) respectively, where $x_i(t) = [x_{i1}(t), x_{i2}(t), x_{i3}(t)]^T$, $u_i(t) = [u_{i1}(t), u_{i2}(t)]^T$, $A = \begin{bmatrix} 0 & 1 & 0 \\ 0 & 0 & 1 \\ -4 & -6 & -4 \end{bmatrix}$, $B = \begin{bmatrix} 0 & 1 \\ 0 & -1 \\ 1 & 0 \end{bmatrix}$. It is assumed that the time-varying input of leader is $u_0(t) = [1, 0.5\sin(t)]^T$ and $u_0(t)$ is unknown for all the followers. The parameter uncertainties and external disturbances of each follower are shown in TABLE 4.1. Moreover, assume that follower 3 has loss of effectiveness fault with $\rho_3(t) = \text{diag}\{0.6 - 0.2\sin(t), 0.4\}$. Follower 4 has both loss of effectiveness and output bias fault with $\rho_5(t) = \text{diag}\{0.5 + 0.2e^{-0.1t}, 0.3 + 0.1\cos(t)\}$ and $f_{b5}(t) = [-2e^{-0.1t}, 1 + 0.5\cos(t)]^T$.

TABLE 4.1: Unknown disturbances for each follower.

No.	Parameter uncertainty	External disturbance
F-1	$\Delta A_1 = \begin{bmatrix} 0 & 0.5\cos(t) & 0 \\ 0 & -0.5\cos(t) & 0 \\ \sin(t) & 0 & 0 \end{bmatrix}$	$d_1=[0.5,-0.5,\sin(t)]^T$
F-2	$\Delta A_2 = 0$	$d_2=[\cos(t),-\cos(t),0]^T$
F-3	$\Delta A_3 = \begin{bmatrix} 0 & 0 & 0 \\ 0 & 0 & 0 \\ 0 & 0 & e^{-0.1t} \end{bmatrix}$	$d_3=[\sin(t),-\sin(t),\cos(t)]^T$
F-4	$\Delta A_4 = 0$	$d_4=[e^{-0.1t},-e^{-0.1t},1]^T$
F-5	$\Delta A_5 = \begin{bmatrix} 0.5\sin(t) & 0 & -\cos(t) \\ -0.5\sin(t) & 0 & \cos(t) \\ 0 & 0 & 0 \end{bmatrix}$	$d_5=[\cos(t),-\cos(t),2e^{-0.2t}]^T$
F-6	$\Delta A_6 = 0$	$d_6=[\sin(t),-\sin(t),-1]^T$
F-7	$\Delta A_7 = \begin{bmatrix} -0.5e^{-0.2t} & 0 & 0 \\ 0.5e^{-0.2t} & 0 & 0 \\ 0 & 0 & -\sin(t) \end{bmatrix}$	$d_7=[0,0,-\cos(t)]^T$
F-8	$\Delta A_8 = 0$	$d_8=[-0.5,0.5,e^{-0.1t}]^T$

The followers need to achieve a time-varying regular octagon formation tracking, and the expected formation vector $h_F(t) = [h_1^T(t), h_2^T(t), \ldots, h_8^T(t)]^T$ is defined as

$$h_i(t) = \begin{bmatrix} r\sin(\varpi t + (i-1)\pi/4) \\ r\varpi\cos(\varpi t + (i-1)\pi/4) \\ -r\varpi^2\sin(\varpi t + (i-1)\pi/4) \end{bmatrix}, \quad i = 1,2,\ldots,8.$$

Let $r = 1$ and $\varpi = 1$. When $h_F(t)$ is achieved, these eight followers locate respectively at the eight vertexes of a regular octagon and rotate at the angular speed of 1 rad/s while tracking the state trajectory of the leader.

Let $\tilde{B} = \begin{bmatrix} 0 & 0 & 1 \\ 1 & 0 & 0 \end{bmatrix}$ and $\bar{B} = [\,1 \;\; 1 \;\; 0\,]$ such that $\tilde{B}B = I_2$ and $\bar{B}B = 0$. It can be verified that the time-varying formation tracking feasibility condition (4.6) is satisfied for all the followers. From (4.7), the formation compensation input is described as $v_i = [-5\cos(t + (i-1)\pi/4),0]^T$, $i = 1,2,\ldots,8$. Let $Q = 0.5I_3$ and solving the Riccati equation (4.8) gives $P = \begin{bmatrix} 0.7722 & 0.5931 & 0.0581 \\ 0.5931 & 1.0268 & 0.1236 \\ 0.0581 & 0.1236 & 0.0918 \end{bmatrix}$. The formation controller in Theorem 4.2 is applied in this simulation. For follower i ($i = 1,2,\ldots,8$), let $\check{\sigma}_i = 0.005$, $\eta_{1i} = 1$, $\eta_{2i} = 1$, $\eta_{3i} = 1$, $\hat{c}_i(0) = 2$, $\hat{\alpha}_i(0) = 2$, and $\hat{\beta}_i(0) = 2$. The initial states of each agent are generated by random numbers within the interval $[-1,1]$.

Fig. 4.2 depicts the state snapshots of the eight followers and the leader at different time instants, where these eight followers are represented by the

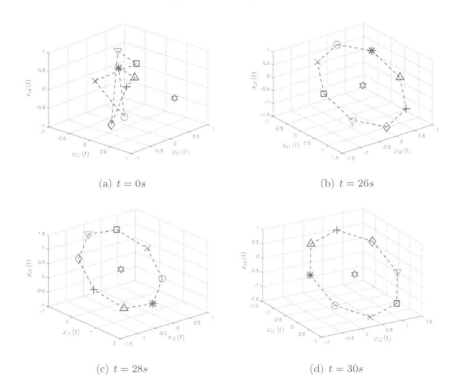

(a) $t = 0s$

(b) $t = 26s$

(c) $t = 28s$

(d) $t = 30s$

FIGURE 4.2: State snapshots of the eight followers and leader at different time instants.

plus sign, diamond, downward-pointing triangle, square, cross, circle, asterisk and upward-pointing triangle respectively, and the leader is denoted by the six-pointed star. Fig. 4.3 shows the state trajectory of the leader within 30s, where the initial state $x_0(0)$ is denoted by the circle and the final state $x_0(30)$ is represented by the five-pointed star. The two norm of the time-varying formation tracking errors of eight followers are shown in Fig. 4.4. The adaptive parameters $\hat{c}_i(t)$, $\hat{\alpha}_i(t)$, and $\hat{\beta}_i(t)$ $(i = 1, 2, \ldots, 8)$ are shown in Fig. 4.5. From Figs. 4.2-4.4, one sees that the followers achieve the regular octagon formation tracking and the leader's state $x_0(t)$ lies in the centre of the octagon. Moreover, the achieved octagon formation rotates around the leader and the edge length is time-varying. As shown in Fig. 4.5, all the adaptive parameters are ultimately bounded. Therefore, under the proposed controller (4.5), practical time-varying regular octagon formation robust tracking with bounded errors is accomplished by weak heterogeneous swarm system.

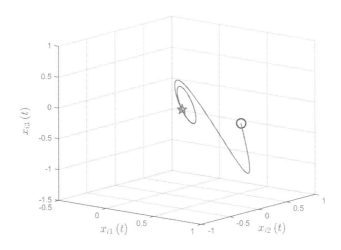

FIGURE 4.3: Trajectory of the leader $x_0(t)$ within $t = 30$s.

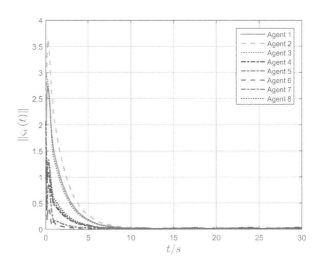

FIGURE 4.4: Time-varying formation tracking errors.

4.3 Finite-time Formation Tracking with Mismatched Disturbances

In the last section, the unknown disturbances for the followers are required to satisfy the given matched condition. However, in some practical scenes,

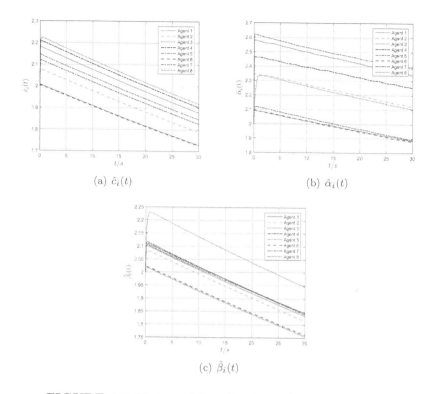

(a) $\hat{c}_i(t)$

(b) $\hat{\alpha}_i(t)$

(c) $\hat{\beta}_i(t)$

FIGURE 4.5: History of the adaptive updating parameters.

there may exist mismatched disturbances which affect the dynamics of each agent different from the channels of control inputs, such as magnetic levitation vehicle systems and missile systems. Since the mismatched disturbances cannot be compensated by the control inputs directly, it is more difficult to realize the expected formation tracking for swarm systems with mismatched disturbances. Moreover, the formation tracking controller in Section 4.2 only focuses on asymptotic stability, which means that the expected formation tracking cannot be realized in finite time. As indicated in [128, 142, 143], finite-time control has faster convergence rate, higher accuracy, and better robust properties. Thus, finite-time formation tracking problems for high-order swarm systems with mismatched disturbances will be further studied in this section. Firstly, a disturbance observer is designed for each follower to estimate the disturbances in finite time. Then, based on the homogeneous finite-time control, the integral sliding mode control, and the super-twisting algorithm, a distributed formation tracking protocol is presented utilizing the neighbouring interaction. Furthermore, it is proved that the desired formation tracking can be realized in finite time by swarm systems in the presence of mismatched disturbances and the leader's unknown input.

4.3.1   Problem Description

Consider a swarm system with one leader and N followers, where the leader is denoted by 0 and the followers are represented by $1, 2, \ldots, N$. The interaction topology of the swarm system is describe by a directed graph $\bar{\mathcal{G}}$.

Assumption 4.5. *The graph $\bar{\mathcal{G}}$ has a spanning tree rooted by the leader.*

Similar to Section 4.2, the Laplacian matrix $\bar{L}$ of graph $\bar{\mathcal{G}}$ can be divided into $\bar{L} = \begin{bmatrix} 0 & 0_{1 \times N} \\ L_2 & L_1 \end{bmatrix}$, where $L_1 \in \mathbb{R}^{N \times N}$ and $L_2 \in \mathbb{R}^{N \times 1}$. According to Lemma 2.2, all eigenvalues of L_1 have positive real parts under Assumption 4.5.

In this section, the followers will be classified into informed ones and uninformed ones as the following definition.

Definition 4.2. *A follower is said to be informed if its neighbour set contains the leader. Let F_I represent the set of the informed followers. The rest of the followers are called the uninformed ones, which are denoted by the set F_U.*

The dynamics of follower i ($i \in \{1, 2, \ldots, N\}$) is described by

$$\begin{cases} \dot{x}_{i,1}(t) = x_{i,2}(t) + d_{i,1}(t), \\ \dot{x}_{i,2}(t) = x_{i,3}(t) + d_{i,2}(t), \\ \qquad \ldots\ldots \\ \dot{x}_{i,n}(t) = u_i(t) + d_{i,n}(t), \\ y_i(t) = x_{i,1}(t), \end{cases} \tag{4.38}$$

where $x_{i,l}(t) \in \mathbb{R}$ ($l = 1, 2, \ldots, n$) is the state, $u_i(t) \in \mathbb{R}$ is the control input, $y_i(t) \in \mathbb{R}$ is the output, $d_{i,1}(t), d_{i,2}(t), \ldots, d_{i,n-1}(t)$ denote the mismatched disturbances, and $d_{i,n}(t)$ represents the matched disturbance. The leader is modeled as

$$\begin{cases} \dot{x}_{0,l}(t) = x_{0,l+1}(t), \ l = 1, 2, \ldots, n-1, \\ \dot{x}_{0,n}(t) = u_0(t), \\ y_0(t) = x_{0,1}(t), \end{cases} \tag{4.39}$$

where $x_{0,l}(t) \in \mathbb{R}$, $u_0(t) \in \mathbb{R}$, and $y_0(t) \in \mathbb{R}$ ($l = 1, 2, \ldots, n$) are the state, control input, and output of the leader respectively.

The control input $u_0(t)$ of the leader is also assumed to be unknown to all the followers in this section. Then, the leader can represent a non-cooperative target whose control input is usually unavailable.

Assumption 4.6. *The leader's control input $u_0(t)$ and its derivative $\dot{u}_0(t)$ are globally bounded, and there exist positive constants δ and ε such that $|u_0(t)| \leqslant \delta$ and $|\dot{u}_0(t)| \leqslant \varepsilon$.*

The desired time-varying output formation of each follower is described by a vector $h_F(t) = [h_1(t), h_2(t), \ldots, h_N(t)]^T$, where $h_i(t) \in \mathbb{R}$ $(i = 1, 2, \ldots, N)$ are n times differentiable. Note that the expected formation is realized by the physical movement of each follower with n-order dynamics. Thus, it is reasonable to have the n times differentiability requirement on $h_i(t)$.

Definition 4.3. *Consider the swarm system composed of (4.38) and (4.39). For any given bounded initial states, if there exists a finite time $T > 0$ such that*

$$\begin{cases} \lim_{t \to T} (y_i(t) - h_i(t) - y_0(t)) = 0, \ i = 1, 2, \ldots, N, \\ y_i(t) - h_i(t) - y_0(t) = 0, \ \forall t \geqslant T, \end{cases} \tag{4.40}$$

then the swarm system is said to achieve the expected finite-time time-varying output formation tracking.

This section aims to present a distributed formation control protocol for high-order swarm system such that the desired time-varying output formation tracking can be accomplished in finite time under the influences of the mismatched disturbances and the leader's unknown input.

The following lemma gives an approach to design finite-time disturbance observer.

Lemma 4.1 ([144]). *Consider the following system:*

$$\dot{x}(t) = u(t) + \vartheta(t), \tag{4.41}$$

where $x(t) \in \mathbb{R}$, $u(t) \in \mathbb{R}$, and $\vartheta(t) \in \mathbb{R}$ are the state, control input, and external disturbance respectively. Suppose that $\vartheta(t)$ is p times differentiable and $\vartheta^{(p)}(t)$ has a known Lipschitz constant l. Construct the following disturbance observer:

$$\begin{aligned} \varphi_0(t) &= -\lambda_0 l^{\frac{1}{p+1}} \operatorname{sig}^{\frac{p}{p+1}} (\xi_0(t) - x(t)) + \xi_1(t), \\ \dot{\xi}_0(t) &= u(t) + \varphi_0(t), \\ \varphi_k(t) &= -\lambda_k l^{\frac{1}{p+1-k}} \operatorname{sig}^{\frac{p-k}{p+1-k}} (\xi_k(t) - \varphi_{k-1}(t)) + \xi_{k+1}(t), \\ \dot{\xi}_k(t) &= \varphi_k(t), k = 1, 2, \ldots, p-1, \\ \dot{\xi}_p(t) &= -\lambda_p l \operatorname{sgn}(\xi_p(t) - \varphi_{p-1}(t)), \end{aligned} \tag{4.42}$$

where $\lambda_0, \lambda_1, \ldots, \lambda_p$ denote positive observer gains. $\xi_0(t) = \hat{x}(t)$, $\xi_1(t) = \hat{\vartheta}(t)$, $\xi_2(t) = \dot{\hat{\vartheta}}(t)$, $\ldots$, $\xi_p(t) = \hat{\vartheta}^{(p-1)}(t)$ are the estimates of $x(t), \vartheta(t), \dot{\vartheta}(t), \ldots, \vartheta^{(p-1)}(t)$. Choose sufficiently large gains $\lambda_0, \lambda_1, \ldots, \lambda_p$. Then, the disturbance observer (4.42) is finite-time convergent.

The following lemma summarizes the results about the super-twisting algorithm (STA) in [145].

Lemma 4.2 ([145]). *Consider the following system:*

$$\begin{cases} \dot{y}_1(t) = -k_1 \text{sig}^{1/2}\left(y_1(t)\right) + y_2(t), \\ \dot{y}_2(t) = -k_2 \,\text{sgn}\left(y_1(t)\right) + \rho(t), \end{cases} \tag{4.43}$$

where $y_1(t) \in \mathbb{R}$ and $y_2(t) \in \mathbb{R}$ stand for the states, k_1 and k_2 are gains to be ascertained, and $\rho(t)$ is the perturbation term. Suppose that $\rho(t)$ is globally bounded and there exists a positive constant γ such that $|\rho(t)| \leqslant \gamma$. For any given γ, there are gains k_1 and k_2 such that the origin is globally finite-time stable. The gains k_1 and k_2 can be chosen by the following algorithm:
i) Select positive constants β_1 and β_2 satisfying $0 < \beta_1 < 1$, $\beta_2 > 1$, $\beta_1\beta_2 > 1$.
ii) Choose appropriate positive constants χ_1 and χ_2 such that

$$\chi_1 - \frac{2}{\beta_2}\chi_2 > \chi_2^2 - \beta_1\left(1 + \chi_1\right)\chi_2 + \frac{1}{4}\left(1 + \chi_1\right)^2.$$

iii) The gains k_1 and k_2 are chosen as $k_1 = \chi_1\sqrt{\frac{2\beta_2}{(1-\beta_1)\chi_2}}\sqrt{\gamma}$, $k_2 = \frac{1+\beta_1}{1-\beta_1}\gamma$.

A homogeneous finite-time control approach for high-order system is summarized as follows.

Lemma 4.3 ([146]). *The following n-order system is considered:*

$$\begin{cases} \dot{z}_l(t) = z_{l+1}(t), \; l = 1, 2, \ldots, n-1, \\ \dot{z}_n(t) = u(t), \end{cases} \tag{4.44}$$

where $z_l(t) \in \mathbb{R}$ ($l = 1, 2, \ldots, n$) is the state, $u(t) \in \mathbb{R}$ is the control input. Construct the following controller:

$$u(t) = -\sum_{l=1}^{n} c_l \text{sig}^{\beta_l}\left(z_l(t)\right), \tag{4.45}$$

where c_l ($l = 1, 2, \ldots, n$) are positive constants to make the polynomial $x^n + c_n x^{n-1} + \cdots + c_2 x + c_1$ Hurwitz, and $\beta_{l-1} = \frac{\beta_l\beta_{l+1}}{2\beta_{l+1}-\beta_l}$ ($l = 2, 3, \ldots, n$), $\beta_{n+1} = 1$. There exists a positive scale $0 < \sigma < 1$ such that the origin of system (4.44) is a globally finite-time stable equilibrium for any $1 - \sigma < \beta_n < 1$ by the controller (4.45).

4.3.2 Finite-time Controller Design and Stability Analysis

It is assumed that the unknown disturbances of each follower satisfy the following condition.

Assumption 4.7. *The disturbance $d_{i,l}(t)$ ($i \in \{1, 2, \ldots, N\}$, $l \in \{1, 2, \ldots, n\}$) is $n - l + 1$ times differentiable, and there is a Lipschitz constant $L_{i,l}$ for $d_{i,l}^{(n-l+1)}(t)$.*

Assumption 4.7 is applied widely to deal with mismatched disturbances in the existing works (see, e.g., [144, 147, 148]). Many types of disturbances, including constant, ramp, and sinusoidal signals, satisfy this assumption.

For follower i ($i \in \{1, 2, \ldots, N\}$), construct the following disturbance observer:

$$
\begin{cases}
\varphi_{i,l}^0(t) = -\lambda_{i,l}^0 (L_{i,l})^{\frac{1}{n-l+2}} \operatorname{sig}^{\frac{n-l+1}{n-l+2}} (\xi_{i,l}^0(t) - x_{i,l}(t)) + \xi_{i,l}^1(t), \\
\dot{\xi}_{i,l}^0(t) = f_{i,l}(t) + \varphi_{i,l}^0(t), \\
\varphi_{i,l}^k(t) = -\lambda_{i,l}^k (L_{i,l})^{\frac{1}{n-l+2-k}} \operatorname{sig}^{\frac{n-l+1-k}{n-l+2-k}} (\xi_{i,l}^k(t) - \varphi_{i,l}^{k-1}(t)) + \xi_{i,l}^{k+1}(t), \\
\dot{\xi}_{i,l}^k(t) = \varphi_{i,l}^k(t), \quad k = 1, 2, \ldots, n-l, \\
\dot{\xi}_{i,l}^{n-l+1}(t) = -\lambda_{i,l}^{n-l+1} L_{i,l} \operatorname{sgn}(\xi_{i,l}^{n-l+1}(t) - \varphi_{i,l}^{n-l}(t)),
\end{cases} \tag{4.46}
$$

where $f_{i,l}(t) = x_{i,l+1}(t)$ ($l = 1, 2, \ldots, n-1$) and $f_{i,n}(t) = u_i(t)$. $\lambda_{i,l}^0$, $\lambda_{i,l}^1$, ..., and $\lambda_{i,l}^{n-l+1}$ ($l = 1, 2, \ldots, n$) denote the positive observer gains. $\xi_{i,l}^0(t) = \hat{x}_{i,l}(t)$, $\xi_{i,l}^1(t) = \hat{d}_{i,l}(t)$, ..., $\xi_{i,l}^{n-l+1}(t) = \hat{d}_{i,l}^{(n-l)}(t)$ are the estimates of $x_{i,l}(t), d_{i,l}(t), \ldots, d_{i,l}^{(n-l)}(t)$.

Lemma 4.4. *Suppose that Assumption 4.7 holds. For high-order system (4.38), the disturbance observer (4.46) is convergent in a finite time T_1 by choosing sufficiently large observer gains $\lambda_{i,l}^0$, $\lambda_{i,l}^1$, ..., $\lambda_{i,l}^{n-l+1}$ ($i = 1, 2, \ldots, N$, $l = 1, 2, \ldots, n$).*

Proof. In light of Lemma 4.1, we can get the finite-time convergence of the observer (4.46) directly. So the detailed proof is omitted here. □

For follower i, let $e_{i,l}(t)$ ($i \in \{1, 2, \ldots, N\}$, $l = 1, 2, \ldots, n$) denote the local formation errors relative to its neighbours in the state components $x_{i,l}(t)$ respectively. $e_{i,l}(t)$ are defined as

$$
\begin{cases}
e_{i,1} = w_{i0}(x_{i,1} - h_i - x_{0,1}) \\
\qquad + \displaystyle\sum_{j=1}^{N} w_{ij}((x_{i,1} - h_i) - (x_{j,1} - h_j)), \\
e_{i,l} = w_{i0}\left(x_{i,l} - h_i^{(l-1)} + \displaystyle\sum_{m=1}^{l-1} \hat{d}_{i,m}^{(l-1-m)} - x_{0,l}\right) \\
\qquad + \displaystyle\sum_{j=1}^{N} w_{ij}\left[\left(x_{i,l} - h_i^{(l-1)} + \displaystyle\sum_{m=1}^{l-1} \hat{d}_{i,m}^{(l-1-m)}\right)\right. \\
\qquad\qquad\left. - \left(x_{j,l} - h_j^{(l-1)} + \displaystyle\sum_{m=1}^{l-1} \hat{d}_{j,m}^{(l-1-m)}\right)\right], \\
l = 2, 3, \ldots, n,
\end{cases} \tag{4.47}
$$

where $w_{i0} > 0$ if $i \in F_I$, and $w_{i0} = 0$ if $i \in F_U$. Each follower is required to interact the state, the desired formation, and the mismatched disturbance estimate information with its neighbours.

For follower i $(i = 1, 2, \ldots, N)$, construct the following time-varying formation tracking protocol:

$$u_i(t) = u_{bi}(t) - \sum_{m=1}^{n-1} \dot{\hat{d}}_{i,m}^{(n-m-1)}(t) - \hat{d}_{i,n}(t) + h_i^{(n)}(t), \qquad (4.48)$$

where $\dot{\hat{d}}_{i,m}^{(n-m-1)}(t) = \varphi_{i,m}^{n-m}(t)$ $(m = 1, 2, \ldots, n-1)$ are calculated by the disturbance observer (4.46). For uninformed followers, $u_{bi}(t)$ $(i \in F_U)$ is designed as

$$u_{bi}(t) = \frac{1}{\displaystyle\sum_{j=0}^{N} w_{ij}} \left(-\sum_{l=1}^{n} c_{i,l}\text{sig}^{\beta_{i,l}}\left(e_{i,l}(t)\right) + \sum_{j=1}^{N} w_{ij}u_{bj}(t) \right).$$

For informed followers, $u_{bi}(t)$ $(i \in F_I)$ is determined by

$$\begin{cases} u_{bi}(t) = \dfrac{1}{\displaystyle\sum_{j=0}^{N} w_{ij}} \left(\sum_{l=1}^{n} c_{i,l}\text{sig}^{\beta_{i,l}}\left(e_{i,l}(t)\right) + \sum_{j=1}^{N} w_{ij}u_{bj}(t) \right) \\ \qquad + \dfrac{w_{i0}}{\displaystyle\sum_{j=0}^{N} w_{ij}} \left(-k_{1i}\text{sig}^{1/2}\left(s_i(t)\right) + g_i(t) \right), \\ \dot{g}_i(t) = -k_{2i}\,\text{sgn}\left(s_i(t)\right), \ i \in F_I. \end{cases}$$

The sliding-mode surface $s_i(t)$ $(i \in F_I)$ is defined as

$$s_i(t) = e_{i,n}(t) + \int_0^t \sum_{l=1}^{n} c_{i,l}\text{sig}^{\beta_{i,l}}\left(e_{i,l}(\tau)\right)\mathrm{d}\tau. \qquad (4.49)$$

The constants $c_{i,l}$ and $\beta_{i,l}$ $(i \in \{1, 2, \ldots, N\},\ l = 1, 2, \ldots, n)$ are determined by Lemma 4.3. The gains k_{1i} and k_{2i} $(i \in F_I)$ of STA are chosen by the algorithm in Lemma 4.2 with $\gamma = w_{i0}\varepsilon$, where ε is the upper bound of $|\dot{u}_0(t)|$.

In the formation tracking protocol (4.48), the first term denotes the control input dependent on the neighbouring interaction to achieve the desired formation tracking, the second and third terms represent the disturbance compensation inputs, and the fourth term stands for the time-varying formation compensation input. Taking advantage of the integral sliding mode in (4.49) and the super-twisting algorithm, a continuous compensation term is constructed in $u_{bi}(t)$ $(i \in F_I)$ for each informed follower to compensate for the leader's unknown input.

The following theorem shows the finite-time stability of the closed-loop system.

Theorem 4.3. *Suppose that Assumptions 4.5-4.7 hold. For the swarm system (4.38) and (4.39), the desired time-varying output formation tracking specified by $h_F(t)$ can be realized in finite time under the distributed protocol (4.48).*

Proof. Let $x_l(t) = [x_{1,l}(t), x_{2,l}(t), \ldots, x_{N,l}(t)]^T$, $d_l(t) = [d_{1,l}(t), d_{2,l}(t), \ldots, d_{N,l}(t)]^T$, $e_l(t) = [e_{1,l}(t), e_{2,l}(t), \ldots, e_{N,l}(t)]^T$ $(l = 1, 2, \ldots, n)$, $u(t) = [u_1(t), u_2(t), \ldots, u_N(t)]^T$, and $y(t) = [y_1(t), y_2(t), \ldots, y_N(t)]^T$. It follows from (4.47) that $e_1(t) = L_1(y(t) - h(t) - \mathbf{1}_N \otimes y_0(t))$. Since L_1 is a nonsingular matrix under Assumption 4.5, we have that $y_i(t) - h_i(t) - y_0(t) = 0$ $(i = 1, 2, \ldots, N)$ if and only if $e_1(t) = 0$. It follows from (4.38), (4.39), and (4.47) that

$$\begin{cases} \dot{e}_1 = e_2 + L_1(d_1 - \hat{d}_1), \\ \dot{e}_l = e_{l+1} + L_1(d_l - \hat{d}_l) + L_1 \sum_{m=1}^{l-1} (\dot{\hat{d}}_m^{(l-m-1)} - \hat{d}_m^{(l-m)}), \; l = 2, 3, \ldots, n-1, \\ \dot{e}_n = L_1(u + d_n - h^{(n)} + \sum_{m=1}^{n-1} \dot{\hat{d}}_m^{(n-m-1)} - \mathbf{1}_N \otimes u_0). \end{cases}$$

$$(4.50)$$

Let $u_b(t) = [u_{b1}(t), u_{b2}(t), \ldots, u_{bN}(t)]^T$. Substituting the controller (4.48) into (4.50) gives $\dot{e}_n(t) = L_1(u_b(t) - \mathbf{1}_N \otimes u_0(t)) + L_1(d_n(t) - \hat{d}_n(t))$. Let $\tilde{q}_1(t) = L_1(d_1(t) - \hat{d}_1(t))$, $\tilde{q}_l(t) = L_1(d_l(t) - \hat{d}_l(t)) + L_1 \sum_{m=1}^{l-1} (\dot{\hat{d}}_m^{(l-m-1)}(t) - \hat{d}_m^{(l-m)}(t))$, $l = 2, 3, \ldots, n-1$, and $\tilde{q}_n(t) = L_1(d_n(t) - \hat{d}_n(t))$. Then, system (4.50) can be transformed to

$$\begin{cases} \dot{e}_l(t) = e_{l+1}(t) + \tilde{q}_l(t), \; l = 1, 2, \ldots, n-1, \\ \dot{e}_n(t) = L_1(u_b(t) - \mathbf{1}_N \otimes u_0(t)) + \tilde{q}_n(t). \end{cases}$$

$$(4.51)$$

From Lemma 4.4, we have that the disturbance observer (4.46) is convergent in a finite time T_1. Thus, $\tilde{q}_l(t)(l = 1, 2, \ldots, n)$ are bounded in $[0, T_1)$.

In the following, we will show the state boundedness of (4.51) in $[0, T_1)$ firstly. Then, it is proved that the desired formation tracking is realized in finite time after the observer (4.46) is convergent.

The proof of the state boundedness of (4.51) in $[0, T_1)$ is divided into the uninformed follower case and the informed follower case.

i) For uninformed follower i $(i \in F_U)$, substituting $u_{bi}(t)$ into (4.51) leads to

$$\begin{cases} \dot{e}_{i,l}(t) = e_{i,l+1}(t) + \tilde{q}_{i,l}(t), \; l = 1, 2, \ldots, n-1, \\ \dot{e}_{i,n}(t) = -\sum_{l=1}^{n} c_{i,l} \mathrm{sig}^{\beta_{i,l}} (e_{i,l}(t)) + \tilde{q}_{i,n}(t). \end{cases}$$

$$(4.52)$$

Construct the Lyapunov function $V_{ai}(t) = \frac{1}{2}\sum_{l=1}^{n} e_{i,l}^2(t)$. Taking the time derivative of $V_{ai}(t)$ along the trajectory of (4.52) gives

$$\dot{V}_{ai}(t) = \sum_{l=1}^{n-1} e_{i,l}(t)\left(e_{i,l+1}(t) + \tilde{q}_{i,l}(t)\right)$$

$$+ e_{i,n}(t)\left(-\sum_{l=1}^{n} c_{i,l}\mathrm{sig}^{\beta_{i,l}}\left(e_{i,l}(t)\right) + \tilde{q}_{i,n}(t)\right)$$

$$\leqslant \sum_{l=1}^{n-1}\left(\frac{1}{2}e_{i,l}^2(t) + \frac{1}{2}e_{i,l+1}^2(t)\right) + \sum_{l=1}^{n}|e_{i,l}(t)|\,|\tilde{q}_{i,l}(t)|$$

$$+ \sum_{l=1}^{n} c_{i,l}\,|e_{i,n}(t)|\,|e_{i,l}(t)|^{\beta_{i,l}}. \tag{4.53}$$

From Lemma 2.4, it can be verified that $|e_{i,l}(t)| \leqslant \frac{1}{2}\left(1 + e_{i,l}^2(t)\right)$, $|e_{i,n}(t)|\,|e_{i,l}(t)|^{\beta_{i,l}} \leqslant \frac{1}{1+\beta_{i,l}}|e_{i,n}(t)|^{1+\beta_{i,l}} + \frac{\beta_{i,l}}{1+\beta_{i,l}}|e_{i,l}(t)|^{1+\beta_{i,l}}$, $|e_{i,n}(t)|^{1+\beta_{i,l}} \leqslant \frac{1+\beta_{i,l}}{2}e_{i,n}^2(t) + \frac{1-\beta_{i,l}}{2}$, and $|e_{i,l}(t)|^{1+\beta_{i,l}} \leqslant \frac{1+\beta_{i,l}}{2}e_{i,l}^2(t) + \frac{1-\beta_{i,l}}{2}$. Note that $\sum_{l=1}^{n-1}\left(\frac{1}{2}e_{i,l}^2 + \frac{1}{2}e_{i,l+1}^2\right) \leqslant \sum_{l=1}^{n} e_{i,l}^2(t)$. Then, it follows from (4.53) that

$$\dot{V}_{ai}(t) \leqslant \sum_{l=1}^{n} e_{i,l}^2(t) + \sum_{l=1}^{n}|\tilde{q}_{i,l}|\frac{1}{2}e_{i,l}^2(t) + \frac{1}{2}e_{i,n}^2(t)\sum_{l=1}^{n} c_{i,l}$$

$$+ \sum_{l=1}^{n}\left(\frac{c_{i,l}\beta_{i,l}}{2}e_{i,l}^2(t)\right) + \sum_{l=1}^{n}\frac{|\tilde{q}_{i,l}|}{2} + \sum_{l=1}^{n}\frac{c_{i,l}(1-\beta_{i,l})}{2}. \tag{4.54}$$

Let $\bar{\tilde{q}}_i(t) = \max_{l=1,2,\dots,n}|\tilde{q}_{i,l}(t)|$, $\bar{c}_i = \max_{l=1,2,\dots,n}\{c_{i,l}\}$, and $\bar{\beta}_i = \max_{l=1,2,\dots,n}\{\beta_{i,l}\}$. From (4.54), we have

$$\dot{V}_{ai}(t) \leqslant \kappa_{ai}V_{ai}(t) + \delta_{ai}, \tag{4.55}$$

where $\kappa_{ai} = 2 + \bar{\tilde{q}}_i(t) + \sum_{l=1}^{n} c_{i,l} + \bar{c}_i\bar{\beta}_i$ and $\delta_{ai} = \sum_{l=1}^{n}\frac{|\tilde{q}_{i,l}(t)|}{2} + \sum_{l=1}^{n}\frac{c_{i,l}(1-\beta_{i,l})}{2}$. Since $|\tilde{q}_{i,l}(t)|$ $(l = 1, 2, \dots, n)$ are bounded in $[0, T_1)$, it holds from (4.55) that $e_{i,l}(t)$ $(i \in F_U, l = 1, 2, \dots, n)$ are bounded in $[0, T_1)$.

ii) For informed follower i $(i \in F_I)$, substituting $u_{bi}(t)$ into (4.51) gives

$$\begin{cases} \dot{e}_{i,l}(t) = e_{i,l+1}(t) + \tilde{q}_{i,l}(t),\ l = 1, 2, \dots, n-1, \\[2mm] \dot{e}_{i,n}(t) = -\sum_{l=1}^{n} c_{i,l}\mathrm{sig}^{\beta_{i,l}}\left(e_{i,l}(t)\right) - w_{i0}k_{1i}\mathrm{sig}^{1/2}\left(s_i(t)\right) \\[2mm] \qquad\quad + w_{i0}\left(g_i(t) - u_0(t)\right) + \tilde{q}_{i,n}(t), \\[2mm] \dot{g}_i(t) = -k_{2i}\,\mathrm{sgn}\left(s_i(t)\right). \end{cases} \tag{4.56}$$

Take the time derivative of $s_i(t)$. It follows from (4.49) that

$$\dot{s}_i(t) = \dot{e}_{i,n}(t) + \sum_{l=1}^{n} c_{i,l} \operatorname{sig}^{\beta_{i,l}} (e_{i,l}(t)). \qquad (4.57)$$

From (4.56) and (4.57), we can get

$$\begin{cases} \dot{s}_i(t) = -w_{i0}k_{1i}\operatorname{sig}^{1/2}(s_i(t)) + w_{i0}(g_i(t) - u_0(t)) + \tilde{q}_{i,n}(t), \\ \dot{g}_i(t) = -k_{2i}\operatorname{sgn}(s_i(t)). \end{cases} \qquad (4.58)$$

Consider the Lyapunov function $V_{bi}(t) = \frac{1}{2}s_i^2(t) + \frac{1}{2}g_i^2(t)$. Taking the time derivative of $V_{bi}(t)$ along the trajectory of (4.58) gives

$$\begin{aligned} \dot{V}_{bi}(t) = & -w_{i0}k_{1i}|s_i(t)|^{3/2} + w_{i0}s_i(t)g_i(t) + s_i(t)(-w_{i0}u_0(t) + \tilde{q}_{i,n}(t)) \\ & - k_{2i}g_i(t)\operatorname{sgn}(s_i(t)). \end{aligned} \qquad (4.59)$$

Based on Lemma 2.4, it holds from (4.59) that

$$\begin{aligned} \dot{V}_{bi}(t) \leqslant & w_{i0}\left(\frac{1}{2}s_i^2(t) + \frac{1}{2}g_i^2(t)\right) + k_{2i}\left(\frac{1}{2}g_i^2(t) + \frac{1}{2}\right) \\ & + |-w_{i0}u_0(t) + \tilde{q}_{i,n}(t)|\left(\frac{1}{2}s_i^2(t) + \frac{1}{2}\right). \end{aligned} \qquad (4.60)$$

Let $\bar{\mu}_i = \max\{|-w_{i0}u_0(t) + \tilde{q}_{i,n}(t)|, k_{2i}\}$. It can be verified from (4.60) that

$$\dot{V}_{bi}(t) \leqslant (w_{i0} + \bar{\mu}_i)V_{bi}(t) + \bar{\mu}_i. \qquad (4.61)$$

Since $u_0(t)$ and $\tilde{q}_{i,n}(t)$ are bounded in $[0, T_1)$, it follows from (4.61) that $V_{bi}(t)$ is bounded in $[0, T_1)$. So are $s_i(t)$ and $g_i(t)$ $(i \in F_I)$.

Let $\tilde{\rho}_i(t) = w_{i0}\left(-k_{1i}\operatorname{sig}^{1/2}(s_i(t)) + g_i(t) - u_0(t)\right) + \tilde{q}_{i,n}(t)$ $(i \in F_I)$. We can obtain that $\tilde{\rho}_i(t)$ is bounded in $[0, T_1)$. Then system (4.56) can be transformed to

$$\begin{cases} \dot{e}_{i,l}(t) = e_{i,l+1}(t) + \tilde{q}_{i,l}(t), \ l = 1, 2, \ldots, n-1, \\ \dot{e}_{i,n}(t) = -\sum_{l=1}^{n} c_{i,l}\operatorname{sig}^{\beta_{i,l}}(e_{i,l}(t)) + \tilde{\rho}_i(t). \end{cases} \qquad (4.62)$$

Consider the Lyapunov function $V_{ci}(t) = \frac{1}{2}\sum_{l=1}^{n} e_{i,l}^2(t)$. Follow the similar steps in (4.53)-(4.55). It can be verified that $e_{i,l}(t)$ $(i \in F_I, \ l = 1, 2, \ldots, n)$ are bounded in $[0, T_1)$.

Therefore, according to the above two cases, the states of the followers in (4.51) are bounded in $[0, T_1)$.

When $t \geqslant T_1$, the disturbance observer (4.46) is convergent. So we get that $\tilde{q}_l(t) = 0$ $(l = 1, 2, \ldots, n)$. Then system (4.51) becomes

$$\begin{cases} \dot{e}_l(t) = e_{l+1}(t), \ l = 1, 2, \ldots, n-1, \\ \dot{e}_n(t) = L_1(u_b(t) - \mathbf{1}_N \otimes u_0(t)). \end{cases} \qquad (4.63)$$

i) For uninformed follower i ($i \in F_U$), it follows from (4.52) that

$$\begin{cases} \dot{e}_{i,l}(t) = e_{i,l+1}(t), \ l = 1, 2, \ldots, n-1, \\ \dot{e}_{i,n}(t) = -\sum_{l=1}^{n} c_{i,l} \mathrm{sig}^{\beta_{i,l}}(e_{i,l}(t)). \end{cases} \tag{4.64}$$

ii) For informed follower i ($i \in F_I$), it can be verified from (4.56) and (4.57) that

$$\begin{cases} \dot{e}_{i,l}(t) = e_{i,l+1}(t), \ l = 1, 2, \ldots, n-1, \\ \dot{e}_{i,n}(t) = -\sum_{l=1}^{n} c_{i,l} \mathrm{sig}^{\beta_{i,l}}(e_{i,l}(t)) + \dot{s}_i(t), \end{cases} \tag{4.65}$$

where $\dot{s}_i(t)$ ($i \in F_I$) is described by

$$\begin{cases} \dot{s}_i(t) = -w_{i0}k_{1i}\mathrm{sig}^{1/2}(s_i(t)) + w_{i0}(g_i(t) - u_0(t)), \\ \dot{g}_i(t) = -k_{2i}\,\mathrm{sgn}(s_i(t)). \end{cases} \tag{4.66}$$

Let $\bar{g}_i(t) = w_{i0}(g_i(t) - u_0(t))$, $\bar{k}_{1i} = w_{i0}k_{1i}$, and $\bar{k}_{2i} = w_{i0}k_{2i}$ ($i \in F_I$). It follows from (4.66) that

$$\begin{cases} \dot{s}_i(t) = -\bar{k}_{1i}\mathrm{sig}^{1/2}(s_i(t)) + \bar{g}_i(t), \\ \dot{\bar{g}}_i(t) = -\bar{k}_{2i}\,\mathrm{sgn}(s_i(t)) - w_{i0}\dot{u}_0(t), \end{cases} \tag{4.67}$$

where $|w_{i0}\dot{u}_0(t)| \leqslant w_{i0}\varepsilon$ under Assumption 4.6. Note that (4.67) is the standard STA. According to Lemma 4.2, the states of all the informed followers will reach the sliding-mode surfaces $s_i(t) = 0$ ($i \in F_I$) in a finite time T_2. On $s_i(t) = 0$ ($i \in F_I$), it can be verified that (4.65) is simplified to (4.64) for each informed follower.

Therefore, when $t \geqslant T_1 + T_2$, we can obtain that all the followers satisfy (4.64). In light of Lemma 4.3, there exist $0 < \sigma_i < 1$ such that $e_{i,l}(t) \to 0$ ($i = 1, 2, \ldots, N$, $l = 1, 2, \ldots, n$) in finite time. Since $e_1(t) = L_1(y(t) - h(t) - \mathbf{1}_N \otimes y_0(t))$ and L_1 is nonsingular, we have that $\tilde{y}_i(t) = y_i(t) - h_i(t) - y_0(t) \to 0$ ($i = 1, 2, \ldots, N$) in finite time. Thus, the finite-time time-varying output formation tracking is realized by the swarm system (4.38) and (4.39) under the proposed control protocol (4.48). This completes the proof of Theorem 4.3. □

Based on the finite-time disturbance observer (4.46), a continuous finite-time time-varying formation tracking protocol is designed as (4.48) for the followers subject to mismatched disturbances and the leader's unknown input. Since the mismatched disturbances affect the dynamics of each follower different from the channels of control inputs, the influences of mismatched disturbances cannot be suppressed by the control input directly. We need to construct the neighbouring errors $e_{i,l}(t)$ as (4.47) based on the interaction of the state, the desired formation, and the mismatched disturbance estimate information, which makes the formation tracking problems considered here more complicated. Moreover, the control approaches in [148] are discontinuous, which will lead to the chattering of control inputs. In light of the homogeneous finite-time control, the integral sliding mode control, and the super-twisting algorithm, the presented controller (4.48) in this section can guarantee both finite-time convergence of the formation tracking errors and large chattering avoidance of the control inputs.

4.3.3 Simulation Example

Consider a third-order swarm system with five followers labelled by $1, 2, \ldots, 5$ and one leader denoted by 0. The directed graph with 0-1 weights is shown in Fig. 4.6, where we can see that $F_I = \{1, 2\}$ and $F_U = \{3, 4, 5\}$. The two dimensional plane (i.e., the XY plane) is considered in this example, and the dynamics of each agent in X-axis and Y-axis can be modeled as (4.38) and (4.39), respectively. Let $x_{i,l}(t) = [x_{i,l}^X(t), x_{i,l}^Y(t)]^T$, $u_i(t) = [u_i^X(t), u_i^Y(t)]^T$, and $y_i(t) = [y_i^X(t), y_i^Y(t)]^T$ $(i = 0, 1, \ldots, 5, \; l = 1, 2, 3)$.

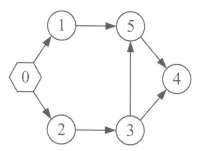

FIGURE 4.6: Directed graph.

The leader's unknown input is set as $u_0^X(t) = 0.5 \sin(t)$ and $u_0^Y(t) = 0.5 \cos(t)$, which satisfies Assumption 4.6. Consider the following disturbances for each follower:

Follower 1: $d_{1,1}^X = -\cos(t)$, $d_{1,2}^X = \sin(2t)$, $d_{1,3}^X = 1$, $d_{1,1}^Y = 1 - \sin(2t)$, $d_{1,2}^Y = 0.5 \cos(2t)$, $d_{1,3}^Y = 1 + \sin(t)$.
Follower 2: $d_{2,1}^X = 2 \sin(0.5t)$, $d_{2,2}^X = -1$, $d_{2,3}^X = 1 + \sin(t)$, $d_{2,1}^Y = 2 \cos(0.5t)$, $d_{2,2}^Y = 1.5$, $d_{2,3}^Y = -1$.
Follower 3: $d_{3,1}^X = 1 + \sin(t)$, $d_{3,2}^X = 1.5$, $d_{3,3}^X = 1$, $d_{3,1}^Y = -1 + 2 \sin(t)$, $d_{3,2}^Y = -\cos(t)$, $d_{3,3}^Y = 1$.
Follower 4: $d_{4,1}^X = -\sin(t)$, $d_{4,2}^X = \cos(2t)$, $d_{4,3}^X = 1 - 0.5 \sin(2t)$, $d_{4,1}^Y = -\cos(1.5t)$, $d_{4,2}^Y = -2$, $d_{4,3}^Y = \sin(0.5t)$.
Follower 5: $d_{5,1}^X = 1 + 2 \cos(t)$, $d_{5,2}^X = -1$, $d_{5,3}^X = \sin(3t)$, $d_{5,1}^Y = 2$, $d_{5,2}^Y = -\sin(2t)$, $d_{5,3}^Y = 1 - \cos(t)$.
It can be verified that $d_{i,l}(t)$ $(i = 1, 2, \ldots, 5, \; l = 1, 2, 3)$ satisfy Assumption 4.7.

The expected time-varying output formation for the followers is described by $h_i(t) = \begin{bmatrix} 2\cos(0.5t + 2(i-1)\pi/5) \\ 2\sin(0.5t + 2(i-1)\pi/5) \end{bmatrix}$, $i = 1, 2, \ldots, 5$. When $h(t)$ is achieved, the outputs of these five followers will form a regular pentagon centred by the leader and rotate around the leader. For follower i, the disturbance observers (4.46) are designed as $L_{1,1}^X = 2$, $L_{1,2}^X = 10$, $L_{1,3}^X = 1$, $L_{1,1}^Y = 16$, $L_{1,2}^Y = 4$, $L_{1,3}^Y = 2$, $L_{2,1}^X = 1$, $L_{2,2}^X = 1$, $L_{2,3}^X = 2$, $L_{2,1}^Y = 1$, $L_{2,2}^Y = 1$, $L_{2,3}^Y = 1$, $L_{3,1}^X = 2$, $L_{3,2}^X = 1$, $L_{3,3}^X = 1$, $L_{3,1}^Y = 4$, $L_{3,2}^Y = 2$, $L_{3,3}^Y = 1$, $L_{4,1}^X = 2$, $L_{4,2}^X = 8$, $L_{4,3}^X = 2$, $L_{4,1}^Y = 10$, $L_{4,2}^Y = 1$, $L_{4,3}^Y = 1$, $L_{5,1}^X = 5$, $L_{5,2}^X = 1$, $L_{5,3}^X = 6$, $L_{5,1}^Y = 1$, $L_{5,2}^Y = 8$, $L_{5,3}^Y = 2$, $\lambda_{i,1}^0 = 10$, $\lambda_{i,1}^1 = 10$, $\lambda_{i,1}^2 = 8$, $\lambda_{i,1}^3 = 5$, $\lambda_{i,2}^0 = 10$, $\lambda_{i,2}^1 = 8$, $\lambda_{i,2}^2 = 5$, $\lambda_{i,3}^0 = 8$, $\lambda_{i,3}^1 = 5$. The control parameters in the formation tracking controller (4.48) are chosen as $c_{i,1} = 8$, $c_{i,2} = 12$, $c_{i,3} = 6$, $\beta_{i,1} = \frac{4}{7}$, $\beta_{i,2} = \frac{2}{3}$, $\beta_{i,3} = \frac{4}{5}$ $(i = 1, 2, \ldots, 5)$, $k_{1j} = 6$, $k_{2j} = 3$ $(j \in F_I)$. For simplicity, the initial states $x_{i,l}(0)$ $(i = 0, 1, \ldots, 5, \ l = 1, 2, 3)$ of each agent are generated by random numbers within the interval $[-2, 2]$, and the initial values of the disturbance observer (4.46) and $g_i(t)$ $(i \in F_I)$ are set to be zero.

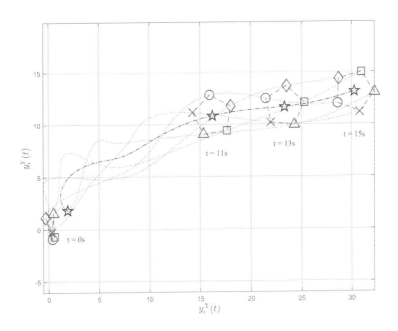

FIGURE 4.7: Output trajectories within 15s and snapshots at $t = 0, 11, 13, 15$s for the leader and the five followers.

Fig. 4.7 gives the output trajectories within 15s and the snapshots at $t = 0, 11, 13, 15$s of the leader and the five followers, in which the leader's

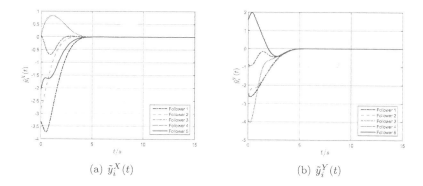

(a) $\tilde{y}_i^X(t)$　　　　　　　　　(b) $\tilde{y}_i^Y(t)$

FIGURE 4.8: Output formation tracking errors of each follower.

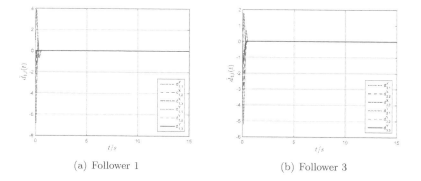

(a) Follower 1　　　　　　　　　(b) Follower 3

FIGURE 4.9: Disturbance estimate errors in the disturbance observer (4.46).

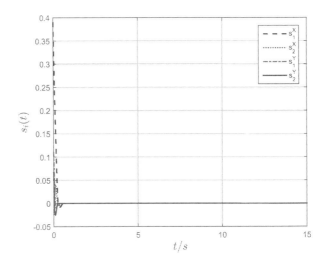

FIGURE 4.10: Sliding-mode surfaces $s_i(t)$ for the informed followers.

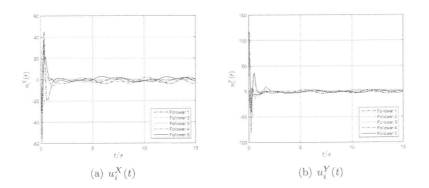

(a) $u_i^X(t)$ (b) $u_i^Y(t)$

FIGURE 4.11: Control inputs of the five followers.

trajectory is denoted by the dash-dot line and the followers' trajectories are displayed by the dotted lines. The leader is depicted by the pentagram, and the five followers are marked by the "□", "◊", "○", "×", and "△", respectively. Fig. 4.8 gives the output formation tracking errors of the followers. Taking followers 1 and 3 as examples, the disturbance estimate errors $\tilde{d}_{i,l}(t)$ ($i \in \{1,3\}$, $l = 1,2,3$) in the disturbance observer (4.46) are shown in Fig. 4.9. The sliding-mode surfaces $s_1(t)$ and $s_2(t)$ for the informed followers are given in Fig. 4.10. The control inputs of the five followers are shown in Fig. 4.11.

From Fig. 4.9, we can see that the disturbance observer (4.46) can provide the accurate estimates of the disturbances fast. As given in Fig. 4.10, the states of the informed followers can reach the sliding-mode surfaces $s_i(t) = 0$ in finite time. Figs. 4.7 and 4.8 indicate that the outputs of the followers form the predefined rotating regular pentagon and track the output trajectory of the leader in finite time. Moreover, as shown in Fig. 4.11, there is no large chattering in the control inputs of all the followers. Thus, the desired finite-time output formation is realized by the third-order swarm system under the continuous protocol (4.48).

4.4 Conclusions

This chapter investigated the time-varying formation tracking problems for high-order weak heterogeneous swarm systems with matched/mismatched disturbances respectively. For the case with matched disturbances, an adaptive time-varying formation robust tracking controller with a totally distributed form was proposed utilizing the neighbourhood state information. Feasible conditions for the followers to accomplish the expected time-varying formation tracking were provided, and the stability analysis was given based on Lyapunov theory. For the case with mismatched disturbances, based on the homogeneous finite-time control, the integral sliding mode control, and the super-twisting algorithm, a distributed continuous formation tracking protocol was proposed utilizing the neighbouring interaction, and it was proved that the desired formation tracking can be realized in finite time. The results in this chapter are mainly based on [51] and [52].

Chapter 5

Formation Tracking Control for Heterogeneous Swarm Systems with a Non-autonomous Leader

5.1   Introduction

Time-varying formation tracking problems for weak heterogeneous swarm systems are studied in Chapter 4, where each agent has identical nominal dynamics with different disturbances. In practical cross-domain mission scenarios, including air-ground cooperative enclosing with multiple UAVs and UGVs and air-sea collaborative detection with multiple UAVs and USVs, there are agents with completely different dynamics. Thus, the swarm systems are fully heterogeneous, which means that the proposed approaches in Chapter 4 are no longer applicable. Formation tracking problems for high-order heterogeneous swarm systems will be further investigated in this chapter, where the dynamics of the agents can be different in the aspects of both matrix parameters and state dimensions.

In general, high-order heterogeneous swarm systems cannot be directly written into a compact form through the Kronecker product, and it is also difficult to construct appropriate Lyapunov functions for the dynamics of each follower itself. Heterogeneous dynamics models would greatly restrict the coordination space for formation control. Thus, there exist great challenges in the analysis and design for the time-varying formation tracking problem of heterogeneous swarm systems. Since formation control and tracking control are the research basis of formation tracking control, time-varying formation control problems for heterogeneous swarm systems are considered firstly in this chapter. Then, based on adaptive control, a distributed time-varying formation tracking controller is proposed for heterogeneous swarm systems with a leader of unknown input. The main contents of this chapter are summarized as follows.

Firstly, time-varying output formation control of high-order heterogeneous swarm systems with layered architecture is proposed. An algorithm to design controller parameters and the time-varying formation feasibility conditions with heterogeneous dynamics are given. Based on the state space decomposition approach and output regulation strategy, it is proved that the high-order

DOI: 10.1201/9781003263470-5

heterogeneous swarm system can achieve the desired time-varying output formation, and the exact expression of the output formation reference function is given.

Furthermore, considering a non-autonomous leader with unknown input, time-varying formation tracking problem for heterogeneous swarm systems is further studied. Based on the output regulation control and the sliding mode control, a hierarchical formation tracking control strategy composed of the distributed observer and the local tracking controller is provided. Using the neighbouring interaction, a distributed formation tracking protocol with the adaptive compensation capability for the unknown input of the leader is proposed. Considering the features of heterogeneous dynamics, the time-varying formation tracking feasible constraints are provided, and a compensation input is applied to expand the feasible formation set. The convergence of formation tracking errors is proved based on the Lyapunov stability theory.

5.2   Time-varying Formation Control for Heterogeneous Swarm Systems

This section focuses on how to design distributed controller such that the outputs of agents can achieve the desired formation shape for leaderless heterogeneous swarm systems. Firstly, a formation reference generator is given to deal with heterogeneous dynamics, and a time-varying formation controller with layered architecture is proposed. Then, an algorithm to design controller parameters is presented, where time-varying formation feasibility conditions with heterogeneous dynamics are given. Furthermore, using the state space decomposition approach and output regulation strategy, it is proved that the high-order heterogeneous swarm system can realize the desired time-varying output formation, and the exact expression of the output formation reference function is also given to describe the macroscopic motion of the swarm system.

5.2.1   Problem Description

Consider a heterogeneous swarm system with N agents. The dynamics of agent i is described by

$$\begin{aligned}
\dot{x}_i(t) &= A_i x_i(t) + B_i u_i(t) + D_i d_i(t), \\
y_i(t) &= C_i x_i(t),
\end{aligned} \tag{5.1}$$

where $i \in \{1, 2, \ldots, N\}$, $x_i(t) \in \mathbb{R}^{n_i}$, $u_i(t) \in \mathbb{R}^{m_i}$, $d_i(t) \in \mathbb{R}^{l_i}$, and $y_i(t) \in \mathbb{R}^p$ are the state, input, disturbance, and output of agent i. Assume that (A_i, B_i) is stabilizable, and (C_i, A_i) is detectable. Since the outputs of all the agents need to achieve a desired formation, the outputs $y_i(t)$ are required to have the same dimensions. The unknown disturbances $d_i(t)$ ($i \in \{1, 2, \ldots, N\}$) are generated by the following exosystem:

$$\dot{d}_i(t) = S_{di} d_i(t). \tag{5.2}$$

By choosing different matrices S_{di}, the exosystem (5.2) can generate different types of disturbances, such as constant, sine, cosine, and ramp signals.

The desired output formation for swarm system (5.1) to achieve is specified by the vector $h_y(t) = [h_{y1}^T(t), h_{y2}^T(t), \ldots, h_{yN}^T(t)]^T \in \mathbb{R}^{Np}$.

Definition 5.1. *For any given bounded initial states, if there exists a function $r_y(t) \in \mathbb{R}^p$ such that*

$$\lim_{t \to \infty} (y_i(t) - h_{yi}(t) - r_y(t)) = 0, \ i = 1, 2, \ldots, N, \tag{5.3}$$

then heterogeneous swarm system (5.1) is said to achieve the time-varying output formation $h_y(t)$, and $r_y(t)$ is called as the formation reference function.

If there are control inputs $u_i(t)$ ($i = 1, 2, \ldots, N$) to drive the outputs of swarm system (5.1) to form the time-varying formation $h_y(t)$, then $h_y(t)$ is feasible for the swarm system (5.1) under the controller $u_i(t)$. From (5.3), one can see that $h_{yi}(t)$ denotes the offset of $y_i(t)$ relative to the common formation reference $r_y(t)$.

5.2.2   Formation Controller Design and Stability Analysis

The interaction topology of swarm system (5.1) is denoted by a directed graph $\mathcal{G}$, and L denotes the Laplacian matrix of the graph $\mathcal{G}$.

Assumption 5.1. *The directed graph $\mathcal{G}$ has a spanning tree.*

Since the agents of swarm system (5.1) have different dynamics, the existing formation control approaches in [32–36] are no longer applicable. In the following, a formation reference generator will be introduced to deal with heterogeneous dynamics, and a time-varying formation controller with layered architecture is proposed.

For agent i ($i = 1, 2, \ldots, N$), construct the following time-varying output formation controller:

$$u_i(t) = K_{1i} x_i(t) + K_{2i} (\varsigma_i(t) + \delta_i(t)) + K_{zi} z_i(t) + v_i(t),$$

$$\dot{\varsigma}_i(t) = S_r \varsigma_i(t) - \alpha K_r \sum_{j=1}^{N} w_{ij} (\xi_i(t) - \xi_j(t)), \tag{5.4}$$

$$\dot{z}_i(t) = \bar{G}_{1i} z_i(t) + \bar{G}_{2i} (y_i(t) - h_{yi}(t) - \xi_i(t)),$$

where $\varsigma_i(t) \in \mathbb{R}^q$ denotes the state of the formation reference generator, $\delta_i(t) \in \mathbb{R}^q$ satisfying $h_{yi}(t) = F \delta_i(t)$ represents the state offset relative to

$\varsigma_i(t)$, $v_i(t) \in \mathbb{R}^{m_i}$ is the time-varying formation compensation input determined by $h_y(t)$, $z_i(t) \in \mathbb{R}^{n_{zi}}$ is the internal state of the controller, $\xi_i(t) \in \mathbb{R}^p$ satisfying $\dot{\xi}_i(t) = F\varsigma_i(t)$ denotes the output of the formation reference generator, S_r and F are constant matrices and (F, S_r) is detectable, K_{1i}, K_{2i}, K_{zi}, K_r, $\bar{G}_{1i}$, and $\bar{G}_{2i}$ are gain matrices to be determined, and α is a positive constant to be designed.

The second equation in the controller (5.4) denotes formation reference generator, which is updated by the local information interaction among neighbouring agents. If we can choose appropriate α and K_r such that $\xi_i(t) - r_y(t) \to 0$ ($i = 1, 2, \ldots, N$), then $\xi_i(t)$ can denote the estimation of agent i for the common formation reference $r_y(t)$. Moreover, the internal state $z_i(t)$ of the controller (5.4) is applied to compensate for the unknown disturbance $d_i(t)$.

Consider the following common assumptions in cooperative control of heterogeneous swarm system [83–94].

Assumption 5.2. *There exist matrices X_i and U_i ($i = 1, 2, \ldots, N$) such that the following regulator equations hold:*

$$X_i S_r = A_i X_i + B_i U_i,$$
$$0 = C_i X_i - F. \tag{5.5}$$

Assumption 5.3. *Let $\sigma(S_{di})$ denote the set of all eigenvalues of matrix S_{di}. For any $\lambda_{di} \in \sigma(S_{di})$, it holds that λ_{di} do not have negative real parts, and*

$$rank\left(\begin{bmatrix} A_i - \lambda_{di} I & B_i \\ C_i & 0 \end{bmatrix}\right) = n_i + p.$$

Definition 5.2 ([149]). *For a square matrix S, if*

$$G_1 = diag\underbrace{\{\beta, \ldots, \beta\}}_{p-tuple}, \; G_2 = diag\underbrace{\{\sigma, \ldots, \sigma\}}_{p-tuple},$$

where β denotes a constant square matrix whose characteristic polynomial equals the minimal polynomial of S, and σ is a constant column vector such that (β, σ) is controllable, then a pair of matrices (G_1, G_2) is said to incorporate the minimum p-copy internal model of the matrix S.

The following algorithm is given to design the control parameters in the formation controller (5.4).

Algorithm 5.1. *For agent i ($i \in \{1, 2, \ldots, N\}$), the time-varying output formation controller (5.4) can be designed by the following steps.*

Step 1. Choose matrices X_i and U_i such that the regulator equation (5.5) holds. For a given time-varying output formation $h_y(t)$, check whether there exists the compensation input $v_i(t)$ such that the following time-varying formation feasibility condition holds:

$$\lim_{t \to \infty} \left(X_i \left(S_r \delta_i - \dot{\delta}_i \right) + B_i v_i \right) = 0. \tag{5.6}$$

If there are $v_i(t)$ $(i = 1, 2, \ldots, N)$ for all the followers to satisfy the condition (5.6), then the algorithm continues. Otherwise, the given formation $h_y(t)$ is not feasible under the proposed controller (5.4), and the algorithm stops.

Step 2. *Choose K_{1i} to make $A_i + B_i K_{1i}$ Hurwitz, and let $K_{2i} = U_i - K_{1i} X_i$.*

Step 3. *Let the gain constant $\alpha > \frac{1}{2\,\mathrm{Re}(\lambda_2)}$, where λ_2 denotes the minimum non-zero eigenvalue of the Laplacian matrix L. Design the gain matrix $K_r = PF^T$, and P denotes a positive definite matrix satisfying the following algebraic Riccati equation (ARE):*

$$PS_r^T + S_r P - PF^T FP + Q = 0, \tag{5.7}$$

where Q is a given positive definite matrix.

Step 4. *Let $\bar{A}_i = A_i + B_i K_{1i}$ and (G_{1i}, G_{2i}) be the minimum p-copy internal model of S_{di}. Select $K_{zi} = \left[K_{zi}^{[1]}, K_{zi}^{[2]} \right]$ such that $\begin{bmatrix} \bar{A}_i + B_i K_{zi}^{[1]} & B_i K_{zi}^{[2]} \\ G_{2i} C_i & G_{1i} \end{bmatrix}$ is Hurwitz, and choose $\bar{G}_{1i} = \begin{bmatrix} \bar{A}_i + B_i K_{zi}^{[1]} & B_i K_{zi}^{[2]} \\ 0 & G_{1i} \end{bmatrix}$ and $\bar{G}_{2i} = \begin{bmatrix} 0 \\ G_{2i} \end{bmatrix}$.*

In the Algorithm 5.1, there is positive definite solution to the ARE (5.7) if and only if (F, S_r) is detectable. Under Assumption 5.3, if (G_{1i}, G_{2i}) incorporates the minimum p-copy internal model of S_{di}, then it follows from Lemma 1.26 in [149] that $\left(\begin{bmatrix} \bar{A}_i & 0 \\ G_{2i} C_i & G_{1i} \end{bmatrix}, \begin{bmatrix} B_i \\ 0 \end{bmatrix} \right)$ is stabilizable.

The following theorem gives a sufficient condition for heterogeneous swarm system to achieve time-varying formation.

Theorem 5.1. *Suppose that Assumptions 5.1-5.3 hold. If the given formation satisfies the feasibility condition (5.6), then high-order heterogeneous swarm system (5.1) can achieve the desired time-varying output formation under the distributed controller (5.4) designed by Algorithm 5.1.*

Proof. Let $\varsigma = [\varsigma_1^T, \varsigma_2^T, \ldots, \varsigma_N^T]^T$. Then, the formation reference generator in (5.4) can be rewritten as

$$\dot{\varsigma} = (I_N \otimes S_r)\varsigma - \alpha(L \otimes K_r F)\varsigma. \tag{5.8}$$

Let λ_i $(i = 1, 2, \ldots, N)$ denote the eigenvalues of L. Under Assumption 5.1, it follows that $\lambda_1 = 0$ and $0 < \mathrm{Re}(\lambda_2) \leqslant \cdots \leqslant \mathrm{Re}(\lambda_N)$. There exists non-singular matrix $T = [\tilde{\tau}_1, \tilde{T}]$, $\tilde{\tau}_1 = \mathbf{1}_N$, $\tilde{T} = [\tilde{\tau}_2, \tilde{\tau}_3, \ldots, \tilde{\tau}_N]$ such that $T^{-1}LT = J = \mathrm{diag}\{0, \bar{J}\}$, where $\bar{J} \in \mathbb{R}^{(N-1) \times (N-1)}$ denotes a Jordan canonical matrix with eigenvalues as λ_i $(i = 2, 3, \ldots, N)$. Let $T^{-1} = [\bar{\tau}_1^T, \bar{T}^T]^T$, $\bar{T} = [\bar{\tau}_2^T, \bar{\tau}_3^T, \ldots, \bar{\tau}_N^T]^T$, $\bar{\tau}_i \in \mathbb{R}^{1 \times N}$ $(i = 1, 2, \ldots, N)$. Define $\vartheta = (T^{-1} \otimes I_q)\varsigma = [\vartheta_1^T, \bar{\vartheta}^T]^T$ and $\bar{\vartheta} = [\vartheta_2^T, \vartheta_3^T, \ldots, \vartheta_N^T]^T$. Then, it holds from (5.8) that

$$\begin{aligned} \dot{\vartheta}_1 &= S_r \vartheta_1, \\ \dot{\bar{\vartheta}} &= \left((I_{N-1} \otimes S_r) - \alpha(\bar{J} \otimes K_r F) \right) \bar{\vartheta}. \end{aligned} \tag{5.9}$$

Let $\varsigma_C = (T \otimes I_q) \begin{bmatrix} \vartheta_1 \\ 0 \end{bmatrix}$ and $\varsigma_{\bar{C}} = (T \otimes I_q) \begin{bmatrix} 0 \\ \bar{\vartheta} \end{bmatrix}$. Then, it holds that $\varsigma = \varsigma_C + \varsigma_{\bar{C}}$, and ς_C and $\varsigma_{\bar{C}}$ are linearly independent. Since $T = [\tilde{\tau}_1, \tilde{T}]$ and $\tilde{\tau}_1 = \mathbf{1}_N$, we have that $\varsigma_C = (T \otimes I_q) \begin{bmatrix} \vartheta_1 \\ 0 \end{bmatrix} = \mathbf{1}_N \otimes \vartheta_1$. Then, it follows that

$$\varsigma_{\bar{C}} = \varsigma - \varsigma_C = \varsigma - \mathbf{1}_N \otimes \vartheta_1. \tag{5.10}$$

Since ς_C and $\varsigma_{\bar{C}}$ are linearly independent and T is nonsingular, it can be verified from (5.10) that $\varsigma - \mathbf{1}_N \otimes \vartheta_1 \to 0$ if and only if $\bar{\vartheta} \to 0$. Therefore, the consensus problem of the formation reference generator (5.8) can be transformed to the stability problem of $\bar{\vartheta}$. We need to design α and K_r such that the matrices $S_r - \alpha \lambda_i K_r F$ ($i = 2, 3, \ldots, N$) are Hurwitz, which leads to $\lim_{t \to \infty} \bar{\vartheta}(t) = 0$.

Consider the system $\dot{\theta}_i = (S_r - \alpha \lambda_i K_r F) \theta_i$, $i = 2, 3, \ldots, N$. Construct the Lyapunov function $V_i = \theta_i^H P^{-1} \theta_i$. Taking the time derivative of V_i gives $\dot{V}_i = \theta_i^H \left(P^{-1} S_r + S_r^T P^{-1} - 2\alpha \operatorname{Re}(\lambda_i) F^T F \right) \theta_i$. Let $\bar{\theta}_i = P^{-1} \theta_i$. Since $\alpha > \frac{1}{2 \operatorname{Re}(\lambda_2)}$, we can obtain from (5.7) that

$$\dot{V}_i = \bar{\theta}_i^H \left(S_r P + P S_r^T - 2\alpha \operatorname{Re}(\lambda_i) P F^T F P \right) \bar{\theta}_i \leqslant -\bar{\theta}_i^H Q \bar{\theta}_i < 0. \tag{5.11}$$

Thus, it follows that $S_r - \alpha \lambda_i K_r F$ ($i = 2, 3, \ldots, N$) are Hurwitz, which implies that the states of formation reference generator (5.8) achieve consensus, i.e., $\lim_{t \to \infty} (\varsigma(t) - \mathbf{1}_N \otimes \vartheta_1(t)) = 0$, where $\vartheta_1 = (\tilde{\tau}_1 \otimes I_q) \varsigma$ denotes the state consensus function of generator (5.8). Furthermore, the output formation reference function r_y can be described as $r_y = F \vartheta_1$.

Let $\tilde{x}_i = x_i - X_i (\varsigma_i + \delta_i)$. From (5.1) and (5.4), we have

$$\dot{\tilde{x}}_i = \bar{A}_i \tilde{x}_i + B_i K_{zi} z_i + D_i d_i + X_i (S_r \delta_i - \dot{\delta}_i) + B_i v_i + \alpha X_i K_{ri} \sum_{j=1}^{N} w_{ij} (\xi_i - \xi_j). \tag{5.12}$$

Since the formation feasibility condition (5.6) holds, it follows that $\tilde{h}_i = X_i (S_r \delta_i - \dot{\delta}_i) + B_i v_i \to 0$. When the states of formation reference generator (5.8) achieve consensus, we can get that $\tilde{\eta}_i = \alpha X_i K_{ri} \sum_{j=1}^{N} w_{ij} (\xi_i - \xi_j) \to 0$. Let $e_i = y_i - \xi_i - h_{yi}$. Then, from $y_i = C_i x_i$, $\xi_i = F \varsigma_i$, $h_{yi} = F \delta_i$, and $C_i X_i = F$, it can be verified that $e_i = C_i \tilde{x}_i$. We can obtain from (5.4) and (5.12) that

$$\begin{aligned}
\dot{\tilde{x}}_i &= \bar{A}_i \tilde{x}_i + B_i K_{zi} z_i + D_i d_i + \tilde{h}_i + \tilde{\eta}_i, \\
\dot{z}_i &= \bar{G}_{1i} z_i + \bar{G}_{2i} C_i \tilde{x}_i, \\
e_i &= C_i \tilde{x}_i.
\end{aligned} \tag{5.13}$$

Let $\varphi_i = [\tilde{x}_i^T, z_i^T]^T$. It holds that

$$\dot{\varphi}_i = A_{ci}\varphi_i + B_{ci}d_i + \tilde{\rho}_i,$$
$$e_i = C_{ci}\varphi_i, \tag{5.14}$$

where $A_{ci} = \begin{bmatrix} \bar{A}_i & B_i K_{zi} \\ \bar{G}_{2i}C_i & \bar{G}_{1i} \end{bmatrix}$, $B_{ci} = \begin{bmatrix} D_i \\ 0 \end{bmatrix}$, $C_{ci} = \begin{bmatrix} C_i & 0 \end{bmatrix}$, and

$\tilde{\rho}_i = \begin{bmatrix} \tilde{h}_i + \tilde{\eta}_i \\ 0 \end{bmatrix}$. Substitute $\bar{G}_{1i}$ and $\bar{G}_{2i}$ into A_{ci}. Based on similarity trans-

formation, we get that $A_{ci} \sim \begin{bmatrix} \bar{A}_i + B_i K_{zi}^{[1]} & B_i K_{zi}^{[2]} & B_i K_{zi}^{[1]} \\ G_{2i}C_i & G_{1i} & 0 \\ 0 & 0 & \bar{A}_i \end{bmatrix}$. Since both

$\bar{A}_i$ and $\begin{bmatrix} \bar{A}_i + B_i K_{zi}^{[1]} & B_i K_{zi}^{[2]} \\ G_{2i}C_i & G_{1i} \end{bmatrix}$ are Hurwitz, it can be verified that A_{ci} is
Hurwitz.

According to Lemma 1.27 in [149], since A_{ci} is Hurwitz and (G_{1i}, G_{2i}) incorporates the minimum p-copy internal model of matrix S_{di}, under Assumption 5.3, there exists an unique solution (X_{di}, Z_{di}) to the following matrix equation:

$$X_{di}S_{di} = \bar{A}_i X_{di} + B_i K_{zi} Z_{di} + D_i,$$
$$Z_{di}S_{di} = \bar{G}_{2i}C_i X_{di} + \bar{G}_{1i}Z_{di}, \tag{5.15}$$
$$0 = C_i X_{di}.$$

Let $\bar{X}_{ci} = \begin{bmatrix} X_{di}^T, Z_{di}^T \end{bmatrix}^T$. It follows from (5.15) that

$$\bar{X}_{ci}S_{di} = A_{ci}\bar{X}_{ci} + B_{ci},$$
$$0 = C_{ci}\bar{X}_{ci}. \tag{5.16}$$

Let $\tilde{\varphi}_i = \varphi_i - \bar{X}_{ci}d_i$. We can obtain from (5.14) and (5.16) that

$$\dot{\tilde{\varphi}}_i = A_{ci}\tilde{\varphi}_i + \tilde{\rho}_i,$$
$$e_i = C_{ci}\tilde{\varphi}_i. \tag{5.17}$$

Based on the input-state stability in Lemma 2.24, since A_{ci} is Hurwitz and $\lim_{t\to\infty} \tilde{\rho}_i = 0$, it can be verified that $\lim_{t\to\infty} \tilde{\varphi}_i = 0$, i.e., $\lim_{t\to\infty} e_i(t) = \lim_{t\to\infty} (y_i(t) - \xi_i(t) - h_{yi}(t)) = 0$. Furthermore, since the formation reference generator (5.8) can achieve consensus, it holds that $\lim_{t\to\infty} (\xi_i(t) - r_y(t)) = 0$. Finally, we can get $\lim_{t\to\infty} (y_i(t) - h_{yi}(t) - r_y(t)) = 0$, which means that heterogeneous swarm system (5.1) achieves the desired time-varying output formation. This completes the proof of Theorem 5.1. □

Based on the proof of Theorem 5.1, the output formation reference function r_y can be described as $r_y = F\vartheta_1$. Furthermore, it follows from (5.9) that the exact expression of r_y is $r_y(t) = Fe^{S_r t}\vartheta_1(0)$. By choosing different system matrices S_r, we can control the motion mode of the whole formation effectively. Let $\tilde{\xi}_i = \xi_i - r_y$ ($i \in \{1, 2, \ldots, N\}$). Then, $\tilde{\xi}_i$ can denote the output error of the formation reference generator (5.8). According to the definition of $e_i = y_i - h_{yi} - \xi_i$ ($i \in \{1, 2, \ldots, N\}$), e_i can be called as output formation local error. If $\tilde{\xi}_i \to 0$ and $e_i \to 0$ hold simultaneously, the desired output formation is achieved by heterogeneous swarm system (5.1).

5.2.3   Simulation Example

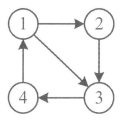

FIGURE 5.1: Directed graph.

Consider a heterogeneous swarm system with 4 agents. The topology of swarm system is shown in Fig. 5.1. Assume that the swarm system moves in the XY plane. The dynamics of each agent is described by (5.1), where $x_i = [x_{Xi}^T, x_{Yi}^T]^T$, $u_i = [u_{Xi}^T, u_{Yi}^T]^T$, and $y_i = [y_{Xi}^T, y_{Yi}^T]^T$. The matrices A_i, B_i, and C_i ($i = 1, 2, 3, 4$) are set as

Agent 1: $A_1 = 0_{2\times2}$, $B_1 = I_2$, $C_1 = I_2$.

Agent 2: $A_2 = I_2 \otimes \begin{bmatrix} 0 & 1 \\ 0 & 0 \end{bmatrix}$, $B_2 = I_2 \otimes \begin{bmatrix} 0 \\ 1 \end{bmatrix}$, $C_2 = I_2 \otimes [\ 1 \quad 0\ ]$.

Agent 3: $A_3 = I_2 \otimes \begin{bmatrix} 0 & 1 \\ -1 & -1 \end{bmatrix}$, $B_3 = I_2 \otimes \begin{bmatrix} 0 \\ 1 \end{bmatrix}$, $C_3 = I_2 \otimes [\ 1 \quad 0\ ]$.

Agent 4: $A_4 = I_2 \otimes \begin{bmatrix} 0 & 1 & 0 \\ 0 & 0 & 1 \\ 2 & -2 & 1 \end{bmatrix}$, $B_4 = I_2 \otimes \begin{bmatrix} 0 & 0 \\ 1 & 0 \\ 0 & 1 \end{bmatrix}$, $C_4 = I_2 \otimes$

$[\ 1 \quad 0 \quad 0\ ]$.

For the formation reference generator, choose $S_r = I_2 \otimes \begin{bmatrix} 0 & 1 \\ 0 & 0 \end{bmatrix}$ and $F = I_2 \otimes [\ 1 \quad 0\ ]$. The disturbances of the four agents are generated by

Agent 1: $S_{d1} = 0_{2\times2}$, $D_1 = I_2$, $d_1(0) = [1, 1]^T$.

Agent 2: $S_{d2} = I_2 \otimes \begin{bmatrix} 0 & 1 \\ -1 & 0 \end{bmatrix}$, $D_2 = I_2 \otimes \begin{bmatrix} 0 & 0 \\ 1 & 0 \end{bmatrix}$, $d_2(0) = [1, 0, 0, 1]^T$.

Agent 3: $S_{d3} = I_2 \otimes \begin{bmatrix} 0 & 1 \\ -4 & 0 \end{bmatrix}$, $D_3 = I_4$, $d_3(0) = [1, 0, 0, 2]^T$.

Agent 4: $S_{d4} = 0_{2\times2}$, $D_4 = I_2 \otimes [1, 1, 1]^T$, $d_4(0) = [-2, -2]^T$.

The outputs of heterogeneous swarm system are required to achieve a spanning square formation in this example, where the desired time-varying output formation vector is described by

$$h_{yi} = \begin{bmatrix} 2\cos\left(t + (i-1)\,\pi/2\right) \\ 2\sin\left(t + (i-1)\,\pi/2\right) \end{bmatrix}, \ i = 1, 2, 3, 4.$$

The corresponding state offset vector is $\delta_{Xi} = \begin{bmatrix} 2\cos\left(t + (i-1)\,\pi/2\right) \\ -2\sin\left(t + (i-1)\,\pi/2\right) \end{bmatrix}$ and

$$\delta_{Yi} = \begin{bmatrix} 2\sin\left(t + (i-1)\,\pi/2\right) \\ 2\cos\left(t + (i-1)\,\pi/2\right) \end{bmatrix}.$$

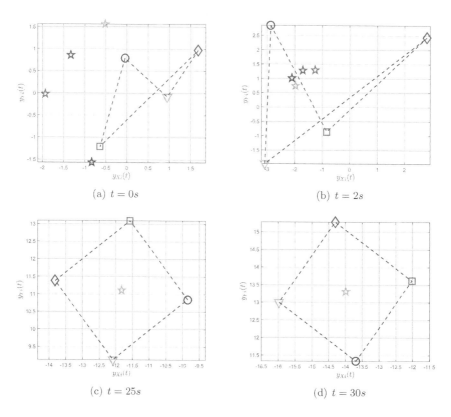

(a) $t = 0s$ (b) $t = 2s$

(c) $t = 25s$ (d) $t = 30s$

FIGURE 5.2: Output snapshots of each agent and formation reference generator at different time instants.

According to Algorithm 5.1, we can design the time-varying output formation controller (5.4) as follows. Choose the following matrices X_i and U_i ($i = 1, 2, 3, 4$) such that the regulator equations (5.5) hold

Agent 1: $X_1 = I_2 \otimes [\,1 \quad 0\,]$, $U_1 = I_2 \otimes [\,0 \quad 1\,]$.
Agent 2: $X_2 = I_4$, $U_2 = 0_{2\times4}$.
Agent 3: $X_3 = I_4$, $U_3 = I_2 \otimes [\,1 \quad 1\,]$.

$$Agent\ 4:\ X_4 = I_2 \otimes \begin{bmatrix} 1 & 0 \\ 0 & 1 \\ 0 & 0 \end{bmatrix},\ U_4 = I_2 \otimes \begin{bmatrix} 0 & 0 \\ -2 & 2 \end{bmatrix}.$$

For the given time-varying output formation $h_y(t)$, it can be verified that the formation feasibility condition (5.6) is satisfied for all the agents and the formation compensation inputs can be described by $v_{X1} = v_{Y1} = 0$, $v_{X2} = -2\cos(t + \pi/2)$, $v_{Y2} = -2\sin(t + \pi/2)$, $v_{X3} = -2\cos(t + \pi)$, $v_{Y3} = -2\sin(t + \pi)$, $v_{X4} = [-2\cos(t + 3\pi/2), 0]^T$, $v_{Y4} = [-2\sin(t + 3\pi/2), 0]^T$.

For each agent, design the gain matrices K_{1i} and K_{2i} as follows:

Agent 1: $K_{11} = -I_2$, $K_{21} = I_2 \otimes [\ 1 \quad 1\ ]$.

Agent 2: $K_{12} = I_2 \otimes [\ -2 \quad -2\ ]$, $K_{22} = I_2 \otimes [\ 2 \quad 2\ ]$.

Agent 3: $K_{13} = I_2 \otimes [\ -1 \quad -1\ ]$, $K_{23} = I_2 \otimes [\ 2 \quad 2\ ]$.

Agent 4: $K_{14} = I_2 \otimes \begin{bmatrix} -2 & -2 & -1 \\ -2 & 2 & -2 \end{bmatrix}$, $K_{24} = I_2 \otimes \begin{bmatrix} 2 & 2 \\ 0 & 0 \end{bmatrix}$.

To make the formation reference generator achieve consensus, choose the positive constant $\alpha = 1$. Let $Q = I_4$, and solving ARE (5.7) gives $P = I_2 \otimes \begin{bmatrix} 1.7321 & 1 \\ 1 & 1.7321 \end{bmatrix}$. Construct the minimum p-copy internal model of S_{di} $(i = 1, 2, 3, 4)$ as follows

Agent 1: $G_{11} = 0_{2\times 2}$, $G_{21} = I_2$.

Agent 2: $G_{12} = I_2 \otimes \begin{bmatrix} 0 & 1 \\ -1 & 0 \end{bmatrix}$, $G_{22} = I_2 \otimes \begin{bmatrix} 0 \\ 1 \end{bmatrix}$.

Agent 3: $G_{13} = I_2 \otimes \begin{bmatrix} 0 & 1 \\ -4 & 0 \end{bmatrix}$, $G_{23} = I_2 \otimes \begin{bmatrix} 0 \\ 1 \end{bmatrix}$.

Agent 4: $G_{14} = 0_{2\times 2}$, $G_{24} = I_2$.

Design gain matrices K_{zi} $(i = 1, 2, 3, 4)$ as

Agent 1: $K_{z1} = I_2 \otimes [\ -1 \quad -2\ ]$, $K_{z1}^{[1]} = -I_2$, $K_{z1}^{[2]} = -2I_2$.

Agent 2: $K_{z2} = I_2 \otimes [\ -7 \quad -3 \quad 5 \quad -5\ ]$, $K_{z2}^{[1]} = I_2 \otimes [\ -7 \quad -3\ ]$, $K_{z2}^{[2]} = I_2 \otimes [\ 5 \quad -5\ ]$.

Agent 3: $K_{z3} = I_2 \otimes [\ -4 \quad -3 \quad 20 \quad 10\ ]$, $K_{z3}^{[1]} = I_2 \otimes [\ -4 \quad -3\ ]$, $K_{z3}^{[2]} = I_2 \otimes [\ 20 \quad 10\ ]$.

Agent 4: $K_{z4} = I_2 \otimes \begin{bmatrix} -2 & -1 & 0 & -2 \\ 0 & 0 & -1 & 0 \end{bmatrix}$, $K_{z4}^{[1]} = I_2 \otimes \begin{bmatrix} -2 & -1 & 0 \\ 0 & 0 & -1 \end{bmatrix}$, $K_{z4}^{[2]} = I_2 \otimes \begin{bmatrix} -2 \\ 0 \end{bmatrix}$.

The initial states of four agents $x_i(0)$, formation reference generator $\varsigma_i(0)$, and internal state of controller $z_i(0)$ $(i = 1, 2, 3, 4)$ are generated by random numbers between -2 and 2.

Fig. 5.2 gives the output snapshots of each agent and formation reference generator at different time instants, where circle, square, diamond, and triangle represent the output of each agent respectively, and the corresponding output of its formation reference generator is represented by pentagram with the same color. The output errors $\tilde{\xi}_i = \xi_i - r_y$ and $e_i = y_i - h_{yi} - \xi_i$ are shown in Fig. 5.3 and Fig. 5.4, respectively. From Figs. 5.2-5.4, we can see that the

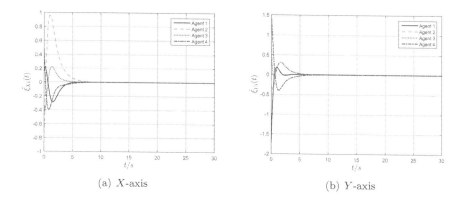

(a) X-axis   (b) Y-axis

FIGURE 5.3: Output error of formation reference generator $\tilde{\xi}_i = \xi_i - r_y$.

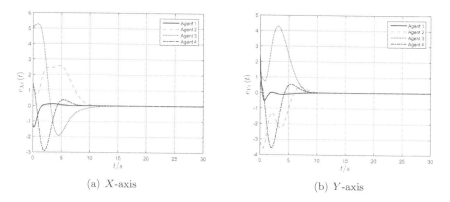

(a) X-axis   (b) Y-axis

FIGURE 5.4: Output formation local error $e_i = y_i - h_{yi} - \xi_i$.

formation reference generators of four agents can achieve consensus, and then the outputs of the agents form the desired spanning square formation centred by the common formation reference. Therefore, heterogeneous swarm system can achieve the desired time-varying output formation under the proposed distributed controller.

5.3   Formation Tracking Control for Heterogeneous Swarm Systems with a Non-Autonomous Leader

Based the time-varying formation control in Section 5.2, this section will further investigate the formation tracking control problems for high-order heterogeneous swarm systems with a non-autonomous leader of unknown input. Using the neighbouring interaction, a distributed formation tracking controller with adaptive compensation capability for the unknown input of the leader is proposed. Considering the features of heterogeneous dynamics, the time-varying formation tracking feasible constraints are provided, and a compensation input is applied to expand the feasible formation set. Based on the Lyapunov stability theory, the convergence of formation tracking errors for heterogeneous swarm systems is proved.

5.3.1   Problem Description

Consider a heterogeneous swarm system with one leader and N followers, and its interaction topology is denoted by a directed graph $\bar{\mathcal{G}}$. The dynamics of follower i ($i \in \{1, 2, \ldots, N\}$) is described by

$$\begin{aligned} \dot{x}_i(t) &= A_i x_i(t) + B_i u_i(t), \\ y_i(t) &= C_i x_i(t), \end{aligned} \tag{5.18}$$

where $x_i(t) \in \mathbb{R}^{n_i}$, $u_i(t) \in \mathbb{R}^{m_i}$, and $y_i(t) \in \mathbb{R}^p$ denote the state, control input, and output of agent i. $A_i \in \mathbb{R}^{n_i \times n_i}$, $B_i \in \mathbb{R}^{n_i \times m_i}$, and $C_i \in \mathbb{R}^{p \times n_i}$ are known constant matrices, where $\mathrm{rank}(B_i) = m_i$. The pairs (A_i, B_i) are stabilizable, and the pairs (C_i, A_i) are detectable. The leader is modeled as

$$\begin{aligned} \dot{v}_0(t) &= S v_0(t) + E r_0(t), \\ y_0(t) &= F v_0(t), \end{aligned} \tag{5.19}$$

where $v_0(t) \in \mathbb{R}^q$ is the state, $r_0(t) \in \mathbb{R}^l$ is the control input, and $y_0(t) \in \mathbb{R}^p$ is the output of the leader. $S \in \mathbb{R}^{q \times q}$, $E \in \mathbb{R}^{q \times l}$ and $F \in \mathbb{R}^{p \times q}$ are constant known matrices.

Similar to Chapter 4, the time-varying input $r_0(t)$ of the leader is assumed to be unknown to all the followers, and only satisfies the following bounded condition.

Assumption 5.4. *The unknown control input of the leader $r_0(t)$ is bounded, and there exists a positive constant γ such that $\|r_0(t)\| \leqslant \gamma$.*

In the existing results for cooperative control on heterogeneous swarm systems [83–94, 101–105], it is assumed that the exosystem has no control input, i.e., $\dot{v}_0(t) = S v_0(t)$. In order to generate more general reference signals and regulate the expected trajectory in real time, the leader is assumed to be nonautonomous with a non-zero input $r_0(t)$ in this section. Furthermore, considering the case where the leader denotes a non-cooperative target, it is assumed that $r_0(t)$ is unknown to all the followers, which is more general than $r_0(t) \equiv 0$. Moreover, if there exist disturbances for the followers, the approach in Section 5.2 can be used to deal with the disturbances similarly. Thus, in this section, the compensation control for unknown disturbances is omitted, and it is assumed that $d_i(t) = 0$ in the follower model (5.18).

In order to specify the expected output formation, a time-varying vector $h_y(t) = [h_{y1}^T(t), h_{y2}^T(t), \cdots, h_{yN}^T(t)]^T$ is introduced, where $h_{yi}(t) \in \mathbb{R}^p$ is piecewise continuously differentiable.

Definition 5.3. *Consider the follower systems (5.18) and the leader system (5.19) on a directed graph $\bar{\mathcal{G}}$. For any bounded initial states, if*

$$\lim_{t\to\infty} (y_i(t) - h_{yi}(t) - y_0(t)) = 0, \ i = 1, 2, \ldots, N, \qquad (5.20)$$

then high-order heterogeneous swarm system (5.18) and (5.19) is said to achieve the desired time-varying output formation tracking.

From Definition 5.3, we can see that the outputs of the followers need to not only accomplish the desired formation but also track the trajectory generated by the leader. The time-varying vectors $h_{yi}(t)$ $(i = 1, 2, \ldots, N)$ can describe both the formation configuration of the followers and the tracking relationship of the whole formation relative to the leader. The formation shape, orientation, and scale of the followers can be determined by choosing $h_{yi}(t)$ appropriately.

5.3.2 Formation Tracking Controller Design and Stability Analysis

Assumption 5.5. *The directed graph $\bar{\mathcal{G}}$ has a spanning tree rooted by the leader.*

Under Assumption 5.5, the Laplacian matrix $\bar{L}$ can be divided as $\bar{L} = \begin{bmatrix} 0 & 0 \\ L_2 & L_1 \end{bmatrix}$, where $L_2 \in \mathbb{R}^{N\times1}$ and $L_1 \in \mathbb{R}^{N\times N}$.

Lemma 5.1 ([150]). *Under Assumption 5.5, all eigenvalues of L_1 have positive real parts. Moreover, there exists a diagonal matrix $D = diag\{d_1, d_2, \cdots, d_N\}$ with $d_i > 0$ $(i = 1, 2, \ldots, N)$ such that $\bar{L}_1 = DL_1 + L_1^T D > 0$.*

Since the state of the leader is only available to a subset of followers, to acquire $v_0(t)$ for each follower, we will construct a distributed observer using the neighbouring information interaction under the influences of the leader's unknown input. Moreover, the upper bound of $r_0(t)$ (i.e., γ) is global information, and a distributed observer for γ is needed. Let $\hat{v}_i(t)$ and $\hat{\eta}_i(t)$ stand for the distributed estimations of $v_0(t)$ and γ for follower i $(i \in \{1, 2, \ldots, N\})$, respectively.

For follower i $(i \in \{1, 2, \ldots, N\})$, consider the following time-varying output formation tracking controller:

$$\dot{\hat{v}}_i(t) = S\hat{v}_i(t) - \left(\hat{\alpha}_i(t) + \varsigma_i^T(t)P\varsigma_i(t)\right)P\varsigma_i(t) - (\hat{\eta}_i(t) + \rho)Ef_i(t), \quad (5.21a)$$

$$\dot{\hat{\eta}}_i(t) = -\frac{1}{\sum_{j=0}^{N} w_{ij}}\left(k\text{sig}^\beta(\tilde{\eta}_{ei}(t)) - \sum_{j=1}^{N} w_{ij}\dot{\hat{\eta}}_j(t)\right), \quad (5.21b)$$

$$\dot{\hat{x}}_i(t) = A_i \hat{x}_i(t) + B_i u_i(t) + L_{oi} \left(C_i \hat{x}_i(t) - y_i(t) \right), \qquad (5.21c)$$

$$u_i(t) = K_{1i} \left(\hat{x}_i(t) - \phi_i(t) \right) + K_{2i} \hat{v}_i(t) - \left(\hat{\eta}_i(t) + \bar{\rho} \right) \delta_i(t) + \tau_i(t), \qquad (5.21d)$$

where $\varsigma_i(t) = w_{i0} \left(\hat{v}_i(t) - v_0(t) \right) + \sum_{j=1}^{N} w_{ij} \left(\hat{v}_i(t) - \hat{v}_j(t) \right)$ and $\tilde{\eta}_{ei}(t) = w_{i0} \left(\hat{\eta}_i(t) - \gamma \right) + \sum_{j=1}^{N} w_{ij} \left(\hat{\eta}_i(t) - \hat{\eta}_j(t) \right)$ denote the local errors of estimated values $\hat{v}_i(t)$ and $\hat{\eta}_i(t)$ for follower i relative to its neighbours, respectively.

In the distributed observer (5.21a) for $v_0(t)$, $\hat{\alpha}_i(t)$ is the adaptive updating gain to avoid the global information of the graph, $f_i(t)$ represents the nonlinear function to suppress the leader's unknown input, ρ denotes any positive constant, and $P \in \mathbb{R}^{q \times q}$ stands for a positive definite matrix to be determined. In the distributed observer (5.21b) for γ, $k > 0$ and $0 < \beta < 1$ represent two positive constants. In the Luenberger state observer (5.21c), $\hat{x}_i(t) \in \mathbb{R}^{n_i}$ is the estimation of $x_i(t)$, and $L_{oi} \in \mathbb{R}^{n_i \times p}$ represents a constant gain matrix. In the observer-based formation tracking controller (5.21d), $\phi_i(t) \in \mathbb{R}^{n_i}$ satisfying $h_{yi}(t) = C_i \phi_i(t)$ is used to drive the outputs of followers to accomplish the expected formation specified by $h_y(t)$, $\delta_i(t) \in \mathbb{R}^{m_i}$ is applied to make up for the leader's unknown input, $\tau_i(t) \in \mathbb{R}^{m_i}$ is the formation compensation input, $\bar{\rho}$ is any positive constant, and $K_{1i} \in \mathbb{R}^{m_i \times n_i}$ and $K_{2i} \in \mathbb{R}^{m_i \times q}$ are two constant gain matrices to be designed.

Assumption 5.6. *The regulator equations*

$$\begin{aligned} X_i S &= A_i X_i + B_i U_i \\ 0 &= C_i X_i - F \end{aligned} \qquad (5.22)$$

have solution pairs (X_i, U_i), $i = 1, 2, \ldots, N$.

Assumption 5.7. *The linear matrix equations*

$$B_i R_i - X_i E = 0 \qquad (5.23)$$

have solutions R_i, $i = 1, 2, \ldots, N$.

The regulator equations (5.22) are given in Assumption 5.6, whose solvability is a necessary condition to achieve output regulation [83–88,149]. Since the formation tracking controller (5.21d) is constructed based on the output regulation theory, Assumption 5.6 is also a necessary condition to solve the formation tracking problems. Assumption 5.7 is similar to the matched condition used in the robust control with disturbances (see, e.g., [139] and [140]), which means that the unknown input of the leader $r_0(t)$ can be compensated by the control input $u_i(t)$ completely for each follower. Moreover, if all $N + 1$ agents have identical dynamics with $A \in \mathbb{R}^{n \times n}$, $B \in \mathbb{R}^{n \times m}$ and $C \in \mathbb{R}^{p \times n}$ as the cases considered in [9,51,151], it can be verified that Assumptions 5.6 and 5.7 hold naturally with $X_i = R_i = I_n$ and $U_i = 0$.

Since $\text{rank}(B_i) = m_i$, there is a non-singular matrix $\Pi_i = [\hat{B}_i^T, \tilde{B}_i^T]^T$ with $\hat{B}_i \in \mathbb{R}^{m_i \times n_i}$ and $\tilde{B}_i \in \mathbb{R}^{(n_i - m_i) \times n_i}$ such that $\hat{B}_i B_i = I_{m_i}$ and $\tilde{B}_i B_i = 0$. Consider the following algorithm to design the formation tracking controller.

Algorithm 5.2. *The output formation tracking protocol (5.21) for follower* i *($i \in \{1, 2, \ldots, N\}$) can be designed by the following six steps.*

Step 1. *For a given time-varying formation vector* $h_y(t) = [h_{y1}^T(t), h_{y2}^T(t),$ $\ldots, h_{yN}^T(t)]^T$ *and a vector* $\phi_i(t)$ *satisfying* $h_{yi}(t) = C_i\phi_i(t)$, *check the following formation tracking feasibility condition:*

$$\lim_{t \to \infty} (\tilde{B}_i A_i \phi_i(t) - \tilde{B}_i \dot{\phi}_i(t)) = 0. \tag{5.24}$$

If there exist $\phi_i(t)$ *($i = 1, 2, \ldots, N$) such that the feasible condition (5.24) holds for each follower, then continue. Otherwise the given formation* $h_y(t)$ *is not feasible under the protocol (5.21) and the algorithm stops.*

Step 2. *Calculate the formation compensation input* $\tau_i(t)$ *as follows:*

$$\tau_i(t) = -\hat{B}_i(A_i \phi_i(t) - \dot{\phi}_i(t)). \tag{5.25}$$

Step 3. *Choose* X_i, U_i, *and* R_i *such that the linear matrix equations (5.22) and (5.23) hold.*

Step 4. *Design* L_{oi} *and* K_{1i} *to make* $A_i + L_{oi}C_i$ *and* $A_i + B_iK_{1i}$ *Hurwitz respectively, and choose* K_{2i} *as* $K_{2i} = U_i - K_{1i}X_i$. *Solve the positive definite matrix* $Q_i \in \mathbb{R}^{n_i \times n_i}$ *from the following Lyapunov equation:*

$$Q_i (A_i + B_iK_{1i}) + (A_i + B_iK_{1i})^T Q_i = -2I_{n_i}. \tag{5.26}$$

Step 5. *To design the distributed observer (5.21a) to get* $\hat{v}_i(t)$, *solve* $P > 0$ *firstly from the algebraic Riccati equation (ARE)*

$$PS + S^T P - P^2 + I_q = 0. \tag{5.27}$$

Since (S, I_q) *is controllable, there exists a unique solution* P *for the ARE (5.27). Then, the gain* $\hat{\alpha}_i(t)$ *is updated by the following adaptive law:*

$$\dot{\hat{\alpha}}_i(t) = \varsigma_i^T(t)P^2\varsigma_i(t), \tag{5.28}$$

where $\hat{\alpha}_i(0) \geqslant 0$. *The non-linear function* $f_i(t)$ *is defined as*

$$f_i(t) = \begin{cases} \dfrac{E^T P \varsigma_i(t)}{\|E^T P \varsigma_i(t)\|}, & \|E^T P \varsigma_i(t)\| \neq 0, \\ 0, & \|E^T P \varsigma_i(t)\| = 0. \end{cases} \tag{5.29}$$

Step 6. *In (5.21d),* $\delta_i(t)$ *is defined as* $\delta_i(t) = R_i g_i(t)$ *with* $g_i(t)$ *denoting the non-linear function*

$$g_i(t) = \begin{cases} \dfrac{H_i^T \hat{\xi}_i(t)}{\|H_i^T \hat{\xi}_i(t)\|}, & \|H_i^T \hat{\xi}_i(t)\| \neq 0, \\ 0, & \|H_i^T \hat{\xi}_i(t)\| = 0, \end{cases} \tag{5.30}$$

where $H_i = Q_i B_i R_i$ *and* $\hat{\xi}_i(t) = \hat{x}_i(t) - \phi_i(t) - X_i\hat{v}_i(t)$.

It should be pointed out that not all time-varying formation vectors can be accomplished even for the homogeneous swarm systems (see, e.g., [9,33]). Considering the heterogeneous dynamics of each agent, the formation tracking feasible condition is proposed as (5.24), which implies that the given formations should be compatible with the dynamic constraints of the swarm systems. From (5.25), we see that the feasible formation set can be expanded by utilizing the compensation input $\tau_i(t)$. To get a more direct form of the feasible condition (5.24), assume that $\mathrm{rank}\,(C_i) = p$ without loss of generality. Then, there is a non-singular matrix $T_i = [\bar{C}_i, \tilde{C}_i]$ with $\bar{C}_i \in \mathbb{R}^{n_i \times p}$ and $\tilde{C}_i \in \mathbb{R}^{n_i \times (n_i - p)}$ such that $C_i \bar{C}_i = I_p$ and $C_i \tilde{C}_i = 0_{p \times (n_i - p)}$. Let $h_{\bar{y}i}(t) \in \mathbb{R}^{n_i - p}$ denote any piecewise continuously differentiable column vector. Then, $\phi_i(t)$ satisfying $h_{yi}(t) = C_i \phi_i(t)$ can be rewritten as $\phi_i(t) = T_i[h_{yi}^T(t), h_{\bar{y}i}^T(t)]^T = \bar{C}_i h_{yi}(t) + \tilde{C}_i h_{\bar{y}i}(t)$. So the feasible condition (5.24) can be transformed to

$$\lim_{t \to \infty} \left(\tilde{h}_{yi}(t) + \tilde{B}_i \left(A_i \tilde{C}_i h_{\bar{y}i}(t) - \tilde{C}_i \dot{h}_{\bar{y}i}(t) \right) \right) = 0, \tag{5.31}$$

where $\tilde{h}_{yi}(t) = \tilde{B}_i (A_i \bar{C}_i h_{yi}(t) - \bar{C}_i \dot{h}_{yi}(t))$. Since the expected output formation vector $h_{yi}(t)$ is predefined, we can calculate $\tilde{h}_{yi}(t)$ directly. Consider the following two cases.
i) If $\lim_{t \to \infty}(\tilde{h}_{yi}(t)) = 0$, then let $h_{\bar{y}i}(t) = 0$. We can obtain that $\phi_i(t) = \bar{C}_i h_{yi}(t)$ and the expected output formation tracking is feasible.
ii) If $\lim_{t \to \infty}(\tilde{h}_{yi}(t)) \neq 0$, then check whether there exists $h_{\bar{y}i}(t)$ such that (5.31) holds. If there is such $h_{\bar{y}i}(t)$ for each follower, then let $\phi_i(t) = \bar{C}_i h_{yi}(t) + \tilde{C}_i h_{\bar{y}i}(t)$ and the given formation tracking is feasible. Otherwise $h_y(t)$ is infeasible under the controller (5.21).

Based on the neighbouring estimation states, the distributed observer (5.21a) is constructed for each follower to asymptotically estimate the leader's state. Different from the existing results in [83–88], there exists a non-linear term (i.e., $-(\hat{\eta}_i(t) + \rho) E f_i(t)$) in (5.21a), which is applied to suppress the influences of the unknown input of the leader in the distributed estimation level. Furthermore, in the observer-based controller (5.21d), by using the formation information $\phi_i(t)$, the local estimation state $\hat{x}_i(t)$, and the distributed estimation state of the leader $\hat{v}_i(t)$, the non-linear term $-(\hat{\eta}_i(t) + \bar{\rho}) R_i g_i(t)$ is constructed to make up for the leader's unknown input in the formation control level.

The following theorem gives sufficient conditions to achieve formation tracking for heterogeneous swarm systems.

Theorem 5.2. *Suppose that Assumptions 5.4-5.7 hold. If the formation tracking feasible condition (5.24) is satisfied, the desired output formation tracking is accomplished by the heterogeneous systems (5.18) and (5.19) with leader's unknown input under the distributed control protocol (5.21) determined by Algorithm 5.2.*

Proof. In the following, we will prove that the distributed observers (5.21a) and (5.21b) are convergent firstly. Then, it is shown that the expected formation tracking can be realized under the controller (5.21d).

Since $\tilde{\eta}_{ei}(t) = \sum_{j=1}^{N} w_{ij}(\hat{\eta}_i(t) - \hat{\eta}_j(t)) + w_{i0}(\hat{\eta}_i(t) - \gamma)$ $(i = 1, 2, \ldots, N)$, the distributed observer (5.21b) can be transformed to $\dot{\hat{\eta}}_{ei}(t) = -k\mathrm{sig}^\beta(\tilde{\eta}_{ei}(t))$. Consider the Lyapunov function $V_{\eta_i}(t) = \tilde{\eta}_{ei}^2(t)$. Then, we can get that $\dot{V}_{\eta_i}(t) = -2k|\tilde{\eta}_{ei}(t)|^{1+\beta} = -2k(V_{\eta_i}(t))^{(1+\beta)/2}$. According to the finite-time stability theory in Lemma 2.27, it can be verified that $\tilde{\eta}_{ei}(t) \to 0$ in a finite time T_η. Let $\tilde{\eta}_i(t) = \hat{\eta}_i(t) - \gamma$, $\tilde{\eta}(t) = [\tilde{\eta}_1(t), \tilde{\eta}_2(t), \ldots, \tilde{\eta}_N(t)]^T$, and $\tilde{\eta}_e(t) = [\tilde{\eta}_{e1}(t), \tilde{\eta}_{e2}(t), \ldots, \tilde{\eta}_{eN}(t)]^T$. It follows that $\tilde{\eta}_e(t) = L_1\tilde{\eta}(t)$, where L_1 is non-singular under Assumption 5.5. Thus, we have that $\tilde{\eta}_i(t) \to 0$ in a finite time T_η.

Let $\tilde{v}_i(t) = \hat{v}_i(t) - v_0(t)$ and $\varphi_i(t) = \hat{\alpha}_i(t) + \varsigma_i^T(t)P\varsigma_i(t)$ $(i = 1, 2, \ldots, N)$. Then, it follows from (5.19) and (5.21a) that

$$\dot{\tilde{v}}_i(t) = S\tilde{v}_i(t) - \varphi_i(t)P\varsigma_i(t) - (\hat{\eta}_i(t) + \rho)\,Ef_i(t) - Er_0(t). \qquad (5.32)$$

Let $\tilde{v}(t) = [\tilde{v}_1^T(t), \tilde{v}_2^T(t), \ldots, \tilde{v}_N^T(t)]^T$, $\varsigma(t) = [\varsigma_1^T(t), \varsigma_2^T(t), \ldots, \varsigma_N^T(t)]^T$, $f(t) = [f_1^T(t), f_2^T(t), \ldots, f_N^T(t)]^T$, and $\bar{r}(t) = \mathbf{1}_N \otimes r_0(t)$. The system (5.32) can be rewritten in the following compact form:

$$\dot{\tilde{v}}(t) = (I_N \otimes S)\,\tilde{v}(t) - (\Phi \otimes P)\,\varsigma(t) - \eta\,(I_N \otimes E)\,f(t)$$
$$- (\tilde{\Pi} \otimes E)f(t) - (I_N \otimes E)\,\bar{r}(t), \qquad (5.33)$$

where $\Phi = \mathrm{diag}\{\varphi_1(t), \varphi_2(t), \ldots, \varphi_N(t)\}$, $\tilde{\Pi} = \mathrm{diag}\{\tilde{\eta}_1(t), \tilde{\eta}_2(t), \ldots, \tilde{\eta}_N(t)\}$, and $\eta = \gamma + \rho$. Since $\varsigma(t) = (L_1 \otimes I_q)\tilde{v}(t)$, it holds from (5.33) that

$$\dot{\varsigma}(t) = (I_N \otimes S)\,\varsigma(t) - (L_1\Phi \otimes P)\,\varsigma(t) - \eta\,(L_1 \otimes E)\,f(t)$$
$$- (L_1\tilde{\Pi} \otimes E)f(t) - (L_1 \otimes E)\,\bar{r}(t). \qquad (5.34)$$

Because L_1 is nonsingular under Assumption 5.5, it can be verified that $\lim_{t\to\infty}\tilde{v}_i(t) = 0$ if and only if $\lim_{t\to\infty}\varsigma_i(t) = 0$.

Construct the Lyapunov functional candidate as follows

$$V_\varsigma(t) = \frac{1}{2}\sum_{i=1}^{N} d_i\,(\varphi_i(t) + \hat{\alpha}_i(t))\,\varsigma_i^T(t)P\varsigma_i(t) + \frac{1}{2}\sum_{i=1}^{N} d_i\tilde{\alpha}_i^2(t), \qquad (5.35)$$

where $d_i > 0$ is defined in Lemma 5.1 and $\tilde{\alpha}_i(t) = \hat{\alpha}_i(t) - \alpha$ with α representing a positive constant to be determined.

Note that $\varsigma_i^T(t)P\varsigma_i(t) = \varphi_i(t) - \hat{\alpha}_i(t)$. Taking the time derivative of $V_\varsigma(t)$ gives $\dot{V}_\varsigma(t) = \Xi_1(t) + \Xi_2(t)$, where $\Xi_1(t) = 2\sum_{i=1}^{N} d_i\varphi_i(t)\varsigma_i^T(t)P\dot{\varsigma}_i(t)$ and $\Xi_2(t) = \sum_{i=1}^{N} d_i\,(\varphi_i(t) - \alpha)\,\dot{\hat{\alpha}}_i(t)$. Substituting (5.34) into $\Xi_1(t)$ gives

$$\Xi_1(t) = \varsigma^T(t)(\Phi D \otimes (PS + S^T P))\varsigma(t) - \varsigma^T(t)(\Phi\bar{L}_1\Phi \otimes P^2)\varsigma(t)$$
$$- 2\eta\varsigma^T(t)(\Phi DL_1 \otimes PE)f(t) - 2\varsigma^T(t)(\Phi DL_1\tilde{\Pi} \otimes PE)f(t)$$
$$- 2\varsigma^T(t)\,(\Phi DL_1 \otimes PE)\,\bar{r}(t), \qquad (5.36)$$

where $\bar{L}_1 = DL_1 + L_1^T D > 0$.

According to the definition of $f_i(t)$ in (5.29), we can obtain that $\varsigma_i^T(t)PEf_i(t) = \|E^T P\varsigma_i(t)\|$ and $\varsigma_i^T(t)PEf_j(t) \leqslant \|E^T P\varsigma_i(t)\|\|f_j(t)\| \leqslant \|E^T P\varsigma_i(t)\|$, $i \neq j$, $i, j = 1, 2, \ldots, N$. Then, it follows that

$$
\begin{aligned}
&- 2\eta\varsigma^T(t)\left(\Phi DL_1 \otimes PE\right)f(t) \\
&= 2\eta \sum_{i=1}^{N} d_i\varphi_i(t)\varsigma_i^T(t)PE\left(\sum_{j=1}^{N} w_{ij}\left(f_j(t) - f_i(t)\right) - w_{i0}f_i(t)\right) \\
&\leqslant -2\eta \sum_{i=1}^{N} d_i w_{i0}\varphi_i(t)\left\|E^T P\varsigma_i(t)\right\|.
\end{aligned}
\tag{5.37}
$$

From Assumption 5.4, we have

$$
\begin{aligned}
&- 2\varsigma^T(t)\left(\Phi DL_1 \otimes PE\right)\bar{r}(t) \\
&= -2 \sum_{i=1}^{N} d_i w_{i0}\varphi_i(t)\varsigma_i^T(t)PEr_0(t) \\
&\leqslant 2\gamma \sum_{i=1}^{N} d_i w_{i0}\varphi_i(t)\|E^T P\varsigma_i(t)\|.
\end{aligned}
\tag{5.38}
$$

Note that $\eta > \gamma$. Substituting (5.37) and (5.38) into (5.36) gives

$$
\begin{aligned}
\Xi_1(t) &\leqslant \varsigma^T(t)(\Phi D \otimes (PS + S^T P))\varsigma(t) - \bar{\lambda}_1\varsigma^T(t)(\Phi^2 \otimes P^2)\varsigma(t) \\
&\quad - 2\varsigma^T(t)(\Phi DL_1\tilde{\Pi} \otimes PE)f(t),
\end{aligned}
\tag{5.39}
$$

where $\bar{\lambda}_1$ denotes the minimum eigenvalue of $\bar{L}_1$.

Since $\dot{\alpha}_i(t) = \varsigma_i^T(t)P^2\varsigma_i(t)$, $\Xi_2(t)$ can be transformed to

$$
\Xi_2(t) = \varsigma^T(t)\left((\Phi - \alpha I_N)D \otimes P^2\right)\varsigma(t).
\tag{5.40}
$$

It holds from (5.39) and (5.40) that

$$
\begin{aligned}
\dot{V}_\varsigma(t) &\leqslant \varsigma^T(t)\left(\Phi D \otimes (PS + S^T P + P^2)\right)\varsigma(t) - \varsigma^T(t)\left((\bar{\lambda}_1\Phi^2 + \alpha D) \otimes P^2\right)\varsigma(t) \\
&\quad - 2\varsigma^T(t)(\Phi DL_1\tilde{\Pi} \otimes PE)f(t).
\end{aligned}
\tag{5.41}
$$

In light of Lemma 2.3, it follows that $-2\varsigma^T(t)(\Phi DL_1\tilde{\Pi} \otimes PE)f(t) \leqslant \bar{c}\varsigma^T(t)(\Phi^2 \otimes P^2)\varsigma(t) + \frac{1}{\bar{c}}\|(DL_1\tilde{\Pi} \otimes E)f(t)\|^2$, where $0 < \bar{c} < \bar{\lambda}_1$. Select α sufficiently large such that $\alpha \geqslant d_m/(\bar{\lambda}_1 - \bar{c})$ with $d_m = \max\{d_1, d_2, \ldots, d_N\}$. Using the Young's inequality in Lemma 2.3, we get $-\varsigma^T(t)(((\bar{\lambda}_1 - \bar{c})\Phi^2 + \alpha D) \otimes P^2)\varsigma(t) \leqslant -2\varsigma^T(t)(\Phi D \otimes P^2)\varsigma(t)$. Recall that $PS + S^T P - P^2 + I_q = 0$. Then, it follows from (5.41) that

$$
\dot{V}_\varsigma(t) \leqslant -\varsigma^T(t)(\Phi D \otimes I_q)\varsigma(t) + \frac{1}{\bar{c}}\|(DL_1\tilde{\Pi} \otimes E)f(t)\|^2.
\tag{5.42}
$$

Note that $\tilde{\Pi}$ is bounded in $[0, T_\eta)$ and $\|f_i(t)\| \leqslant 1$. We can obtain that $\frac{1}{c}\|(DL_1\tilde{\Pi} \otimes E)f(t)\|^2$ is bounded in $[0, T_\eta)$, and its upper bound is assumed to be $\bar{f}_m$. It follows from (5.42) that

$$\dot{V}_\varsigma(t) \leqslant -\varsigma^T(t)\,(\Phi D \otimes I_q)\,\varsigma(t) + \bar{f}_m \leqslant \bar{f}_m. \tag{5.43}$$

Thus, $V_\varsigma(t)$ is bounded in $[0, T_\eta)$, which implies that $\varsigma_i(t)$ and $\hat{\alpha}_i(t)$ are bounded in $[0, T_\eta)$.

When $t \geqslant T_\eta$, we get that $\tilde{\Pi} = 0$. Then, it can be verified from (5.42) that

$$\dot{V}_\varsigma(t) \leqslant -\varsigma^T(t)(\Phi D \otimes I_q)\varsigma(t) \leqslant 0. \tag{5.44}$$

It holds from (5.44) that $V_\varsigma(t)$ is bounded and so is $\hat{\alpha}_i(t)$. Since $\hat{\alpha}_i(t)$ updated by (5.28) is monotonically increasing, it can be verified that $\hat{\alpha}_i(t)$ converges to some positive constant. Note that $\dot{V}_\varsigma(t) \equiv 0$ leads to $\varsigma(t) = 0$. Thus, based on the LaSalle's Invariance principle in [127], it holds that $\lim_{t\to\infty}\varsigma(t) = 0$. Since L_1 is nonsingular, we have $\lim_{t\to\infty}\tilde{v}_i(t) = 0$, i.e., $\lim_{t\to\infty}(\hat{v}_i(t) - v_0(t)) = 0$.

Based on the convergence of the distributed observers (5.21a) and (5.21b), we will prove that the expected output formation tracking can be achieved under the proposed controller (5.21d). Let $\tilde{x}_i(t) = \hat{x}_i(t) - x_i(t)$ $(i = 1, 2, \ldots, N)$. Then, it holds from (5.18) and (5.21c) that $\dot{\tilde{x}}_i(t) = (A_i + L_{oi}C_i)\tilde{x}_i(t)$. Since L_{oi} is selected to make $A_i + L_{oi}C_i$ Hurwitz, we have $\lim_{t\to\infty}\tilde{x}_i(t) = 0$. For follower i $(i = 1, 2, \ldots, N)$, substituting (5.21d) into (5.18) gives

$$\dot{x}_i(t) = (A_i + B_iK_{1i})x_i(t) + B_iK_{1i}\tilde{x}_i(t) + B_iK_{2i}v_0(t) + B_iK_{2i}\tilde{v}_i(t)$$
$$- (\hat{\eta}_i(t) + \bar{\rho})\,B_i\delta_i(t) - B_iK_{1i}\phi_i(t) + B_i\tau_i(t). \tag{5.45}$$

Let $\xi_i(t) = x_i(t) - \phi_i(t) - X_iv_0(t)$ and $\bar{\eta} = \gamma + \bar{\rho}$. Note that $X_iS = A_iX_i + B_iU_i$, $B_iR_i - X_iE = 0$, and $\delta_i(t) = R_ig_i(t)$. It follows from (5.19) and (5.45) that

$$\dot{\xi}_i(t) = (A_i + B_iK_{1i})\xi_i(t) - (\bar{\eta} + \tilde{\eta}_i(t))B_iR_ig_i(t) - B_iR_ir_0(t) + B_iK_{1i}\tilde{x}_i(t)$$
$$+ B_iK_{2i}\tilde{v}_i(t) + A_i\phi_i(t) - \dot{\phi}_i(t) + B_i\tau_i(t). \tag{5.46}$$

Consider the following Lyapunov functional candidate for each follower: $V_{\xi i}(t) = \xi_i^T(t)Q_i\xi_i(t)$, $i = 1, 2, \ldots, N$, where $Q_i > 0$ satisfies (5.26). The time derivative of $V_{\xi i}(t)$ along the trajectory of (5.46) is described by

$$\dot{V}_{\xi i}(t) = \xi_i^T(t)(Q_i(A_i + B_iK_{1i}) + (A_i + B_iK_{1i})^TQ_i)\xi_i(t)$$
$$- 2(\bar{\eta} + \tilde{\eta}_i(t))\xi_i^T(t)Q_iB_iR_ig_i(t)$$
$$- 2\xi_i^T(t)Q_iB_iR_ir_0(t) + 2\xi_i^T(t)Q_i(B_iK_{1i}\tilde{x}_i(t) + B_iK_{2i}\tilde{v}_i(t))$$
$$+ 2\xi_i^T(t)Q_i(A_i\phi_i(t) - \dot{\phi}_i(t) + B_i\tau_i(t)). \tag{5.47}$$

Let $\tilde{\xi}_i(t) = \hat{\xi}_i(t) - \xi_i(t) = \tilde{x}_i(t) - X_i\tilde{v}_i(t)$ $(i = 1, 2, \ldots, N)$. Since $\lim_{t\to\infty}\tilde{x}_i(t) = 0$ and $\lim_{t\to\infty}\tilde{v}_i(t) = 0$, it can be verified that $\lim_{t\to\infty}\tilde{\xi}_i(t) = 0$.

Note that $H_i = Q_i B_i R_i$ and $\bar{\eta} = \gamma + \bar{\rho} > \gamma$. From Assumption 5.4 and (5.30), we have

$$
\begin{aligned}
&- 2\bar{\eta}\xi_i^T(t)H_i g_i(t) - 2\xi_i^T(t)H_i r_0(t) \\
&= -2\bar{\eta}\hat{\xi}_i^T(t)H_i g_i(t) - 2\hat{\xi}_i^T(t)H_i r_0(t) + 2\bar{\eta}\tilde{\xi}_i^T(t)H_i g_i(t) + 2\tilde{\xi}_i^T(t)H_i r_0(t) \\
&\leqslant -2(\bar{\eta} - \gamma)\|H_i^T \hat{\xi}_i(t)\| + 2(\bar{\eta} + \gamma)\|H_i^T \tilde{\xi}_i(t)\| \\
&\leqslant 2(\bar{\eta} + \gamma)\|H_i^T \tilde{\xi}_i(t)\|.
\end{aligned} \tag{5.48}
$$

Substituting (5.26) and (5.48) into (5.47) gives

$$
\begin{aligned}
\dot{V}_{\xi i}(t) \leqslant &- 2\|\xi_i(t)\|^2 - 2\tilde{\eta}_i(t)\xi_i^T(t)H_i g_i(t) + 2(\bar{\eta} + \gamma)\|H_i^T \tilde{\xi}_i(t)\| \\
&+ 2\xi_i^T(t)Q_i(B_i K_{1i}\tilde{x}_i(t) + B_i K_{2i}\tilde{v}_i(t)) \\
&+ 2\xi_i^T(t)Q_i(A_i\phi_i(t) - \dot{\phi}_i(t) + B_i\tau_i(t)).
\end{aligned} \tag{5.49}
$$

Since the formation tracking feasibility condition (5.24) is satisfied, it follows that $\lim_{t\to\infty}(\tilde{B}_i(A_i\phi_i(t) - \dot{\phi}_i(t)) + \tilde{B}_i B_i\tau_i(t)) = 0$. From (5.25), we get

$$
\hat{B}_i(A_i\phi_i(t) - \dot{\phi}_i(t)) + \hat{B}_i B_i\tau_i(t) = 0.
$$

Note that $\mathrm{II}_i = [\hat{B}_i^T, \tilde{B}_i^T]^T$ is a non-singular matrix. Then, it holds that

$$
\lim_{t\to\infty}(A_i\phi_i(t) - \dot{\phi}_i(t) + B_i\tau_i(t)) = 0.
$$

Let $\tilde{\phi}_i(t) = Q_i(A_i\phi_i(t) - \dot{\phi}_i(t) + B_i\tau_i(t))$ $(i = 1, 2, \ldots, N)$. It follows that $\lim_{t\to\infty}\tilde{\phi}_i(t) = 0$. Let $\tilde{\vartheta}_i(t) = Q_i(B_i K_{1i}\tilde{x}_i(t) + B_i K_{2i}\tilde{v}_i(t))$. Since $\lim_{t\to\infty}\tilde{x}_i(t) = 0$ and $\lim_{t\to\infty}\tilde{v}_i(t) = 0$, we can obtain $\lim_{t\to\infty}\tilde{\vartheta}_i(t) = 0$. Using the Young's inequality, it can be verified that

$$
2\xi_i^T(t)\tilde{\phi}_i(t) \leqslant \frac{1}{2}\xi_i^T(t)\xi_i(t) + 2\tilde{\phi}_i^T(t)\tilde{\phi}_i(t),
$$

$$
2\xi_i^T(t)\tilde{\vartheta}_i(t) \leqslant \frac{1}{2}\xi_i^T(t)\xi_i(t) + 2\tilde{\vartheta}_i^T(t)\tilde{\vartheta}_i(t),
$$

$$
-2\tilde{\eta}_i(t)\xi_i^T(t)H_i g_i(t) \leqslant \frac{1}{2}\xi_i^T(t)\xi_i(t) + 2\|\tilde{\eta}_i(t)H_i g_i(t)\|^2.
$$

Substituting the above three inequalities into (5.49) yields

$$
\begin{aligned}
\dot{V}_{\xi i}(t) \leqslant &- \frac{1}{2\lambda_{\max}(Q_i)}V_{\xi i}(t) + 2(\bar{\eta} + \gamma)\|H_i^T \tilde{\xi}_i(t)\| + 2\tilde{\vartheta}_i^T(t)\tilde{\vartheta}_i(t) \\
&+ 2\tilde{\phi}_i^T(t)\tilde{\phi}_i(t) + 2\|\tilde{\eta}_i(t)H_i g_i(t)\|^2.
\end{aligned} \tag{5.50}
$$

Since $\tilde{\eta}_i(t)$ is bounded in $[0, T_\eta)$ and $\|g_i(t)\| \leqslant 1$, we have that $2\|\tilde{\eta}_i(t)H_i g_i(t)\|^2$ is bounded in $[0, T_\eta)$. Note that $\tilde{\xi}_i(t)$, $\tilde{\vartheta}_i(t)$ and $\tilde{\phi}_i(t)$ are bounded. Thus, it holds from (5.50) that $V_{\xi i}(t)$ is bounded in $[0, T_\eta)$, so is $\xi_i(t)$.

When $t \geqslant T_\eta$, we have that $\tilde{\eta}_i(t) = 0$. Recall that $\lim_{t\to\infty}\tilde{\xi}_i(t) = 0$, $\lim_{t\to\infty}\tilde{\vartheta}_i(t) = 0$, and $\lim_{t\to\infty}\tilde{\phi}_i(t) = 0$. Then, by Lemma 2.19 in [150], it holds from (5.50) that $\lim_{t\to\infty}V_{\xi i}(t) = 0$, which implies that $\lim_{t\to\infty}\xi_i(t) = 0$ ($i = 1, 2, \ldots, N$). Let $e_i(t) = y_i(t) - h_{yi}(t) - Fv_0(t)$ be the output formation tracking error. Since $h_{yi}(t) - C_i\phi_i(t) = 0$ and $C_iX_i - F = 0$, we have $\lim_{t\to\infty}e_i(t) = 0$ ($i = 1, 2, \ldots, N$). Therefore, the expected time-varying output formation tracking is achieved by systems (5.18) and (5.19). This completes the proof of Theorem 5.2.   □

From the proof of Theorem 5.2, we can see that the unknown input of the leader has influences on the design of both the distributed observer (5.21a) and the formation controller (5.21d). Besides $f_i(t)$ in the distributed observer (5.21a), we also need to construct an appropriate compensation term in the formation controller (5.21d) for each follower using its local information. Note that we cannot obtain $\xi_i(t)$ directly for all the followers. Combining the estimation states $\hat{v}_i(t)$, $\hat{x}_i(t)$ and the formation information $\phi_i(t)$, a novel compensation term $-(\hat{\eta}_i(t) + \bar{\rho})\delta_i(t)$ is constructed in (5.21d) to suppress $r_0(t)$ in the formation control level. Thus, only the output regulation strategy with $r_0(t) \equiv 0$ in [83–88] cannot solve the formation tracking problems considered in this section. Moreover, since the swarm system is homogeneous in [51], the influences of $r_0(t)$ can be compensated directly in the formation controller using the neighbouring relative state. There is no need to construct a distributed robust observer and combine a compensation term with output regulation strategy for homogeneous case in [51].

In light of the adaptive updating gain $\hat{a}_i(t)$, the proposed observer (5.21a) is distributed without needing any global information of the directed topology. Note that the existing approaches for undirected graph in [51, 152, 153] cannot be applied to solve the adaptive time-varying formation tracking problems on directed graph by a simple modification. Since the Laplacian matrix of directed graph is asymmetric, we have to construct a different adaptive formation tracking protocol and select an appropriate Lyapunov function for directed graph. Compared with the protocol for undirected graph in [51], there exist two distinct differences for the proposed distributed observer (5.21a) in this paper. Firstly, we add a smooth function $\varsigma_i^T(t)P\varsigma_i(t)$ to $\hat{a}_i(t)$ to accelerate the convergence of the estimation error and give extra freedom for stability analysis. Secondly, we construct an extra distributed observer (5.21b) to get γ for each follower using the neighbouring interaction, so the complicated interrelations between the non-linear functions to suppress $r_0(t)$ and the directed topology can be handled. Moreover, the Lyapunov function candidate (5.35) in this section has a quite different and more complicated structure than the one in Section 4.2.

Since the non-linear functions $f_i(t)$ and $g_i(t)$ ($i \in \{1, 2, \ldots, N\}$) in (5.29) and (5.30) are discontinuous, the formation tracking protocol (5.21) will cause chattering to the control input $u_i(t)$. In the following, we will replace $f_i(t)$ and $g_i(t)$ with continuous functions to avoid the large chattering of $u_i(t)$. The continuous non-linear functions $\bar{f}_i(t)$ and $\bar{g}_i(t)$ are defined as

$$\bar{f}_i(t) = \frac{E^T P \varsigma_i(t)}{\|E^T P \varsigma_i(t)\| + \varepsilon_{fi}}, \tag{5.51}$$

$$\bar{g}_i(t) = \frac{H_i^T \hat{\xi}_i(t)}{\|H_i^T \hat{\xi}_i(t)\| + \varepsilon_{gi}}, \tag{5.52}$$

where ε_{fi} and ε_{gi} are two small positive constants. Moreover, motivated by the σ-modification adaptive approach in [139], the adaptive law for $\hat{\alpha}_i(t)$ in (5.28) is redesigned as

$$\dot{\hat{\alpha}}_i(t) = -\sigma_i \hat{\alpha}_i(t) + \varsigma_i^T(t) P^2 \varsigma_i(t), \tag{5.53}$$

where $\hat{\alpha}_i(0) \geqslant 0$ and σ_i denotes a positive constant. As shown in [139], the σ-modification adaptive law (5.53) has robustness to bounded noise.

Theorem 5.3. *Suppose that Assumptions 5.4-5.7 hold. Using the continuous functions $\bar{f}_i(t)$, $\bar{g}_i(t)$, and the σ-modification adaptive law (5.53) in the protocol (5.21), if the formation tracking feasible condition (5.24) is satisfied, then the formation tracking error $e_i(t)$ is uniformly ultimately bounded.*

Proof. Construct the same Lyapunov functional candidate $V_\varsigma(t)$ as (5.35). Considering the structure of $\bar{f}_i(t)$, we get $\varsigma_i^T(t) P E \bar{f}_i(t) = \frac{\|E^T P \varsigma_i(t)\|^2}{\|E^T P \varsigma_i(t)\| + \varepsilon_{fi}}$, $\varsigma_i^T(t) P E \bar{f}_j(t) \leqslant \|E^T P \varsigma_i(t)\|$, $i \neq j$, $i, j = 1, 2, \ldots, N$. Then, it follows that

$$\varsigma_i^T(t) P E \left(\sum_{j=1}^{N} w_{ij}(\bar{f}_j(t) - \bar{f}_i(t)) - w_{i0}\bar{f}_i(t) \right)$$

$$\leqslant -w_{i0} \frac{\|E^T P \varsigma_i(t)\|^2}{\|E^T P \varsigma_i(t)\| + \varepsilon_{fi}} + \varepsilon_{fi} \sum_{j=1}^{N} w_{ij}. \tag{5.54}$$

From (5.38) and (5.54), we can obtain

$$-2\eta \varsigma^T(t)(\Phi D L_1 \otimes P E)\bar{f}(t) - 2\varsigma^T(t)(\Phi D L_1 \otimes P E)\bar{r}(t)$$

$$\leqslant 2\eta \sum_{i=1}^{N} d_i \varphi_i(t) \varepsilon_{fi} \sum_{j=0}^{N} w_{ij}. \tag{5.55}$$

Under the σ-modification adaptive law (5.53), $\Xi_2(t)$ can be transformed to $\Xi_2(t) = \varsigma^T(t)((\Phi - \alpha I_N)D \otimes P^2)\varsigma(t) - \sum_{i=1}^{N} d_i \sigma_i (\varphi_i(t) - \alpha)\hat{\alpha}_i(t)$. Note that $\varphi_i(t) \geqslant \hat{\alpha}_i(t) \geqslant 0$, $-\frac{1}{2}(\hat{\alpha}_i(t) - \alpha)\hat{\alpha}_i(t) \leqslant -\frac{1}{4}\hat{\alpha}_i^2(t) + \frac{1}{4}\alpha^2$, $-\frac{1}{2}\tilde{\alpha}_i(t)(\tilde{\alpha}_i(t) + \alpha) \leqslant -\frac{1}{4}\tilde{\alpha}_i^2(t) + \frac{1}{4}\alpha^2$. Then, it follows that

$$\Xi_2(t) \leqslant \varsigma^T(t) \left((\Phi - \alpha I_N)D \otimes P^2 \right) \varsigma(t) - \frac{1}{4} \sum_{i=1}^{N} d_i \sigma_i \left(\hat{\alpha}_i^2(t) + \tilde{\alpha}_i^2(t) \right)$$

$$+ \frac{1}{2} \sum_{i=1}^{N} d_i \sigma_i \alpha^2. \tag{5.56}$$

Follow the similar steps in the proof of Theorem 5.2. It can be verified from (5.55) and (5.56) that

$$\dot{V}_\varsigma(t) \leqslant -\varsigma^T(t)(\Phi D \otimes I_q)\varsigma(t) + \frac{1}{\bar{c}}\|(DL_1\tilde{\Pi}\otimes E)\bar{f}(t)\|^2 + 2\eta\sum_{i=1}^{N} d_i\varphi_i(t)\varepsilon_{fi}\sum_{j=0}^{N} w_{ij}$$

$$-\frac{1}{4}\sum_{i=1}^{N} d_i\sigma_i(\hat{\alpha}_i^2(t) + \tilde{\alpha}_i^2(t)) + \frac{1}{2}\sum_{i=1}^{N} d_i\sigma_i\alpha^2. \tag{5.57}$$

Let $\bar{\varepsilon}_{fi} = 2\eta\varepsilon_{fi}\sum_{j=0}^{N} w_{ij}$. Then, it can be verified that

$$2\eta\sum_{i=1}^{N} d_i\varphi_i(t)\varepsilon_{fi}\sum_{j=0}^{N} w_{ij}$$

$$\leqslant \frac{1}{4}\sum_{i=1}^{N} d_i\sigma_i\hat{\alpha}_i^2(t) + \frac{1}{2\lambda_{\max}(P)}\sum_{i=1}^{N} d_i(\varsigma_i^T(t)P\varsigma_i(t))^2$$

$$+\sum_{i=1}^{N} d_i\left(\frac{1}{\sigma_i} + \frac{\lambda_{\max}(P)}{2}\right)\bar{\varepsilon}_{fi}^2.$$

Since $\frac{1}{2\lambda_{\max}(P)}\sum_{i=1}^{N} d_i(\varsigma_i^T(t)P\varsigma_i(t))^2 \leqslant \frac{1}{2}\varsigma^T(t)(\Phi D \otimes I_q)\varsigma(t)$, it holds from (5.57) that

$$\dot{V}_\varsigma(t) \leqslant -\frac{1}{2}\varsigma^T(t)(\Phi D \otimes I_q)\varsigma(t) - \frac{1}{4}\sum_{i=1}^{N} d_i\sigma_i\tilde{\alpha}_i^2(t)$$

$$+\frac{1}{\bar{c}}\|(DL_1\tilde{\Pi}\otimes E)\bar{f}(t)\|^2 + \varpi,$$

where $\varpi = \sum_{i=1}^{N} d_i\left(\frac{1}{\sigma_i} + \frac{\lambda_{\max}(P)}{2}\right)\bar{\varepsilon}_{fi}^2 + \frac{1}{2}\sum_{i=1}^{N} d_i\sigma_i\alpha^2$. It can be verified that $V_\varsigma(t)$ is bounded in $[0, T_\eta)$. When $t \geqslant T_\eta$, since $\tilde{\Pi} = 0$, it holds that

$$\dot{V}_\varsigma(t) \leqslant -\frac{1}{2}\varsigma^T(t)(\Phi D \otimes I_q)\varsigma(t) - \frac{1}{4}\sum_{i=1}^{N} d_i\sigma_i\tilde{\alpha}_i^2(t) + \varpi. \tag{5.58}$$

Because $\varphi_i(t) \geqslant \hat{\alpha}_i(t) \geqslant 0$, we can obtain from (5.35) that

$$V_\varsigma(t) \leqslant \varsigma^T(t)(\Phi D \otimes P)\varsigma(t) + \frac{1}{2}\sum_{i=1}^{N} d_i\tilde{\alpha}_i^2(t).$$

Then, we get from (5.58) that

$$\dot{V}_\varsigma(t) \leqslant -\mu V_\varsigma(t) - \left(\frac{1}{2} - \mu\lambda_{\max}(P)\right)\varsigma^T(t)(\Phi D \otimes I_q)\varsigma(t)$$

$$-\frac{1}{4}\sum_{i=1}^{N}(\sigma_i - 2\mu)d_i\tilde{\alpha}_i^2(t) + \varpi, \tag{5.59}$$

where μ is a positive constant. Let μ be sufficiently small such that $\mu < \frac{1}{2\lambda_{\max}(P)}$ and $\mu < \frac{1}{2}\sigma_i$ ($i = 1, 2, \ldots, N$). Then, it holds from (5.59) that

$$\dot{V}_\varsigma(t) \leqslant -\mu V_\varsigma(t) + \varpi. \tag{5.60}$$

According to the Comparison lemma in Lemma 2.25, it can be verified that $V_\varsigma(t) \leqslant (V_\varsigma(0) - \varpi/\mu)e^{-\mu t} + \varpi/\mu$. Thus, $V_\varsigma(t)$ is ultimately bounded with the upper bound ϖ/μ, which implies that $\varsigma_i(t)$ and $\hat{\alpha}_i(t)$ ($i = 1, 2, \ldots, N$) are ultimately bounded. Since $\varsigma(t) = (L_1 \otimes I_q)\tilde{v}(t)$ and L_1 is nonsingular, we can obtain that $\tilde{v}_i(t)$ ($i = 1, 2, \ldots, N$) are ultimately bounded.

Construct the same Lyapunov functional candidate $V_{\xi i}(t) = \xi_i^T(t)Q_i\xi_i(t)$ as the one in Theorem 5.2. From (5.52), we can obtain that $-2\bar{\eta}\hat{\xi}_i^T(t)H_i\bar{g}_i(t) - 2\hat{\xi}_i^T(t)H_i r_0(t) \leqslant 2\bar{\eta}\varepsilon_{gi}$. Follow the similar steps in the proof of Theorem 5.2. It can be verified that $V_{\xi i}(t)$ is bounded in $[0, T_\eta)$. When $t \geqslant T_\eta$, it holds that

$$\dot{V}_{\xi i}(t) \leqslant -\frac{1}{2\lambda_{\max}(Q_i)}V_{\xi i}(t) + 2\bar{\eta}\varepsilon_{gi} + 2(\bar{\eta} + \gamma)\|H_i^T\tilde{\xi}_i(t)\| \\ + 2\tilde{\vartheta}_i^T(t)\tilde{\vartheta}_i(t) + 2\tilde{\phi}_i^T(t)\tilde{\phi}_i(t). \tag{5.61}$$

Since $\tilde{v}_i(t)$ is ultimately bounded and $\lim_{t\to\infty}\tilde{x}_i(t) = 0$, we can obtain that $\tilde{\xi}_i(t)$ and $\tilde{\vartheta}_i(t)$ are ultimately bounded. Note that $\lim_{t\to\infty}\tilde{\phi}_i(t) = 0$. According to the Comparison lemma and the input-to-state stability, it can be verified that $V_{\xi i}(t)$ is ultimately bounded, which implies that $e_i(t) = C_i\xi_i(t)$ is ultimately bounded. This completes the proof of Theorem 5.3. $\qquad\square$

In Theorem 5.3, the continuous output formation tracking protocol using $\bar{f}_i(t)$ and $\bar{g}_i(t)$ ($i \in \{1, 2, \ldots, N\}$) can avoid the chattering of the control inputs effectively. From (5.60) and (5.61), we see that the formation tracking error $e_i(t)$ can converge to a small bounded set if the positive constants ε_{fi}, ε_{gi}, and σ_i are selected to be relatively small. If the formation tracking error is small enough to meet the requirements of practical applications, the heterogeneous swarm system is said to achieve *practical* time-varying output formation tracking.

5.3.3 Simulation Example

Consider a heterogeneous swarm system with 4 followers and one leader, which moves in the XY plane. The directed topology of the swarm system is shown in Fig. 5.5. The model of each follower is described by (5.18), where $x_i(t) = [x_{Xi}^T(t), x_{Yi}^T(t)]^T \in \mathbb{R}^{2n_i}$ ($n_1 = 1$, $n_2 = n_3 = 2$, $n_4 = 3$), $u_i(t) = [u_{Xi}^T(t), u_{Yi}^T(t)]^T \in \mathbb{R}^{2m_i}$ ($m_1 = m_2 = m_3 = 1$, $m_4 = 2$), and $y_i(t) = [y_{Xi}^T(t), y_{Yi}^T(t)]^T \in \mathbb{R}^2$. The matrices A_i, B_i, and C_i ($i = 1, 2, 3, 4$) are selected the same as the simulation example in Section 5.2. The dynamics of the leader is described by (5.19), where $v_0(t) = [v_{X0}^T(t), v_{Y0}^T(t)]^T \in \mathbb{R}^4$, $r_0(t) = [r_{X0}^T(t), r_{Y0}^T(t)]^T \in \mathbb{R}^2$, $y_0(t) = [y_{X0}^T(t), y_{Y0}^T(t)]^T \in \mathbb{R}^2$, $S = I_2 \otimes \begin{bmatrix} 0 & 1 \\ 0 & 0 \end{bmatrix}$,

$E = I_2 \otimes \begin{bmatrix} 0 \\ 1 \end{bmatrix}$, $F = I_2 \otimes [\, 1 \quad 0 \,]$. Choose the unknown input of the leader as $r_0(t) = [0.5\cos(t), 0.5\sin(t)]^T$, where we can see that $r_0(t)$ satisfies Assumption 5.4.

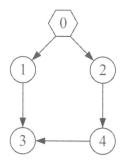

FIGURE 5.5: Directed graph.

The followers are required to achieve time-varying square formation tracking, and the desired formation vector is chosen as $h_{yi}(t) = [3\cos(2t + (i-1)\pi/2), 3\sin(2t + (i-1)\pi/2)]^T$, $i = 1, 2, 3, 4$. When $h_y(t)$ is realized, the outputs of the four followers will form a square and rotate around the leader. Algorithm 5.2 is applied to design the time-varying formation tracking controller (5.21). For each follower, there exist $\phi_i(t)$ ($i = 1, 2, 3, 4$) such that the formation tracking feasibility condition (5.24) holds, and $\phi_i(t)$ can be designed as

$$\phi_1(t) = [3\cos(2t), 3\sin(2t)]^T,$$
$$\phi_2(t) = [3\cos(2t + \pi/2), -6\sin(2t + \pi/2), 3\sin(2t + \pi/2), 6\cos(2t + \pi/2)]^T,$$
$$\phi_3(t) = [3\cos(2t + \pi), -6\sin(2t + \pi), 3\sin(2t + \pi), 6\cos(2t + \pi)]^T,$$
$$\phi_4(t) = [3\cos(2t + 3\pi/2), -6\sin(2t + 3\pi/2), -12\cos(2t + 3\pi/2),$$
$$3\sin(2t + 3\pi/2), 6\cos(2t + 3\pi/2), -12\sin(2t + 3\pi/2)]^T.$$

It follows from (5.25) that

$$\tau_1(t) = [-6\sin(2t), 6\cos(2t)]^T,$$
$$\tau_2(t) = [-12\cos(2t + \pi/2), -12\sin(2t + \pi/2)]^T,$$
$$\tau_3(t) = [-6\sin(2t + \pi) - 9\cos(2t + \pi), -9\sin(2t + \pi) + 6\cos(2t + \pi)]^T,$$
$$\tau_4(t) = [0, 12\sin(2t + 3\pi/2) + 6\cos(2t + 3\pi/2), 0,$$
$$6\sin(2t + 3\pi/2) - 12\cos(2t + 3\pi/2)]^T.$$

Choose the following matrices X_i, U_i, and R_i ($i = 1, 2, 3, 4$) such that the linear matrix equations (5.22) and (5.23) hold: $X_1 = I_2 \otimes [\, 1 \quad 0 \,]$, $U_1 = I_2 \otimes [\, 0 \quad 1 \,]$, $R_1 = 0_{2 \times 2}$, $X_2 = I_4$, $U_2 = 0_{2 \times 4}$, $R_2 = I_2$, $X_3 = I_4$, $U_3 =$

$I_2 \otimes [\ 1 \quad 1\ ]$, $R_3 = I_2$, $X_4 = I_2 \otimes \begin{bmatrix} 1 & 0 \\ 0 & 1 \\ 0 & 0 \end{bmatrix}$, $U_4 = I_2 \otimes \begin{bmatrix} 0 & 0 \\ -2 & 2 \end{bmatrix}$, $R_4 =$

$I_2 \otimes \begin{bmatrix} 1 \\ 0 \end{bmatrix}$. Design the gain matrices L_{oi}, K_{1i}, and K_{2i} $(i = 1, 2, 3, 4)$ as
$L_{o1} = -5I_2$, $K_{11} = -I_2$, $K_{21} = I_2 \otimes [\ 1 \quad 1\ ]$, $L_{o2} = I_2 \otimes [-10, -50]^T$,
$K_{12} = I_2 \otimes [\ -2 \quad -2\ ]$, $K_{22} = I_2 \otimes [\ 2 \quad 2\ ]$, $L_{o3} = I_2 \otimes [-9, -40]^T$, $K_{13} =$
$I_2 \otimes [\ -1 \quad -1\ ]$, $K_{23} = I_2 \otimes [\ 2 \quad 2\ ]$, $L_{o4} = I_2 \otimes [-16, -114, -334]^T$, $K_{14} =$
$I_2 \otimes \begin{bmatrix} -2 & -2 & -1 \\ -2 & 2 & -2 \end{bmatrix}$, $K_{24} = I_2 \otimes \begin{bmatrix} 2 & 2 \\ 0 & 0 \end{bmatrix}$. In this example, the continuous
functions $\bar{f}_i(t)$ and $\bar{g}_i(t)$ and the σ-modification adaptive law (5.53) are used,
where $\varepsilon_{fi} = 0.02$, $\varepsilon_{gi} = 0.02$, and $\sigma_i = 0.1$. Let $k = 1$, $\beta = \frac{1}{2}$, $\rho = \bar{\rho} = 0.5$,
and $\hat{\alpha}_{Xi}(0) = \hat{\alpha}_{Yi}(0) = 0.5$. The initial states of $x_i(t)$, $v_0(t)$,$\hat{v}_i(t)$, $\hat{\eta}_i(t)$, and
$\hat{x}_i(t)$ are generated by random numbers between -2 and 2.

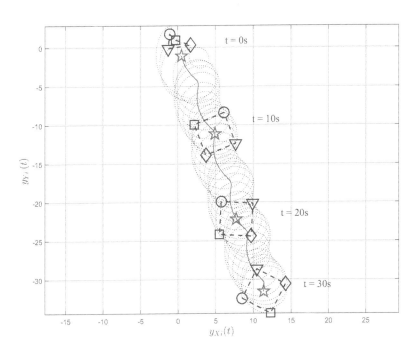

FIGURE 5.6: Output snapshots at different time instants and output trajectories within $t = 30$s of swarm system.

The output snapshots of heterogeneous swarm system at different times
($t = 0$, 10, 20, 30s) and the output trajectories within $t = 30$s are shown
in Fig. 5.6, where the four followers are denoted by circle, square, diamond,
and triangle respectively, and the leader is marked by pentagram. Fig. 5.7
shows the time-varying output formation tracking errors for the followers.

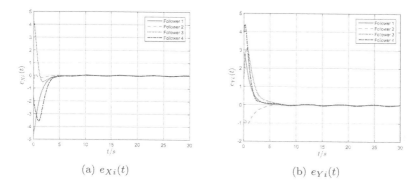

(a) $e_{Xi}(t)$ (b) $e_{Yi}(t)$

FIGURE 5.7: Output formation tracking errors for four followers.

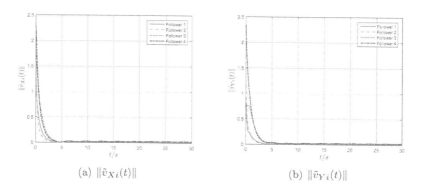

(a) $\|\tilde{v}_{Xi}(t)\|$ (b) $\|\tilde{v}_{Yi}(t)\|$

FIGURE 5.8: Estimation errors for the distributed observer (5.21a).

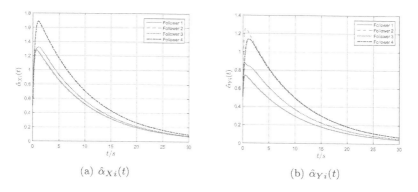

(a) $\hat{\alpha}_{Xi}(t)$ (b) $\hat{\alpha}_{Yi}(t)$

FIGURE 5.9: Adaptive gains in the distributed observer (5.21a).

The estimation errors $\|\tilde{v}_i(t)\|$ and the adaptive gains $\hat{\alpha}_i(t)$ are given in Figs. 5.8 and 5.9. From Figs. 5.6-5.9, we can see that both $\tilde{v}_i(t)$ and $\hat{\alpha}_i(t)$ are ultimately bounded, and the outputs of four followers can form a spanning square and tracking the leader with a relatively small error. Therefore, the desired formation tracking is achieved approximately by heterogeneous swarm system with a leader of unknown input.

5.4   Conclusions

High-order heterogeneous swarm systems were considered in this chapter, where each agent could have different dynamics matrices and state dimensions. Leaderless time-varying output formation control problems were investigated firstly. A distributed formation controller with hierarchical architecture and an algorithm to design control parameters were proposed, where the feasibility conditions of time-varying formation under heterogeneous dynamics were analyzed. In light of the state space decomposition approach and output regulation strategy, it was proved that the high-order heterogeneous swarm system can achieve the desired time-varying output formation, and the exact expression of the output formation reference function was also given. Furthermore, based on the formation control, the time-varying output formation tracking problem of high-order heterogeneous swarm systems with a leader of unknown input was further studied. A distributed adaptive observer using neighbouring interaction was constructed for each follower to estimate the leader state under the influence of its unknown input. Based on the distributed observer, the local state observer, and the desired formation vector, a time-varying output formation tracking controller and its corresponding design algorithm were presented using output regulation strategy and sliding-mode control theory. The feasibility conditions of time-varying output formation tracking for heterogeneous swarm systems and the approach to expand feasible formation set were studied. Using Lyapunov stability theory, the convergence of time-varying output formation tracking error was proved. The results in this chapter are mainly based on [106].

Chapter 6

Formation Tracking for Heterogeneous Swarm Systems with Multiple Leaders

6.1 Introduction

In Chapters 4 and 5, only one real/virtual leader was considered. However, in several applications, there are multiple leaders for the followers to track [9, 54, 108]. For example, in the cooperative reconnaissance application with a group of manned ground vehicles and UAVs, the UAVs are regarded as followers which fly around the ground vehicles with a desired formation. It is more robust for follower UAVs to track the convex combination of multiple leader vehicles rather than a specified leader. Since there are multiple leaders in the swarm system that need to be tracked, how to effectively describe the relationship between formation reference and multiple leaders and specify the macro motion of the whole formation is a challenging problem to be solved. Furthermore, in practical applications, each follower can only receive directly from part of the leaders by itself, and multiple leaders' information is incomplete for all the followers. How to design a time-varying formation tracking controller for high-order heterogeneous swarm system without relying on the well-informed follower assumption is more practical and challenging.

Formation tracking problems for homogeneous swarm systems with multiple identical leaders were firstly formalized in [9], where the achieved formation can track the state convex combination of the leaders. Based on [9], more constraints of swarm systems were considered, such as communication delays [54] and switching graphs [108]. In [9, 54, 108], it is assumed that each follower is well-informed or uninformed, where a well-informed follower can communicate with all the leaders and an uninformed follower has no leaders as its neighbour. This assumption is too restrictive since it forces some followers to receive from all the leaders directly. In practice, it is more possible and realistic that a follower only contains a subset of leaders as its neighbour. Thus, it is significant to further study the formation tracking approaches without depending on the well-informed follower assumption. In this case, all the followers cannot acquire the complete information of the multiple leaders, which makes the design and analysis more complicated. Besides, in containment control [110–112], all

DOI: 10.1201/9781003263470-6

the leaders were assumed to share identical dynamics, and this assumption is crucial to make the output regulation strategy applicable to multiple leaders. In practice, not only the followers but also the leaders can be heterogeneous. How to design formation tracking protocol for heterogeneous swarm systems with multiple non-identical leaders independent of the well-informed follower assumption is still open.

Inspired by the above facts and challenges, this chapter focuses on the time-varying formation tracking problems for heterogeneous swarm systems with multiple leaders, where the followers are required to achieve the desired time-varying formation and track the state convex combination of the multiple leaders simultaneously. The main contents of this chapter are summarized as follows.

Firstly, time-varying output formation tracking problems with switching directed topologies are studied based on the well-informed follower assumption. The followers are classified into the well-informed ones and the uninformed ones. Using the neighbouring relative information, a distributed observer is constructed for each follower to estimate the convex combination of multiple leaders' states under the influences of switching directed topologies. An output formation tracking protocol based on the distributed observer and an algorithm to determine the control parameters of the protocol are presented. Sufficient conditions for the heterogeneous swarm systems with multiple leaders and switching directed topologies to achieve the desired time-varying output formation tracking are proposed.

Furthermore, formation tracking problems for heterogeneous swarm systems are investigated without assuming that each follower is well-informed or uninformed. Besides the followers are heterogeneous, the multiple leaders can also have non-identical dynamics. Both the output and the dynamical matrices of each leader are only available to the followers which contain this leader as neighbour. Based on the local estimation and the interaction with neighbouring followers, a novel distributed observer is designed for each follower to estimate the dynamical matrices and the states of multiple leaders without requiring the well-informed follower assumption. Using the finite-time stability theory and the adaptive updating gains, the proposed observer can be designed in a totally distributed form by each follower with no need for the eigenvalue information of both the leaders' system matrices and the Laplacian matrix. An adaptive algorithm is proposed to solve the regulator equations in finite time based on the estimation of the leaders' matrices. Then, a time-varying formation tracking protocol is presented using the estimated states of multiple leaders, the online solutions of the regulator equations, and the desired formation vector generated by the local exosystem. It is proved that the outputs of the followers can not only realize the expected formation shape but also track the predefined convex combination of multiple leaders.

6.2 Time-varying Formation Tracking with Switching Directed Graphs

In practical applications, due to the formation transformation, complex terrain environment, electromagnetic interference, and so on, the communication links of swarm system would be vulnerable, and the communication edges between agents may be disconnected or reconnected, which leads to the switching of interaction topology and then changes the connectivity of swarm system. This section studies the time-varying output formation tracking problems for heterogeneous linear swarm systems with multiple leaders in the presence of switching directed topologies. The followers are classified into the well-informed ones and the uninformed ones. Firstly, using the neighbouring relative information, a distributed observer is constructed for each follower to estimate the convex combination of multiple leaders' states under the influences of switching directed topologies. Then, an output formation tracking protocol based on the distributed observer and an algorithm to determine the control parameters of the protocol are presented. Considering the features of heterogeneous dynamics, the time-varying formation tracking feasible constraints are provided, and a compensation input is applied to expand the feasible formation set. Finally, sufficient conditions for heterogeneous swarm systems with multiple leaders and switching directed topologies to achieve the desired time-varying output formation tracking are proposed.

6.2.1 Problem Description

The swarm system is composed of M ($M \geqslant 1$) followers and N ($N \geqslant 1$) leaders. If an agent has no neighbour, it is called a leader. The follower and leader sets are denoted by $\mathcal{F} = \{1, 2, \ldots, M\}$ and $\mathcal{E} = \{M+1, M+2, \ldots, M+N\}$, respectively. The followers are classified into well-informed and uninformed ones.

Definition 6.1 ([9]). *Follower i ($i \in \mathcal{F}$) is said to be well-informed if its neighbour set N_i contains all leaders. Follower j ($j \in \mathcal{F}$) is said to be uninformed if N_j contains no leaders. Let $\mathcal{F}_w$ be the well-informed follower set and $\mathcal{F}_u$ denote the uninformed follower set.*

It is assumed that the directed topology of the swarm system is switching. There is an infinite sequence of uniformly bounded non-overlapping time intervals $[t_\kappa, t_{\kappa+1})$ ($\kappa = 1, 2, 3 \ldots$) with $\tau_0 \leqslant t_{\kappa+1} - t_\kappa \leqslant \tau_1$ and $t_1 = 0$, where the positive constant τ_0 is called the dwell time. The interaction topology changes at the switching sequence $t_{\kappa+1}$ ($\kappa = 1, 2, 3 \ldots$). The index set of all possible graphs is described by $\{1, 2, \ldots, z\}$. Let $\sigma(t): [0, \infty) \to \{1, 2, \ldots, z\}$ denote the switching signal, then its value is the index of the graph. The directed graph and Laplacian matrix at $\sigma(t)$ are defined as $\mathcal{G}_{\sigma(t)}$ and $L_{\sigma(t)}$, respectively. For any $i \in \mathcal{F}_w$ and $j \in \mathcal{E}$, the weights from the leaders to the well-informed followers are defined as $w_{ij}^{\sigma(t)} = b_j$ with b_j being a positive constant, which means that the leader j has the same weight b_j to different well-informed followers at any $\sigma(t)$. For a given switching signal $\sigma(t)$, let $N_\sigma(T_1, T_2)$ denote the number of switching of $\sigma(t)$ over the interval (T_1, T_2).

Definition 6.2 ([108]). *For any $T_2 > T_1 \geqslant 0$, if the inequality $N_\sigma(T_1, T_2) \leqslant N_0 + \frac{T_2 - T_1}{\tau_a}$ holds for constants $N_0 \geqslant 0$ and $\tau_a > 0$, then τ_a is called the average dwell time.*

The dynamics of follower i ($i \in \mathcal{F}$) is described by

$$\begin{cases} \dot{x}_i(t) = A_i x_i(t) + B_i u_i(t), \\ y_i(t) = C_i x_i(t), \end{cases} \tag{6.1A}$$

where $x_i(t) \in \mathbb{R}^{n_i}$, $u_i(t) \in \mathbb{R}^{m_i}$ and $y_i(t) \in \mathbb{R}^p$ are the state, control input and output of the follower i, respectively. $A_i \in \mathbb{R}^{n_i \times n_i}$, $B_i \in \mathbb{R}^{n_i \times m_i}$ and $C_i \in \mathbb{R}^{p \times n_i}$ are constant known matrices with $\text{rank}(B_i) = m_i$. The pairs (A_i, B_i) are stabilizable and the pairs (C_i, A_i) are detectable. The dynamics of leader j ($j \in \mathcal{E}$) is described by

$$\begin{cases} \dot{v}_j(t) = S v_j(t), \\ y_j(t) = F v_j(t), \end{cases} \tag{6.1B}$$

where $v_j(t) \in \mathbb{R}^q$ and $y_j(t) \in \mathbb{R}^p$ denote the state and output of the leader j, respectively. $S \in \mathbb{R}^{q \times q}$ and $F \in \mathbb{R}^{p \times q}$ are constant known matrices, where S has no eigenvalues with negative real parts. The pair (F, S) is observable.

Remark 6.1. *The swarm system (6.1) consists of M followers and N leaders. The followers can have both heterogeneous system matrices (i.e., A_i, B_i and C_i) and different system dimensions (i.e., n_i and m_i). Since the swarm system is required to achieve a desired output formation tracking, all followers and leaders need to have the same output dimension p. If $N = 1$, the leader's dynamics (6.1B) is used widely in the cooperative output regulation works [84–86]. In the case where all agents have the same dynamics with $A \in \mathbb{R}^{n \times n}$, $B \in \mathbb{R}^{n \times m}$ and $C = I_n$, the swarm system model (6.1) becomes the one studied in [9]. Plenty of robot systems can be represented by linear models approximately in the formation control level. For example, using the feedback linearization technique in [41], the unicycle robots can be modeled as second-order integrator for formation control.*

Assumption 6.1. *The following regulator equations*

$$\begin{cases} X_i S = A_i X_i + B_i U_i, \\ 0 = C_i X_i - F, \end{cases} \tag{6.2}$$

have solution pairs (X_i, U_i), $i = 1, 2, \ldots, M$.

Assumption 6.1 is standard for cooperative control of heterogeneous swarm systems (see, e.g., [83–88]). The solvability of regulator equations (6.2) is a necessary condition for the output regulation problems. As shown in [149], the regulator equations (6.2) are solvable if

$$\text{rank} \begin{bmatrix} A_i - \lambda I_{n_i} & B_i \\ C_i & 0 \end{bmatrix} = n_i + p, \ i = 1, 2, \dots, M,$$

hold for all $\lambda \in \sigma(S)$.

The expected time-varying formation is specified by a vector $h_F(t) = [h_1^T(t), h_2^T(t), \dots, h_M^T(t)]^T$, where $h_i(t) \in \mathbb{R}^q$ is required to be piecewise continuously differentiable. The desired output formation vector is defined as $h_{yi}(t) = F h_i(t)$ $(i \in \mathcal{F})$.

Definition 6.3. *For any given bounded initial states and follower i ($i \in \mathcal{F}$), if there are positive constants β_j ($j = M + 1, M + 2, \dots, M + N$) with $\sum_{j=M+1}^{M+N} \beta_j = 1$ such that*

$$\lim_{t \to \infty} \left(y_i(t) - h_{yi}(t) - \sum_{j=M+1}^{M+N} \beta_j y_j(t) \right) = 0, \tag{6.3}$$

then swarm system (6.1) is said to accomplish time-varying output formation tracking with multiple leaders.

To explain Definition 6.3 more intuitively, consider the following example for formation tracking with multiple leaders. As shown in Fig. 6.1, during the cooperative flying of multiple heterogeneous manned/unmanned combat aerial vehicles (CAVs), a fleet of unmanned CAVs can keep desired time-varying tactical formation centred by the convex combination of the positions of all the available manned CAVs to enclose them. In this configuration, those dangerous tasks such as reconnaissance and attack can be accomplished by the unmanned CAVs, and the safety of the manned CAVs can be guaranteed due to that the defense system of the opponent will be triggered and consumed by the unmanned CAVs flying outside. It should be pointed out that the tracking target for the formation is the convex combination of all the available manned CAVs instead of one of the specific manned CAVs. If some of the manned CAVs fail or crash, the tracking target becomes the convex combination of the remaining available manned CAVs. If the tracking target for the formation is one specific manned CAV, the failure or crash of this manned CAV may destroy the whole mission since no tracking target is available for the formation. Therefore, tracking the convex combination of all the available manned CAVs is more robust than tracking one of the specific manned CAVs.

Remark 6.2. *In Definition 6.3, $\sum_{j=M+1}^{M+N} \beta_j y_j(t)$ stands for the convex combination of the leaders' outputs. When the output formation tracking is accomplished, all followers regard $\sum_{j=M+1}^{M+N} \beta_j y_j(t)$ as the formation reference*

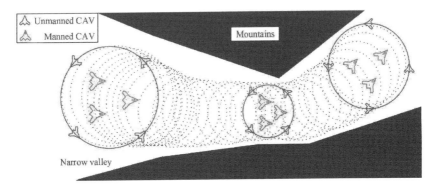

FIGURE 6.1: An example for formation tracking with multiple leaders.

and maintain the time-varying offsets $h_{yi}(t)$ $(i = 1, 2, \ldots, M)$ relative to it. Note that $\sum_{j=M+1}^{M+N} \beta_j y_j(t)$ only determines the common formation reference of the followers, while the desired formation pattern of the followers is specified by time-varying vectors $h_{yi}(t)$ $(i = 1, 2, \ldots, M)$. If $N = 1$, Definition 3 can be transformed into $\lim_{t \to \infty}(y_i(t) - h_{yi}(t) - y_{M+1}(t)) = 0$, which means that the expected output formation tracking with one leader is accomplished in Chapters 4 and 5. If $N = 1$ and $\sum_{i=1}^{M} h_{yi}(t) = 0$, it follows from (6.3) that $\lim_{t \to \infty}(y_{M+1}(t) - \sum_{i=1}^{M} y_i(t)/M) = 0$, then the swarm systems achieve the target enclosing with one target [6].

6.2.2 Formation Tracking Controller Design and Stability Analysis with Switching Graphs

In this section, in order to acquire the convex combination of leaders' states, a distributed observer with neighbouring relative information is presented for each follower firstly. The convergence of the observer is proved based on the piecewise Lyapunov theory and the threshold for the average dwell time of the switching topologies is derived. Then, a formation tracking protocol and an algorithm to design the control parameters are given. Finally, it is proved that the output formation tracking errors of followers can converge to zero using the proposed approach.

Assumption 6.2. *Each follower is well-informed or uninformed. For each directed graph $\mathcal{G}_{\sigma(t)}$ and uninformed follower i $(i \in \mathcal{F}_u)$, there is at least one well-informed follower which has a directed path to this uninformed one.*

Remark 6.3. *The requirement about the well-informed followers is to make all followers acquire the same convex combination of the leaders' states, which means that the followers can reach an agreement on the formation reference. In practical applications, the followers with better communication and detection*

*capabilities are considered to be well-informed, and other followers with rela-
tively weak capabilities are regarded as uninformed ones.*

Since the leaders have no neighbour, the Laplacian matrix $L_{\sigma(t)}$ has the
following form

$$L_{\sigma(t)} = \begin{bmatrix} L_{1\sigma(t)} & L_{2\sigma(t)} \\ 0_{N \times M} & 0_{N \times N} \end{bmatrix},$$

where $L_{1\sigma(t)} \in \mathbb{R}^{M \times M}$ and $L_{2\sigma(t)} \in \mathbb{R}^{M \times N}$. Under Assumption 2, the following two lemmas are given.

Lemma 6.1 ([154]). *Under Assumption 6.2, all the eigenvalues of $L_{1\sigma(t)}$
have positive real parts, and there exists a diagonal matrix $G_{\sigma(t)} =
diag\{g_{1\sigma(t)}, g_{2\sigma(t)}, \ldots, g_{M\sigma(t)}\}$ with $g_{i\sigma(t)} > 0$ such that $\Xi_{\sigma(t)} = G_{\sigma(t)} L_{1\sigma(t)} +
L_{1\sigma(t)}^T G_{\sigma(t)} > 0$. Moreover, one such $G_{\sigma(t)}$ can be determined by*

$$[g_{1\sigma(t)}, g_{2\sigma(t)}, \ldots, g_{M\sigma(t)}]^T = (L_{1\sigma(t)}^T)^{-1} \mathbf{1}_M.$$

Lemma 6.2 ([9]). *Under Assumption 6.2, each entry of $-L_{1\sigma(t)}^{-1} L_{2\sigma(t)}$ is
nonnegative, and each row of $-L_{1\sigma(t)}^{-1} L_{2\sigma(t)}$ is identical and has a sum equal
to one. $-L_{1\sigma(t)}^{-1} L_{2\sigma(t)}$ has the following form:*

$$-L_{1\sigma(t)}^{-1} L_{2\sigma(t)} = \mathbf{1}_M \otimes \frac{[b_{M+1}, b_{M+2}, \ldots, b_{M+N}]}{\sum_{k=M+1}^{M+N} b_k}.$$

Since not all followers can obtain the leaders' outputs directly, it is necessary to construct a distributed observer by using the neighbouring relative
information. Consider the following distributed observer:

$$\dot{\hat{\zeta}}_i(t) = S\hat{\zeta}_i(t) - \mu K \left(\sum_{j=1}^{M} w_{ij}(\hat{y}_i(t) - \hat{y}_j(t)) + \sum_{k=M+1}^{M+N} w_{ik}(\hat{y}_i(t) - y_k(t)) \right), \quad (6.4)$$

where $\hat{\zeta}_i(t) \in \mathbb{R}^q$ and $\hat{y}_i(t) = F\hat{\zeta}_i(t)$, $i \in \mathcal{F}$. $y_k(t) = Fv_k(t)$ denotes the
output of the leader k ($k \in \mathcal{E}$). μ is a positive constant to be determined
later. The constant gain matrix $K \in \mathbb{R}^{q \times p}$ is designed as $K = P^{-1} F^T$, and P
is a positive definite matrix satisfying the following linear matrix inequality
(LMI):

$$S^T P + PS - F^T F + \alpha P < 0, \qquad (6.5)$$

where α is a given positive constant.

Remark 6.4. *The outputs of the leaders $y_k(t)$ ($k \in \mathcal{E}$) are applied to construct
the distributed observer (6.4), while the virtual leader's state information is
used in [83, 86, 87]. The proposed observer (6.4) is dependent on its neighbouring relative output information, rather than the relative state information, which can reduce communication burden effectively. As shown in [155],
for any given positive constant α, the LMI (6.5) is solvable if the pair (F, S)
is observable.*

The following theorem guarantees that $\hat{\zeta}_i(t)$ $(i \in \mathcal{F})$ will converge to the same convex combination of the leaders' states with the switching directed topologies.

Theorem 6.1. *Suppose that Assumption 6.2 holds. The distributed observer (6.4) will converge to the same convex combination of the leaders' states, i.e.,*

$$\lim_{t \to \infty} \left(\hat{\zeta}_i(t) - \frac{1}{\sum_{k=M+1}^{M+N} b_k} \sum_{j=M+1}^{M+N} b_j v_j(t) \right) = 0, \ i \in \mathcal{F},$$

if μ in (6.4) is chosen as $\mu \geqslant \max \left\{ \frac{\lambda_{\max}(G_{\sigma(t)})}{\lambda_{\min}(\Xi_{\sigma(t)})} \right\}$ $(\sigma(t) \in \{1, 2, \ldots, z\})$, and the average dwell time of the switching topologies satisfies

$$\tau_a > \frac{\ln \varphi}{\alpha}, \tag{6.6}$$

where $\varphi = \max \left\{ \frac{\lambda_{\max}(G_i)}{\lambda_{\min}(G_j)} \right\}$ $(i, j \in \{1, 2, \ldots, z\})$.

Proof. Let $\hat{\zeta}_F(t) = [\hat{\zeta}_1^T(t), \hat{\zeta}_2^T(t), \ldots, \hat{\zeta}_M^T(t)]^T$ and $v_L(t) = [v_{M+1}^T(t), v_{M+2}^T(t), \ldots, v_{M+N}^T(t)]^T$. Then, the distributed observer (6.4) can be rewritten in the following compact form:

$$\dot{\hat{\zeta}}_F(t) = (I_M \otimes S) \hat{\zeta}_F(t) - \mu \left(L_{1\sigma(t)} \otimes P^{-1} F^T F \right) \hat{\zeta}_F(t)$$
$$- \mu \left(L_{2\sigma(t)} \otimes P^{-1} F^T F \right) v_L(t). \tag{6.7}$$

Let $\tilde{\zeta}_F(t) = \hat{\zeta}_F(t) - (-L_{1\sigma(t)}^{-1} L_{2\sigma(t)} \otimes I_q) v_L(t)$. It follows from (6.7) that

$$\dot{\tilde{\zeta}}_F(t) = \left(I_M \otimes S - \mu \left(L_{1\sigma(t)} \otimes P^{-1} F^T F \right) \right) \tilde{\zeta}_F(t). \tag{6.8}$$

In order to check the stability of the switched linear system (6.8), consider the following piecewise Lyapunov functional candidate:

$$V(t) = \tilde{\zeta}_F^T(t) \left(G_{\sigma(t)} \otimes P \right) \tilde{\zeta}_F(t). \tag{6.9}$$

The graph $\mathcal{G}_{\sigma(t)}$ keeps fixed in each interval $[t_\kappa, t_{\kappa+1})$ $(\kappa = 1, 2, 3 \ldots)$. For any $t \in [t_\kappa, t_{\kappa+1})$, the time derivative of $V(t)$ along (6.8) can be obtained as

$$\dot{V}(t) = \tilde{\zeta}_F^T(t) \left(G_{\sigma(t)} \otimes (PS + S^T P) \right) \tilde{\zeta}_F(t) - \mu \tilde{\zeta}_F^T(t) \left(\Xi_{\sigma(t)} \otimes F^T F \right) \tilde{\zeta}_F(t), \tag{6.10}$$

where $\Xi_{\sigma(t)} = G_{\sigma(t)} L_{1\sigma(t)} + L_{1\sigma(t)}^T G_{\sigma(t)} > 0$.

Let μ be sufficiently large such that $\mu \geqslant \max \left\{ \frac{\lambda_{\max}(G_{\sigma(t)})}{\lambda_{\min}(\Xi_{\sigma(t)})} \right\}$ $(\sigma(t) \in \{1, 2, \ldots, z\})$. It holds from (6.10) that

$$\dot{V}(t) \leqslant \tilde{\zeta}_F^T(t) \left(G_{\sigma(t)} \otimes \left(PS + S^T P - F^T F \right) \right) \tilde{\zeta}_F(t). \tag{6.11}$$

Substituting the LMI (6.5) into (6.11) yields

$$\dot{V}(t) \leqslant -\alpha \tilde{\zeta}_F^T(t)\left(G_{\sigma(t)} \otimes P\right)\tilde{\zeta}_F(t) = -\alpha V(t). \tag{6.12}$$

From (6.12), one gets

$$V\left(t_{\kappa+1}^-\right) \leqslant e^{-\alpha(t_{\kappa+1}-t_\kappa)}V(t_\kappa). \tag{6.13}$$

Note that

$$\lambda_{\min}(G_{\sigma(t)})\tilde{\zeta}_F^T(t)(I_M \otimes P)\tilde{\zeta}_F(t) \leqslant V(t) \leqslant \lambda_{\max}(G_{\sigma(t)})\tilde{\zeta}_F^T(t)(I_M \otimes P)\tilde{\zeta}_F(t).$$

It follows from Lemma 6.2 that $\tilde{\zeta}_F(t)$ is continuous. Since the system (6.8) switches at the time instant $t_{\kappa+1}$, one can obtain

$$V(t_{\kappa+1}) \leqslant \varphi V\left(t_{\kappa+1}^-\right), \tag{6.14}$$

where $\varphi = \max\left\{\frac{\lambda_{\max}(G_i)}{\lambda_{\min}(G_j)}\right\}$ $(i, j \in \{1, 2, \ldots, z\})$. It follows from (6.13) and (6.14) that

$$V(t_{\kappa+1}) \leqslant \varphi e^{-\alpha(t_{\kappa+1}-t_\kappa)}V(t_\kappa). \tag{6.15}$$

For any $t \in [t_{\kappa+1}, t_{\kappa+2})$, with recursion approach, one can obtain from (6.15) that

$$\begin{aligned} V(t) &\leqslant e^{-\alpha(t-t_{\kappa+1})}V(t_{\kappa+1}) \\ &\leqslant \varphi e^{-\alpha(t-t_\kappa)}V(t_\kappa) \\ &\leqslant \varphi^\kappa e^{-\alpha(t-t_1)}V(t_1). \end{aligned} \tag{6.16}$$

From Definition 6.2, one gets $\kappa \leqslant N_0 + \frac{t-t_1}{\tau_a}$. Then it follows from (6.16) that

$$\begin{aligned} V(t) &\leqslant \varphi^{(t-t_1)/\tau_a}e^{-\alpha(t-t_1)}\varphi^{N_0}V(t_1) \\ &= e^{-\left(\alpha-\frac{\ln\varphi}{\tau_a}\right)(t-t_1)}\varphi^{N_0}V(t_1). \end{aligned} \tag{6.17}$$

Since the average dwell time $\tau_a > \frac{\ln\varphi}{\alpha}$, one can obtain from (6.17) that $\lim_{t\to\infty}\tilde{\zeta}_F(t) = 0$, i.e., $\lim_{t\to\infty}(\hat{\zeta}_F(t) - (-L_{1\sigma(t)}^{-1}L_{2\sigma(t)} \otimes I_q)v_L(t)) = 0$. It holds from Lemma 6.2 that $\hat{\zeta}_i(t)$ $(i \in \mathcal{F})$ in the distributed observer (6.4) converges to the same convex combination of the leaders' states. This completes the proof of Theorem 6.1. $\qquad\square$

Remark 6.5. *From Theorem 6.1, we see that all followers can obtain the same convex combination of the multiple leaders under the influences of switching directed topologies. The dwell time for each time interval is required to be larger than some positive threshold in [33], while only the lower bound of the average dwell time is needed in the current section. By using the average dwell time,*

Theorem 6.1 can be applied to deal with more general switching directed graphs than the cases in [33]. The jointly connected graphs were studied in [86] and [156] under the assumption that S has no eigenvalues with positive real parts, where the leader's state information is used and there exists only one leader. However, in this section, only the output information of the multiple leaders is available and S may have eigenvalues with positive real parts. Therefore, the approaches in [86] and [156] are not directly applicable to the formation tracking problems discussed in this section.

Based on the distributed observer (6.4), for follower i $(i \in \mathcal{F})$, consider the following output formation tracking protocol:

$$
\begin{aligned}
u_i(t) &= K_{1i}\hat{x}_i(t) + K_{2i}\left(\hat{\zeta}_i(t) + h_i(t)\right) + r_i(t), \\
\dot{\hat{x}}_i(t) &= A_i\hat{x}_i(t) + B_iu_i(t) + L_{oi}\left(C_i\hat{x}_i(t) - y_i(t)\right),
\end{aligned}
\tag{6.18}
$$

where $\hat{x}_i(t) \in \mathbb{R}^{n_i}$ is the estimate of $x_i(t)$, $r_i(t) \in \mathbb{R}^{m_i}$ denotes the time-varying formation tracking compensation input, and K_{1i}, K_{2i} and L_{oi} are constant gain matrices to be determined later.

Since B_i is required to be of full column rank, there is a non-singular matrix $\Gamma_i = [\hat{B}_i^T, \tilde{B}_i^T]^T$ such that $\hat{B}_iB_i = I_{m_i}$ and $\tilde{B}_iB_i = 0$, where $\hat{B}_i \in \mathbb{R}^{m_i \times n_i}$ and $\tilde{B}_i \in \mathbb{R}^{(n_i-m_i) \times n_i}$ $(i \in \mathcal{F})$. In the following, an algorithm to determine the control parameters in (6.18) is given.

Algorithm 6.1. *For follower i $(i \in \mathcal{F})$, the time-varying output formation tracking protocol (6.18) can be designed with the following five steps.*
Step 1: *Solve the regulator equation (6.2) for the pair (X_i, U_i).*
Step 2: *For a given time-varying formation vector $h_F(t)$, check the following formation tracking feasible condition:*

$$
\lim_{t \to \infty}\left(\tilde{B}_iX_i\left(Sh_i(t) - \dot{h}_i(t)\right)\right) = 0.
\tag{6.19}
$$

If (6.19) is satisfied, then continue, otherwise $h_F(t)$ is not feasible under the formation tracking protocol (6.18) and the algorithm stops.
Step 3: *The compensation input $r_i(t)$ is described by*

$$
r_i(t) = -\hat{B}_iX_i\left(Sh_i(t) - \dot{h}_i(t)\right).
\tag{6.20}
$$

Step 4: *Based on Theorem 6.1, construct the distributed observer (6.4) for the swarm system (6.1) with switching topologies to get the estimate of the convex combination of the leaders' states (i.e., $\hat{\zeta}_i(t)$).*
Step 5: *Choose K_{1i} such that $A_i + B_iK_{1i}$ is Hurwitz, and K_{2i} is designed as $K_{2i} = U_i - K_{1i}X_i$. L_{oi} is selected such that $A_i + L_{oi}C_i$ is Hurwitz.*

Remark 6.6. *Similar to Chapter 5, considering the influences of hetero-geneous dynamics of followers and leaders, the formation tracking feasible condition is presented as (6.19), and the compensation input $r_i(t)$ $(i \in \mathcal{F})$*

calculated by (6.20) is used to extend the feasible time-varying formation set. If all the agents have identical dynamics with state dimension n, it can be verified that $X_i = I_n$. Then the feasible condition (6.19) can be simplified to the constraints in [9]. In [118, 119], the desired time-varying formation is required to have the same dynamics as the virtual leader (i.e., $\dot{h}_i(t) = Sh_i(t)$), which can be viewed as a special case of the condition (6.19).

The following theorem shows that the output formation tracking can be accomplished by swarm system (6.1).

Theorem 6.2. *Suppose that Assumptions 6.1 and 6.2 hold. If the expected time-varying formation $h_F(t)$ satisfies the feasible condition (6.19) and the average dwell time of the switching topologies satisfies (6.6), then swarm system (6.1) with multiple leaders and switching topologies can accomplish the time-varying output formation tracking under the designed control protocol (6.18).*

Proof. Let $\tilde{x}_i(t) = \hat{x}_i(t) - x_i(t)$ ($i \in \mathcal{F}$). Then, it follows from (6.18) that

$$\dot{\tilde{x}}_i(t) = (A_i + L_{oi}C_i)\,\tilde{x}_i(t). \tag{6.21}$$

Since L_{oi} is chosen such that $A_i + L_{oi}C_i$ is Hurwitz, one can obtain $\lim_{t\to\infty}\tilde{x}_i(t) = 0$. Substituting (6.18) into (6.1A) leads to

$$\dot{x}_i(t) = (A_i + B_iK_{1i})\,x_i(t) + B_iK_{1i}\tilde{x}_i(t) + B_i\,(U_i - K_{1i}X_i)\left(\hat{\zeta}_i(t) + h_i(t)\right)$$
$$+ B_ir_i(t). \tag{6.22}$$

Let $\zeta(t) = \frac{1}{\sum_{k=M+1}^{M+N} b_k}\sum_{j=M+1}^{M+N} b_jv_j(t)$, $\tilde{\zeta}_i(t) = \hat{\zeta}_i(t) - \zeta(t)$ and $\vartheta_i(t) = x_i(t) - X_i(\zeta(t) + h_i(t))$ ($i = 1, 2, \ldots, M$). It holds from (6.22) that

$$\dot{\vartheta}_i(t) = (A_i + B_iK_{1i})\,\vartheta_i(t) + B_iK_{1i}\tilde{x}_i(t) + B_iK_{2i}\tilde{\zeta}_i(t)$$
$$+ (A_iX_i + B_iU_i)\,\zeta(t) - X_i\dot{\zeta}(t)$$
$$+ (A_iX_i + B_iU_i)\,h_i(t) + B_ir_i(t) - X_i\dot{h}_i(t). \tag{6.23}$$

Note that $\dot{\zeta}(t) = S\zeta(t)$. It follows from (6.2) and (6.23) that

$$\dot{\vartheta}_i(t) = (A_i + B_iK_{1i})\,\vartheta_i(t) + B_iK_{1i}\tilde{x}_i(t) + B_iK_{2i}\tilde{\zeta}_i(t)$$
$$+ X_i\left(Sh_i(t) - \dot{h}_i(t)\right) + B_ir_i(t). \tag{6.24}$$

Since the formation tracking feasible condition (6.19) is satisfied, one can obtain

$$\lim_{t\to\infty}\left(\tilde{B}_iX_i\left(Sh_i(t) - \dot{h}_i(t)\right) + \tilde{B}_iB_ir_i(t)\right) = 0. \tag{6.25}$$

It follows from (6.20) that

$$\hat{B}_i X_i \left(Sh_i(t) - \dot{h}_i(t) \right) + \hat{B}_i B_i r_i(t) = 0. \tag{6.26}$$

Note that the matrix $\Gamma_i = [\hat{B}_i^T, \tilde{B}_i^T]^T$ is nonsingular. Then it holds from (6.25) and (6.26) that

$$\lim_{t \to \infty} \left(X_i \left(Sh_i(t) - \dot{h}_i(t) \right) + B_i r_i(t) \right) = 0. \tag{6.27}$$

If the average dwell time of the switching topologies satisfies (6.6), one gets from Theorem 6.1 that $\lim_{t \to \infty} \tilde{\zeta}_i(t) = 0$. Since K_{1i} is selected to make $A_i + B_i K_{1i}$ Hurwitz in Algorithm 6.1, it follows from (6.24) and (6.27) that $\lim_{t \to \infty} \vartheta_i(t) = 0$, i.e.,

$$\lim_{t \to \infty} (x_i(t) - X_i(\zeta(t) + h_i(t))) = 0. \tag{6.28}$$

Let $e_i(t)$ represent the output formation tracking error for follower i ($i \in \mathcal{F}$) and $e_i(t) = y_i(t) - h_{yi}(t) - \sum_{j=M+1}^{M+N} \beta_j y_j(t)$. One can obtain from (6.28) that

$$\lim_{t \to \infty} (C_i x_i(t) - C_i X_i(\zeta(t) + h_i(t))) = 0. \tag{6.29}$$

Substituting the regulator equation (6.2) into (6.29) gives

$$\lim_{t \to \infty} (y_i(t) - h_{yi}(t) - F\zeta(t)) = 0. \tag{6.30}$$

According to the definition of $\zeta(t)$, it follows from (6.30) that

$$\lim_{t \to \infty} \left(y_i(t) - h_{yi}(t) - \frac{1}{\sum_{k=M+1}^{M+N} b_k} \sum_{j=M+1}^{M+N} b_j y_j(t) \right) = 0. \tag{6.31}$$

Let $\beta_j = b_j / (\sum_{k=M+1}^{M+N} b_k)$, $j = M+1, M+2, \ldots, M+N$. Then it can be verified that $\lim_{t \to \infty} e_i(t) = 0$ ($i \in \mathcal{F}$), which means that the expected time-varying output formation tracking is accomplished by swarm system (6.1) with multiple leaders. This completes the proof of Theorem 6.2. $\qquad\square$

Remark 6.7. *From (6.31), it can be obtained that all followers regard the same convex combination of multiple leaders' outputs as the formation reference and maintain time-varying offsets $h_{yi}(t)$ relative to it. In the case where $N = 1$, Theorem 6.2 can be applied directly to deal with the time-varying formation tracking problems for heterogeneous swarm systems with one leader. By choosing the formation vector $h_F(t)$ properly, certain time-invariant formation tracking problems, target enclosing problems and cooperative output regulation problems for a group of heterogeneous agents can be solved by the proposed approach. Moreover, if all agents have the same dynamics with $A \in \mathbb{R}^{n \times n}$, $B \in \mathbb{R}^{n \times m}$ and $C = I_n$ and the interaction topologies are fixed, Theorem 2 in [9] can be regarded as a special case of Theorem 6.2 in this section.*

6.2.3 Numerical Simulations

In order to verify the effectiveness of the theoretical results, two simulation examples are provided in this section.

Example 6.1. Consider a swarm system consisting of six heterogeneous followers and three leaders, where the follower set $\mathcal{F} = \{1, 2, \ldots, 6\}$ and the leader set $\mathcal{E} = \{7, 8, 9\}$. These agents move in the XY plane. As shown in Fig. 6.2, it is assumed that there exist two possible directed topologies $\mathcal{G}_1$ and $\mathcal{G}_2$ with 0-1 weights.

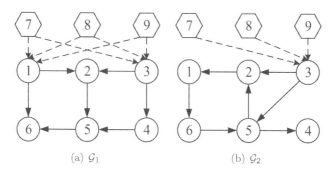

(a) $\mathcal{G}_1$ (b) $\mathcal{G}_2$

FIGURE 6.2: Switching directed graphs.

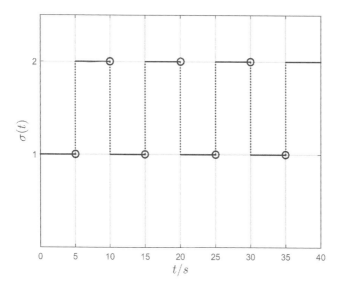

FIGURE 6.3: Switching signal.

These three leaders are assumed to have the dynamics (6.1B) with $S = I_2 \otimes \begin{bmatrix} 0 & 1 \\ 0 & 0 \end{bmatrix}$, $F = I_2 \otimes \begin{bmatrix} 1 & 0 \end{bmatrix}$ and $v_j(t) = [\chi_{Xj}(t), \upsilon_{Xj}(t), \chi_{Yj}(t), \upsilon_{Yj}(t)]^T$

$(j = 7, 8, 9)$, where $\chi_{Xj}(t)$, $\upsilon_{Xj}(t)$, $\chi_{Yj}(t)$ and $\upsilon_{Yj}(t)$ represent the position and velocity along the X-axis and Y-axis, respectively. Consider the following heterogeneous dynamics of the followers:

$$A_1 = A_2 = I_2 \otimes \begin{bmatrix} 0 & 1 \\ 0 & 0 \end{bmatrix}, \; B_1 = B_2 = I_2 \otimes \begin{bmatrix} 0 \\ 1 \end{bmatrix}, \; C_1 = C_2 = I_2 \otimes [\, 1 \quad 0 \,],$$

$$A_3 = 0_{2\times2}, \; B_3 = I_2, \; C_3 = I_2, \; A_4 = A_5 = I_2 \otimes \begin{bmatrix} 0 & 1 & 0 \\ 0 & 0 & 1 \\ -1 & -2 & -1 \end{bmatrix}, \; B_4 =$$

$$B_5 = I_2 \otimes \begin{bmatrix} 0 & 0 \\ 0 & 1 \\ 1 & 0 \end{bmatrix}, \; A_6 = I_2 \otimes \begin{bmatrix} 0 & 1 & 0 \\ 0 & 0 & 1 \\ 2 & -2 & 1 \end{bmatrix}, \; B_6 = I_2 \otimes \begin{bmatrix} 0 & 0 \\ 1 & 0 \\ 0 & 1 \end{bmatrix}, \; \text{and}$$

$C_4 = C_5 = C_6 = I_2 \otimes [\, 1 \quad 0 \quad 0 \,]$.

According to the characteristics of F and C_i $(i = 1, 2, \ldots, 6)$, one can obtain that the output of each agent is its position in the XY plane. Since output formation tracking problems are considered in this section, only the figures of position are provided in the following. Note that follower 3 is assumed to have first-order dynamics, which means that it is impossible to plot the velocity for the heterogeneous swarm systems.

The six heterogeneous followers are required to accomplish a rotating circular formation tracking, where the expected time-varying formation vector is specified by

$$h_i(t) = \begin{bmatrix} 4\cos(t + (i-1)\pi/3) \\ -4\sin(t + (i-1)\pi/3) \\ 4\sin(t + (i-1)\pi/3) \\ 4\cos(t + (i-1)\pi/3) \end{bmatrix}, \; i = 1, 2, \ldots, 6.$$

When the desired formation tracking is accomplished, the positions of the six followers will form a regular hexagon and rotate around the convex combination of the three leaders' positions with $r = 4$ m and $\omega = 1$ rad/s. Since the possible graphs have 0-1 weights, it can be verified from Lemma 6.2 that the convex combination of the leaders' states is calculated by $\zeta(t) = (v_7(t) + v_8(t) + v_9(t))/3$. Follow the steps in Algorithm 6.1 to determine the formation tracking protocol (6.18) and the distributed observer (6.4). For each follower, solving the regulator equation (6.2) gives $X_1 = X_2 = I_2 \otimes \begin{bmatrix} 1 & 0 \\ 0 & 1 \end{bmatrix}$,

$U_1 = U_2 = I_2 \otimes [\, 0 \quad 0 \,]$, $X_3 = I_2 \otimes [\, 1 \quad 0 \,]$, $U_3 = I_2 \otimes [\, 0 \quad 1 \,]$,

$X_4 = X_5 = I_2 \otimes \begin{bmatrix} 1 & 0 \\ 0 & 1 \\ 0 & 0 \end{bmatrix}$, $U_4 = U_5 = I_2 \otimes \begin{bmatrix} 1 & 2 \\ 0 & 0 \end{bmatrix}$, $X_6 = I_2 \otimes \begin{bmatrix} 1 & 0 \\ 0 & 1 \\ 0 & 0 \end{bmatrix}$,

and $U_6 = I_2 \otimes \begin{bmatrix} 0 & 0 \\ -2 & 2 \end{bmatrix}$. Let $\hat{B}_1 = I_2 \otimes [\, 0 \quad 1 \,]$ and $\tilde{B}_1 = I_2 \otimes [\, 1 \quad 0 \,]$

such that $\hat{B}_1 B_1 = I_2$ and $\tilde{B}_1 B_1 = 0_{2\times2}$. It can be verified that follower 1 satisfies the formation tracking feasible condition (6.19). Then one can obtain from (6.20) that $r_1(t) = [-4\cos(t), -4\sin(t)]^T$. Similarly, the condition

(6.19) is satisfied for all the other followers and the compensation inputs $r_i(t)$ $(i = 2, 3, \ldots, 6)$ are described by $r_2(t) = [-4\cos(t + \pi/3), -4\sin(t + \pi/3)]^T$, $r_3(t) = 0_{2\times 1}$, $r_4(t) = [0, -4\cos(t + \pi), 0, -4\sin(t + \pi)]^T$, $r_5(t) = [0, -4\cos(t + 4\pi/3), 0, -4\sin(t + 4\pi/3)]^T$, and $r_6(t) = [-4\cos(t + 5\pi/3), 0, -4\sin(t + 5\pi/3), 0]^T$.

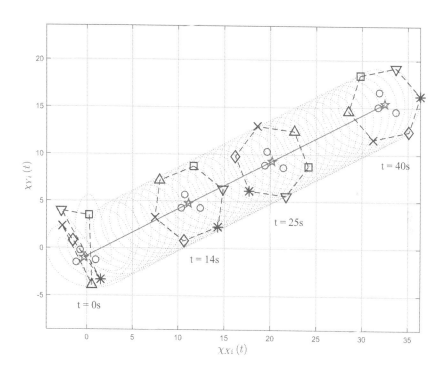

FIGURE 6.4: Position trajectories within $t = 40$s and snapshots at different time instants of the swarm system.

Let $\alpha = 1$. Solving the LMI (6.5) gives $P = I_2 \otimes \begin{bmatrix} 0.5927 & -0.4947 \\ -0.4947 & 0.6974 \end{bmatrix}$. From Theorem 6.1, one can obtain that $\mu \geqslant 3.03$ and $\tau_a > 2.16$. Therefore, μ in the distributed observer (6.4) is chosen as $\mu = 5$. The switching period between $\mathcal{G}_1$ and $\mathcal{G}_2$ is set to be 5 s, and the switching signal is shown in Fig. 6.3. In order to make $A_i + B_i K_{1i}$ and $A_i + L_{oi} C_i$ $(i = 1, 2, \ldots, 6)$ Hurwitz, the matrices K_{1i} and L_{oi} are selected as $K_{11} = K_{12} = I_2 \otimes [-2 \quad -2]$, $L_{o1} = L_{o2} = I_2 \otimes [-9, -20]^T$, $K_{13} = -I_2$, $L_{o3} = -5I_2$, $K_{14} = K_{15} = I_2 \otimes \begin{bmatrix} 1 & 2 & -1 \\ -2 & -2 & -1 \end{bmatrix}$, $L_{o4} = L_{o5} = I_2 \otimes [-14, -58, -33]^T$, $K_{16} = I_2 \otimes \begin{bmatrix} -2 & -2 & -1 \\ -2 & 2 & -3 \end{bmatrix}$, and $L_{o6} = I_2 \otimes [-16, -88, -178]^T$. The initial states of the nine agents in the XY plane are generated by $x_{1,2}(0) = [8\varpi, \varpi, 8\varpi, \varpi]^T$, $x_3(0) = [8\varpi, 8\varpi]^T$, $x_{4,5,6}(0) = [8\varpi, \varpi, \varpi, 8\varpi, \varpi, \varpi]^T$, and $v_{7,8,9}(0) = [3\varpi, 0.8 + 0.05\varpi, 3\varpi, 0.4 + 0.05\varpi]^T$. The initial values of $\hat{\zeta}_i(t)$ and $\hat{x}_i(t)$ $(i = 1, 2, \ldots, 6)$ are set to be ϖ with appropriate sizes, where ϖ denotes a random number between -0.5 and 0.5.

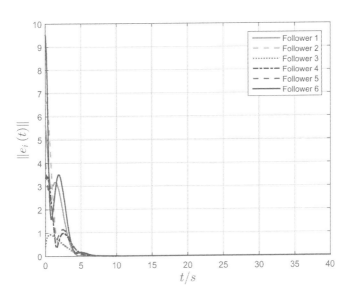

FIGURE 6.5: Curves of the output formation tracking errors for each follower.

Fig. 6.4 shows the position trajectories within $t = 40$s and the position snapshots at different time instants $(t = 0, 14, 25, 40$s$)$ of the swarm system, where the six followers are represented by square, upward-pointing triangle, cross, diamond, asterisk, and downward-pointing triangle respectively, the three leaders are denoted by circles, and the convex combination of the leaders is marked by five-pointed star. The trajectories of the six followers and the convex combination are denoted by the dotted lines and solid line, respectively. Fig. 6.5 gives the Euclidean norm of the output formation tracking error $\|e_i(t)\|$ $(i = 1, 2, \ldots, 6)$ for each follower. The estimate errors of the distributed observer (6.4) and the state observer in (6.18) are depicted in Figs. 6.6 and 6.7 respectively, where one sees that the distributed observer (6.4) of each follower can converge to the same convex combination of the leaders' states, and all the followers can estimate their own states using the state observer in (6.18). From Figs. 6.4 and 6.5, one can obtain that the positions of the six followers form a regular hexagon formation, and the convex combination of the three leaders' positions locates in the centre of the formation. Moreover, the achieved formation keeps rotating around the leaders. Therefore, the expected time-varying output formation tracking is accomplished by the heterogeneous swarm system with multiple leaders and switching directed topologies.

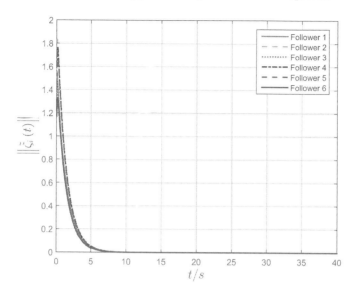

FIGURE 6.6: Estimate errors of the distributed observer (6.4) for each follower.

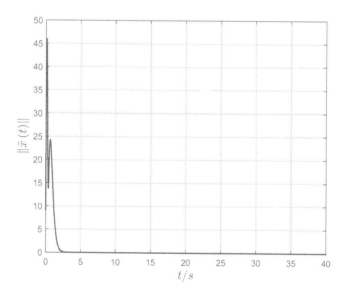

FIGURE 6.7: Estimate error of the state observer in (6.18).

Example 6.2. Consider a heterogeneous multi-robot system composed of a group of UGVs and UAVs. These robots are required to perform an air-ground cooperative surveillance task in a predefined formation pattern. Assume that there are eight followers $\mathcal{F} = \{1, 2, \ldots, 8\}$ and two leaders $\mathcal{E} = \{9, 10\}$, where all the possible graphs are shown in Fig. 6.8.

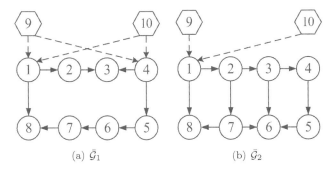

(a) $\bar{\mathcal{G}}_1$ (b) $\bar{\mathcal{G}}_2$

FIGURE 6.8: Switching directed graphs.

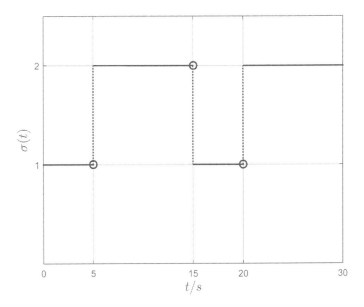

FIGURE 6.9: Switching signal.

The two leaders are selected as quadrotor UAVs, and the eight followers are set as four AmigoBot UGVs ($\mathcal{F}_G = \{1, 2, 3, 4\}$) and four quadrotor UAVs ($\mathcal{F}_A = \{5, 6, 7, 8\}$). It is assumed that all the quadrotor UAVs fly at a predefined constant height, and there is no formation controller along Z-axis for each UAV. Thus, the formation tracking problems for the heterogeneous multi-robot system can be considered in the XY plane. Based on the outer/inner loop framework in [8], the dynamics of a quadrotor UAV in outer loop can be described by

$$\begin{cases} \dot{\chi}_i(t) = \nu_i(t), \\ \dot{\nu}_i(t) = \alpha_{xi}\chi_i(t) + \alpha_{vi}\nu_i(t) + u_i(t), \end{cases}$$

where $\chi_i(t) \in \mathbb{R}^2$, $\nu_i(t) \in \mathbb{R}^2$, and $u_i(t) \in \mathbb{R}^2$ $(i = 5, 6, \ldots, 10)$ denote the position, velocity, and control input respectively, and α_{xi} and α_{vi} are two damping constants. Using the feedback linearization technique in [45], the kinematic equation of the j-th AmigoBot UGV $(j \in \mathcal{F}_G)$ is described by

$$\dot{\chi}_j(t) = u_j(t),$$

where $\chi_j(t) \in \mathbb{R}^2$ and $u_j(t) \in \mathbb{R}^2$ are the position and control input, respectively. Choose $\alpha_{x9} = \alpha_{x10} = -1$, $\alpha_{v9} = \alpha_{v10} = 0$, and $u_9 = u_{10} = 0$. Then, the leader UAVs are modeled as (6.1B) with $S = I_2 \otimes \begin{bmatrix} 0 & 1 \\ -1 & 0 \end{bmatrix}$ and $F = I_2 \otimes \begin{bmatrix} 1 & 0 \end{bmatrix}$. Let $\alpha_{xk} = \alpha_{vk} = 0$ $(k \in \mathcal{F}_A)$, and one has that $A_k = I_2 \otimes \begin{bmatrix} 0 & 1 \\ 0 & 0 \end{bmatrix}$, $B_k = I_2 \otimes \begin{bmatrix} 0 \\ 1 \end{bmatrix}$, and $C_k = I_2 \otimes \begin{bmatrix} 1 & 0 \end{bmatrix}$ $(k \in \mathcal{F}_A)$. In the formation control level, the UGVs are modeled as $A_j = 0_{2\times2}$, $B_j = I_2$, and $C_j = I_2$ $(j \in \mathcal{F}_G)$.

All the UAVs fly at a predefined height $h_Z = 5$m. The follower UAVs need to achieve a rotating circular formation tracking, which is specified by

$$h_k(t) = \begin{bmatrix} 3\cos(t + (k-5)\pi/2) \\ -3\sin(t + (k-5)\pi/2) \\ 3\sin(t + (k-5)\pi/2) \\ 3\cos(t + (k-5)\pi/2) \end{bmatrix}, \, k = 5, 6, 7, 8.$$

The UGVs are required to realize the same formation shape on the ground, i.e., the position of j-th follower UGV $(j = 1, 2, 3, 4)$ in the XY plane is equal to the position of the k-th follower UAV $(k = j + 4)$. Note that all the possible graphs have 0-1 weights. From Lemma 6.2, one has that the convex combination of the leaders' states is calculated by $\zeta(t) = (v_9(t) + v_{10}(t))/2$. Follow the steps in Algorithm 6.1 to determine the formation tracking protocol (6.18). Choose $X_j = I_2 \otimes \begin{bmatrix} 1 & 0 \end{bmatrix}$, $U_j = I_2 \otimes \begin{bmatrix} 0 & 1 \end{bmatrix}$ $(j = 1, 2, 3, 4)$, $X_k = I_4$, $U_k = I_2 \otimes \begin{bmatrix} -1 & 0 \end{bmatrix}$ $(k = 5, 6, 7, 8)$ such that the regulator equations (2) hold. It can be verified that all the followers satisfy the feasible condition (6.19) and the compensation inputs $r_i(t) = [0, 0]^T$ $(i = 1, 2, \ldots, 8)$. Let $\alpha = 1$. Solving the LMI (5) gives $P = I_2 \otimes \begin{bmatrix} 0.3333 & -0.1667 \\ -0.1667 & 0.1667 \end{bmatrix}$. From Theorem 6.1, one can obtain that $\mu \geqslant 5.05$ and $\tau_a > 2.56$. Therefore, μ in the distributed observer (4) is chosen as $\mu = 6$, and the switching signal among $\bar{\mathcal{G}}_1$ and $\bar{\mathcal{G}}_2$ is given in Fig. 6.9. The gain matrices K_{1i} and L_{oi} $(i = 1, 2, \ldots, 8)$ are selected as $K_{1j} = -I_2$, $L_{oj} = -5I_2$ $(j = 1, 2, 3, 4)$, $K_{1k} = I_2 \otimes \begin{bmatrix} -2 & -2 \end{bmatrix}$, $L_{ok} = I_2 \otimes \begin{bmatrix} -9 & -20 \end{bmatrix}^T$ $(k = 5, 6, 7, 8)$. The initial states of the leader UAVs

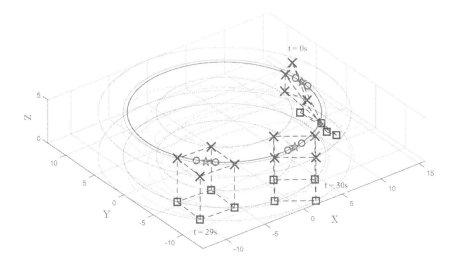

FIGURE 6.10: Position trajectories within $t = 30$s and snapshots at $t = 0, 29, 30$s of the multi-robot system.

are set as $v_9 = [10, -1, 1, 10]^T$ and $v_{10} = [10, 1, -1, 10]^T$, and the initial states of the followers are generated by random numbers. Choose the initial values $\hat{\zeta}(0) = 0$ and $\hat{x}(0) = 0$.

The position trajectories within $t = 30$s and the position snapshots at different time instants ($t = 0, 29, 30$s) of the multi-root system are shown in Fig. 6.10, where the four follower UGVs are represented by squares, the four follower UAVs are depicted by crosses, the two leader UAVs are denoted by circles, and the convex combination of the leaders is marked by five-pointed star. The trajectories of the followers and the convex combination are denoted by the dotted lines and solid line, respectively. Fig. 6.11 shows the Euclidean norm of the output formation tracking error $\|e_i(t)\|$ ($i = 1, 2, \ldots, 8$) for each follower. From Figs. 6.10 and 6.11, one can obtain that the positions of the followers form a circular formation, and the achieved formation keeps rotating around the two leader UAVs. Therefore, the expected time-varying output formation tracking is realized by the heterogeneous multi-robot system.

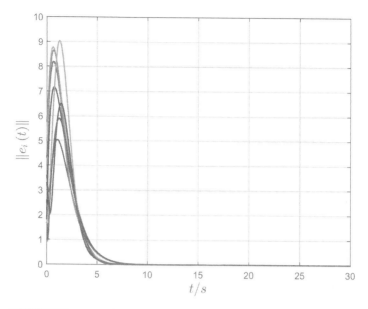

FIGURE 6.11: Curves of the output formation tracking errors.

6.3 Time-varying Formation Tracking without Well-Informed Follower

In the last section, the proposed formation approach relies on the well-informed follower in Assumption 6.2, and all the leaders are required to have the same dynamics model (6.1B) with the known matrices S and F. This section will further investigate the time-varying formation tracking problems for heterogeneous swarm systems with multiple leaders without assuming that each follower is well-informed or uninformed. Besides the followers are heterogeneous, the multiple leaders can also have non-identical dynamics. Both the output and the dynamical matrices of each leader are only available to the followers which contain this leader as neighbour. Firstly, based on the local estimation and the interaction with neighbouring followers, a novel distributed observer is designed for each follower to estimate the dynamical matrices and the states of multiple leaders without requiring the well-informed follower assumption. Using the finite-time stability theory and the adaptive updating gains, the proposed observer can be designed in a totally distributed form by each follower with no need for the eigenvalue information of both the leaders' system matrices and the Laplacian matrix, even if the leaders' system matrices have eigenvalues with positive real parts. Then, an adaptive algorithm is proposed to solve the regulator equations in finite time based on the estimation of the leaders' matrices. By enhancement of the exponential convergence in the existing works to the finite-time convergence, the control gain of the proposed algorithm can be chosen independent of the eigenvalue information of the leaders' system matrices. Furthermore, the desired time-varying output formation of each follower is generated by local active exosystem, where an external input is applied to generate more time-varying formation types. Finally, a time-varying formation tracking protocol is presented using the estimated states of multiple leaders, the online solutions of the regulator equations, and the desired formation vector generated by the local exosystem. It is proved that the outputs of the followers can not only realize the expected formation shape but also track the predefined convex combination of multiple leaders.

6.3.1 Problem Formulation

Consider a heterogeneous swarm system with M leaders and N followers, where the leader set is represented by $\mathcal{T}_L = \{1, 2, \ldots, M\}$ and the follower set is denoted by $\mathcal{T}_F = \{M + 1, M + 2, \ldots, M + N\}$.

The non-identical dynamics of leaders are described by

$$\begin{cases} \dot{z}_j(t) = S_j z_j(t), \\ q_j(t) = F_j z_j(t), \end{cases} \tag{6.32}$$

where $z_j(t) \in \mathbb{R}^{n_j}$ and $q_j(t) \in \mathbb{R}^p$ ($j \in \mathcal{T}_L$) are the state and the output, respectively. The pair (F_j, S_j) is detectable. As pointed out in [87, 157], it is restrictive to require that each follower needs to know the matrices S_j and F_j. Therefore, both the output $q_j(t)$ and the matrices S_j and F_j are assumed to be available to only the followers which contain the leader j as neighbour in this section.

The dynamics of follower i ($i \in \mathcal{T}_F$) is described by

$$\begin{cases} \dot{x}_i(t) = A_i x_i(t) + B_i u_i(t), \\ y_i(t) = C_i x_i(t), \end{cases} \tag{6.33}$$

where $x_i(t) \in \mathbb{R}^{n_i}$, $u_i(t) \in \mathbb{R}^{m_i}$, and $y_i(t) \in \mathbb{R}^p$ are the state, the control input, and the output respectively, and $A_i \in \mathbb{R}^{n_i \times n_i}$, $B_i \in \mathbb{R}^{n_i \times m_i}$, and $C_i \in \mathbb{R}^{p \times n_i}$ are constant matrices with rank$(B_i) = m_i$. The pair (A_i, B_i) is stabilizable, and the pair (C_i, A_i) is detectable.

As shown in (6.32) and (6.33), both the leaders and the followers can be heterogeneous. Only homogeneous swarm systems were considered in [9] and [54]. In [108, 110–112], although the followers were different, all the leaders were assumed to share identical dynamics.

The existing works on formation tracking with multiple leaders in [9, 54, 108] all required that each follower is well-informed or uninformed, where a well-informed follower can communicate with all the leaders and an uninformed follower has no leaders as its neighbour. However, this assumption is too restrictive since it forces some followers to receive from all the leaders directly. In practice, it is more possible and realistic to require that a follower only contains a subset of leaders as its neighbour. Thus, the well-informed follower requirement is removed in this section, and the following assumption on interaction topology is introduced.

Assumption 6.3. *There exist no isolated leaders, i.e., $\mathcal{T}_L \subset \cup_{i=M+1}^{M+N}\mathcal{N}_i$. Moreover, the topology $\mathcal{G}_F$ among the followers is undirected and connected.*

Remark 6.8. *Without well-informed follower assumption, the containment-based control strategy in [9, 54, 108] cannot reach an agreement on the same formation reference, and then the followers would not achieve the desired formation tracking. Moreover, as is known to all, in the distributed tracking of swarm systems with one leader, a basic requirement for the topology is that at least one follower can receive from the leader directly. However, considering the multiple leaders case without the well-informed follower assumption, all the followers cannot acquire the complete information of multiple leaders directly, which makes the design and analysis more complicated. Each follower has to combine the local estimation and the interaction with neighbouring followers together to gather sufficient information for the multiple leaders.*

A time-varying vector $h_y(t) = [h_{y,M+1}^T(t), h_{y,M+2}^T(t), \ldots, h_{y,M+N}^T(t)]^T$ is applied to specify the expected formation shape for the followers.

Definition 6.4. *Consider a heterogeneous swarm system composed of (6.32) and (6.33). For any bounded initial states, if there are constants $0 \leqslant \rho_j \leqslant 1$ $(j = 1, 2, \ldots, M)$ with $\sum_{j=1}^{M} \rho_j = 1$ such that*

$$\lim_{t\to\infty} \left(y_i(t) - h_{y,i}(t) - \sum_{j=1}^{M} \rho_j q_j(t) \right) = 0, \; i \in \mathcal{T}_F, \tag{6.34}$$

then the heterogeneous swarm system realizes the expected time-varying output formation tracking with multiple non-identical leaders.

Remark 6.9. *The physical meaning of Definition 6.4 is explained as follows. In (6.34), $\sum_{j=1}^{M} \rho_j q_j(t)$ represents the convex combination of the multiple leaders' outputs. If the expected formation tracking is realized, all the followers will take $\sum_{j=1}^{M} \rho_j q_j(t)$ as the common formation reference and maintain the time-varying offsets $h_{y,i}(t)$ $(i \in \mathcal{T}_F)$ relative to it. By choosing different values of ρ_j $(j = 1, 2, \ldots, M)$, any point in the convex hull formed by multiple leaders can be designed as the predefined formation reference.*

This section mainly pours attention to how to design a fully distributed output formation tracking protocol for heterogeneous swarm systems with multiple non-identical leaders without requiring the well-informed follower assumption.

6.3.2   Distributed Observer Design with Multiple Leaders

In this subsection, a novel distributed observer is designed for each follower to estimate the dynamical matrices and the states of multiple leaders without requiring the well-informed follower assumption. Using the finite-time stability theory and the adaptive updating gains, the proposed observer can be designed in a distributed form by each follower with no need for the eigenvalue information of both the leaders' system matrices and the Laplacian matrix.

Let $\bar{S} = \text{diag}\{S_1, S_2, \ldots, S_M\}$, $\bar{F} = \text{diag}\{F_1, F_2, \ldots, F_M\}$, $\bar{z}(t) = [z_1^T(t), z_2^T(t), \ldots, z_M^T(t)]^T$, and $\bar{q}(t) = [q_1^T(t), q_2^T(t), \ldots, q_M^T(t)]^T$. Then, system (6.32) can be rewritten in the following compact form:

$$\begin{cases} \dot{\bar{z}}(t) = \bar{S}\bar{z}(t), \\ \bar{q}(t) = \bar{F}\bar{z}(t), \end{cases} \tag{6.35}$$

where $\bar{z}(t) \in \mathbb{R}^{\bar{n} \times 1}$ $(\bar{n} = n_1 + n_2 + \cdots + n_M)$ and $\bar{q}(t) \in \mathbb{R}^{Mp \times 1}$.

Note that there may not exist well-informed followers under Assumption 6.3. Then, all the followers cannot estimate the state $\bar{z}(t)$ only by using their own measurements, and no followers can acquire the complete information of the matrices $\bar{S}$ and $\bar{F}$. Based on the local estimation and the communication with neighbouring observers, a novel distributed observer is designed for each follower to estimate the matrices $\bar{S}$, $\bar{F}$ and the state $\bar{z}(t)$ without requiring the well-informed follower assumption in the following.

Let $s_j = \text{vec}(S_j)$ and $f_j = \text{vec}(F_j)$, $j = 1, 2, \ldots, M$. Then, define $\bar{s} = [s_1^T, s_2^T, \ldots, s_M^T]^T \in \mathbb{R}^{n_s \times 1}$ $(n_s = n_1^2 + n_2^2 + \cdots + n_M^2)$ and $\bar{f} = [f_1^T, f_2^T, \ldots, f_M^T]^T \in \mathbb{R}^{n_f \times 1}$ $(n_f = p(n_1 + n_2 + \cdots + n_M))$. For follower i $(i \in \mathcal{T}_F)$, let $\hat{s}_i = [\hat{s}_{i,1}^T, \hat{s}_{i,2}^T, \ldots, \hat{s}_{i,M}^T]^T$ and $\hat{f}_i = [\hat{f}_{i,1}^T, \hat{f}_{i,2}^T, \ldots, \hat{f}_{i,M}^T]^T$ denote the estimations of $\bar{s}$ and $\bar{f}$, respectively. Let $\hat{S}_i = \text{diag}\{\hat{S}_{i,1}, \hat{S}_{i,2}, \ldots, \hat{S}_{i,M}\}$ and $\hat{F}_i = \text{diag}\{\hat{F}_{i,1}, \hat{F}_{i,2}, \ldots, \hat{F}_{i,M}\}$, where $\hat{S}_{i,j} = \text{mtx}_{n_j}^{n_j}(\hat{s}_{i,j})$ and $\hat{F}_{i,j} = \text{mtx}_{n_j}^{p}(\hat{f}_{i,j})$, $j = 1, 2, \ldots, M$. Then, $\hat{S}_i$ and $\hat{F}_i$ denote the estimated matrices of $\bar{S}$ and $\bar{F}$ for follower i $(i \in \mathcal{T}_F)$, respectively. Since the matrices $\bar{S}$ and $\bar{F}$ can be estimated by the same approach, for simplification, only distributed observer for $\bar{S}$ is provided in detail in the following. By replacing $\hat{s}_i$ with $\hat{f}_i$, all the results can be applied to estimate the matrix $\bar{F}$ directly.

For follower i $(i \in \mathcal{T}_F)$, let $\hat{\xi}_i = [\hat{\xi}_{i,1}^T, \hat{\xi}_{i,2}^T, \ldots, \hat{\xi}_{i,M}^T]^T \in \mathbb{R}^{\bar{n} \times 1}$, where $\hat{\xi}_{i,j} \in \mathbb{R}^{n_j \times 1}$ denotes the i-th follower's estimation of z_j $(j = 1, 2, \ldots, M)$. Construct the following distributed observer:

$$\dot{\hat{s}}_i = -\eta \text{sig}^\gamma \left(\begin{bmatrix} w_{i1}(\hat{s}_{i,1} - s_1) \\ w_{i2}(\hat{s}_{i,2} - s_2) \\ \vdots \\ w_{iM}(\hat{s}_{i,M} - s_M) \end{bmatrix} + \sum_{k=M+1}^{M+N} w_{ik}(\hat{s}_i - \hat{s}_k) \right), \tag{6.36}$$

$$\dot{\hat{\xi}}_i = \hat{S}_i \hat{\xi}_i - \begin{bmatrix} \hat{\alpha}_{i1} w_{i1} K_1 (F_1 \hat{\xi}_{i,1} - q_1) \\ \hat{\alpha}_{i2} w_{i2} K_2 (F_2 \hat{\xi}_{i,2} - q_2) \\ \vdots \\ \hat{\alpha}_{iM} w_{iM} K_M (F_M \hat{\xi}_{i,M} - q_M) \end{bmatrix} - \sum_{k=M+1}^{M+N} \hat{\beta}_{ik} w_{ik} (\hat{\xi}_i - \hat{\xi}_k), \quad (6.37)$$

where $\eta > 0$ and $0 < \gamma < 1$ are two positive constants. The gain matrix K_j $(j = 1, 2, \ldots, M)$ is determined by $K_j = \begin{cases} Q_j F_j^T, & w_{ij} > 0 \\ 0, & w_{ij} = 0 \end{cases}$, where Q_j $(j \in \mathcal{T}_L \cap \mathcal{N}_i)$ is a positive definite matrix calculated by the following algebraic Riccati equation (ARE):

$$S_j Q_j + Q_j S_j^T - Q_j F_j^T F_j Q_j + I_{n_j} = 0. \qquad (6.38)$$

The adaptive gains $\hat{\alpha}_{ij}$ and $\hat{\beta}_{ik}$ are generated by the following updating laws:

$$\dot{\hat{\alpha}}_{ij} = w_{ij} \left\| F_j \hat{\xi}_{i,j} - q_j \right\|^2, \ j = 1, 2, \ldots, M, \qquad (6.39)$$

$$\dot{\hat{\beta}}_{ik} = w_{ik} \left\| \hat{\xi}_i - \hat{\xi}_k \right\|^2, \ k = M + 1, M + 2, \ldots, M + N, \qquad (6.40)$$

where the initial values $\hat{\alpha}_{ij}(0) \geqslant 0$, $\hat{\beta}_{ik}(0) \geqslant 0$, and $\hat{\beta}_{ik}(0) = \hat{\beta}_{ki}(0)$.

To interpret the information interaction among the followers more clearly, considering the following illustrative example.

Illustrative example 6.1. *Consider a heterogeneous swarm system with two leaders and three followers, where $\mathcal{T}_L = \{1, 2\}$ and $\mathcal{T}_F = \{3, 4, 5\}$. The information sent and received by each follower is given in Fig. 6.12. Taking follower 3 (F3) as an example, it is required to send the estimated information $\hat{s}_3 = [\hat{s}_{3,1}^T, \hat{s}_{3,2}^T]^T$, $\hat{f}_3 = [\hat{f}_{3,1}^T, \hat{f}_{3,2}^T]^T$, and $\hat{\xi}_3 = [\hat{\xi}_{3,1}^T, \hat{\xi}_{3,2}^T]^T$ to its neighbour follower 4 (F4), where $\hat{s}_{3,j}$, $\hat{f}_{3,j}$, and $\hat{\xi}_{3,j}$ $(j = 1, 2)$ denote the F3's estimations for the dynamics s_j, f_j, and the state z_j of leader j respectively. The received information of F3 includes s_1, f_1, and q_1 from leader 1 (L1) and the estimated information $\hat{s}_4 = [\hat{s}_{4,1}^T, \hat{s}_{4,2}^T]^T$, $\hat{f}_4 = [\hat{f}_{4,1}^T, \hat{f}_{4,2}^T]^T$, and $\hat{\xi}_4 = [\hat{\xi}_{4,1}^T, \hat{\xi}_{4,2}^T]^T$ from F4. As shown in Fig. 6.12, to estimate both the dynamics matrices and the states of multiple leaders in a distributed form, more information interaction among followers is required in this section than [9, 54, 108].*

Remark 6.10. *In (6.36) and (6.37), the weight $w_{il} > 0$ if and only if $l \in \mathcal{N}_i$, otherwise $w_{il} = 0$ $(l = 1, 2, \ldots, M + N)$. Note that the matrices S_j and F_j $(j \in \mathcal{T}_L)$ are available to the followers which contain the leader j as neighbour. Thus, for $j \in \mathcal{T}_L \cap \mathcal{N}_i$, the i-th follower can solve Q_j from (6.38) directly. Since (F_j, S_j) is detectable, it can be verified that the ARE (6.38) has a unique solution Q_j. Motivated by [158], two adaptive updating gains $\hat{\alpha}_{ij}$ and $\hat{\beta}_{ik}$ are*

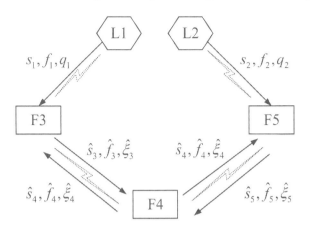

FIGURE 6.12: Information sent and received by each follower in Illustrative example 6.1.

applied in (6.37) to avoid using the eigenvalue information of both the Laplacian matrix and the leaders' system matrices such that the observer (6.37) can be performed by each follower in a fully distributed form. Moreover, if $\hat{s}_i$ ($i \in \mathcal{T}_F$) and s_j ($j \in \mathcal{T}_L$) in (6.36) are replaced by $\hat{f}_i$ and f_j respectively, then the distributed observer (6.36) can be modified to estimate $\bar{F}$ directly.

Let $W_i = \mathrm{diag}_{j=1,\ldots,M}\{w_{ij}I_{n_j^2}\}$ and $\hat{W}_i = \mathrm{diag}_{j=1,\ldots,M}\{\hat{\alpha}_{ij}w_{ij}K_jF_j\}$ ($i \in \mathcal{T}_F$). Then, it follows from (6.36) and (6.37) that

$$\dot{\hat{s}}_i = -\eta\mathrm{sig}^\gamma\left(W_i\left(\hat{s}_i - \bar{s}\right) + \sum_{k=M+1}^{M+N} w_{ik}\left(\hat{s}_i - \hat{s}_k\right)\right), \qquad (6.41)$$

$$\dot{\hat{\xi}}_i = \hat{S}_i\hat{\xi}_i - \hat{W}_i\left(\hat{\xi}_i - \bar{z}\right) - \sum_{k=M+1}^{M+N} \hat{\beta}_{ik}w_{ik}\left(\hat{\xi}_i - \hat{\xi}_k\right). \qquad (6.42)$$

Let $\tilde{s}_i = \hat{s}_i - \bar{s}$, $\tilde{\xi}_i = \hat{\xi}_i - \bar{z}$, and $\tilde{S}_i = \hat{S}_i - \bar{S}$ denote the estimation errors of follower i ($i \in \mathcal{T}_F$). From (6.41) and (6.42), one can obtain that

$$\dot{\tilde{s}}_i = -\eta\mathrm{sig}^\gamma\left(W_i\tilde{s}_i + \sum_{k=M+1}^{M+N} w_{ik}\left(\tilde{s}_i - \tilde{s}_k\right)\right), \qquad (6.43)$$

$$\dot{\tilde{\xi}}_i = \bar{S}\tilde{\xi}_i - \hat{W}_i\tilde{\xi}_i - \sum_{k=M+1}^{M+N} \hat{\beta}_{ik}w_{ik}\left(\tilde{\xi}_i - \tilde{\xi}_k\right) + \tilde{S}_i\left(\tilde{\xi}_i + \bar{z}\right). \qquad (6.44)$$

Let $\bar{D} = \bar{W} + (L_F \otimes I_{n_s})$, where $\bar{W} = \text{diag}\{W_{M+1}, W_{M+2}, \ldots, W_{M+N}\}$ and L_F denotes the Laplacian matrix of the graph $\mathcal{G}_F$ among the followers. The following lemma shows that $\bar{D} > 0$.

Lemma 6.3. *Under Assumption 6.3, $\bar{D} = \bar{W} + (L_F \otimes I_{n_s})$ is a positive definite matrix.*

Proof. Since the graph $\mathcal{G}_F$ is connected, it follows from Lemma 2.1 that $L_F \geqslant 0$, and there exists an orthogonal matrix $U = [\mathbf{1}_N/\sqrt{N}, H_1]$ with $H_1 \in \mathbb{R}^{N \times (N-1)}$ such that $U^T L_F U = \Lambda = \text{diag}\{0, \lambda_2, \lambda_3, \ldots, \lambda_N\}$.

For any non-zero vector $y \in \mathbb{R}^N$, it will be proved that $y^T L_F y = 0$ if and only if $y = c\mathbf{1}_N$ firstly. i) If $y = c\mathbf{1}_N$, one can obtain that $y^T L_F y = 0$ directly since $L_F \mathbf{1}_N = 0$. ii) If $y^T L_F y = 0$, let $\bar{y} = U^T y$. Then it holds that $y^T L_F y = \bar{y}^T \Lambda \bar{y} = 0$. Note that $\Lambda = \text{diag}\{0, \lambda_2, \lambda_3, \ldots, \lambda_N\}$. It can be verified that $\bar{y}$ should have the following form: $\bar{y} = [\bar{c}, 0, \ldots, 0]^T$. Thus, one has that $y = U\bar{y} = [\mathbf{1}_N/\sqrt{N}, H_1][\bar{c}, 0, \ldots, 0]^T = \frac{\bar{c}}{\sqrt{N}}\mathbf{1}_N$. Let $c = \frac{\bar{c}}{\sqrt{N}}$. So y should have the form $y = c\mathbf{1}_N$. Thus, one gets that $y^T L_F y = 0$ if and only if $y = c\mathbf{1}_N$.

Let $x = [x_1^T, x_2^T, \ldots, x_N^T]^T$ denote a non-zero vector with $x_i \in \mathbb{R}^{n_s}$ ($i = 1, 2, \ldots, N$). Since $\bar{W}$ and L_F are positive semidefinite, one has that $x^T \bar{D} x \geqslant 0$. Consider the following two cases.

i) If $x \neq \mathbf{1}_N \otimes \bar{x}$, $\bar{x} \in \mathbb{R}^{n_s}$, it follows that $x^T \bar{D} x \geqslant x^T (L_F \otimes I_{n_s}) x > 0$, where the property that $y^T L_F y = 0$ if and only if $y = c\mathbf{1}_N$ has been used.

ii) If $x = \mathbf{1}_N \otimes \bar{x}$, one has that $x^T (L_F \otimes I_{n_s}) x = 0$. Then it holds that $x^T \bar{D} x = x^T \bar{W} x = \bar{x}^T (\mathbf{1}_N^T \otimes I_{n_s}) \bar{W} (\mathbf{1}_N \otimes I_{n_s}) \bar{x}$. Note that $\bar{W} = \text{diag}\{W_{M+1}, W_{M+2}, \ldots, W_{M+N}\}$. It follows that $(\mathbf{1}_N^T \otimes I_{n_s}) \bar{W} (\mathbf{1}_N \otimes I_{n_s}) = \sum_{i=M+1}^{M+N} W_i$. Since there exist no isolated leaders under Assumption 6.3, for leader j ($j = 1, 2, \ldots, M$), one can obtain that $w_{M+1,j} + w_{M+2,j} + \cdots + w_{M+N,j} > 0$. Recall that $W_i = \text{diag}_{j=1,2,\ldots,M}\{w_{ij} I_{n_2}\}$ ($i = M+1, M+2, \ldots, M+N$). It can be verified that $\sum_{i=M+1}^{M+N} W_i > 0$. So it holds that $x^T \bar{D} x > 0$.

According to the above two cases, one can obtain that $\bar{D} = \bar{W} + (L_F \otimes I_{n_s}) > 0$. This completes the proof of Lemma 6.3. $\square$

The following theorem shows the convergence of the distributed observer.

Theorem 6.3. *Suppose that Assumption 6.3 holds. For follower i ($i \in \mathcal{T}_F$), considering the system composed of (6.32), (6.36) and (6.37), it can be verified that (i) $\lim_{t \to T_s} \tilde{s}_i(t) = 0$ and $\tilde{s}_i(t) = 0$ ($t \geqslant T_s$), i.e., $\hat{S}_i(t) \to \bar{S}$ in a finite time T_s, and (ii) $\lim_{t \to \infty}(\hat{\xi}_i(t) - \bar{z}(t)) = 0$ and the adaptive gains $\hat{\alpha}_{ij}(t)$ ($j = 1, 2, \ldots, M$) and $\hat{\beta}_{ik}(t)$ ($k = M+1, M+2, \ldots, M+N$) converge to some positive constants.*

Proof. The proof is divided into three steps. Firstly, it is proved that $\hat{S}_i(t) \to \bar{S}$ in a finite time T_s. Secondly, the boundedness of $\tilde{\xi}_i(t)$, $\hat{\alpha}_{ij}(t)$ and $\hat{\beta}_{ik}(t)$ in $t \in [0, T_s)$ is analyzed. Thirdly, when $t \geqslant T_s$, we will show that $\lim_{t \to \infty} \tilde{\xi}_i(t) = 0$, and $\hat{\alpha}_{ij}(t)$ and $\hat{\beta}_{ik}(t)$ converge to some positive constants.

Step 1: Finite-time convergence of $\tilde{s}_i(t)$.

Let $\tilde{s} = [\tilde{s}_{M+1}^T, \tilde{s}_{M+2}^T, \ldots, \tilde{s}_{M+N}^T]^T$. Then, it follows from (6.43) that

$$\dot{\tilde{s}} = -\eta \mathrm{sig}^\beta\left(\left(\bar{W} + (L_F \otimes I_{n_s})\right)\tilde{s}\right) = -\eta \mathrm{sig}^\beta\left(\bar{D}\tilde{s}\right). \tag{6.45}$$

According to Lemma 6.3, one has that $\bar{D}$ is a positive definite matrix. Consider the following Lyapunov functional candidate:

$$V_1 = \tilde{s}^T \bar{D}\tilde{s}. \tag{6.46}$$

The time derivative of V_1 along (6.45) can be obtained as

$$\dot{V}_1 = -2\eta \tilde{s}^T \bar{D}\mathrm{sig}^\gamma\left(\bar{D}\tilde{s}\right). \tag{6.47}$$

Let $\tilde{\varsigma} = \bar{D}\tilde{s}$. Based on Lemma 2.4, for a vector $x = [x_1, x_2, \ldots, x_n]^T \in \mathbb{R}^n$, one gets that $x^T \mathrm{sig}^\gamma(x) = \sum_{i=1}^n |x_i|^{1+\gamma} \geqslant (\sum_{i=1}^n |x_i|^2)^{(1+\gamma)/2} = (\|x\|^2)^{(1+\gamma)/2}$. Note that $\tilde{\varsigma}^T \tilde{\varsigma} = \tilde{s}^T \bar{D}^2 \tilde{s} \geqslant \lambda_{\min}(\bar{D})\tilde{s}^T \bar{D}\tilde{s} = \lambda_{\min}(\bar{D})V_1$. Then, it follows from (6.47) that

$$\dot{V}_1 = -2\eta \tilde{\varsigma}^T \mathrm{sig}^\gamma\left(\tilde{\varsigma}\right)$$

$$\leqslant -2\eta\left(\|\tilde{\varsigma}\|^2\right)^{\frac{1+\gamma}{2}}$$

$$\leqslant -2\eta\left[\lambda_{\min}(\bar{D})\right]^{\frac{1+\gamma}{2}} V_1^{\frac{1+\gamma}{2}}. \tag{6.48}$$

From Lemma 2.27, one can obtain that $\tilde{s} \to 0$ in a finite time T_s, where T_s can be estimated by $T_s \leqslant \frac{2V_1^{(1-\gamma)/2}(\tilde{s}(0))}{c(1-\gamma)}$ with $c = 2\eta[\lambda_{\min}(\bar{D})]^{\frac{1+\gamma}{2}}$. Note that $\hat{S}_{i,j} = \mathrm{mtx}_{n_j}^{n_j}(\hat{s}_{i,j})$ and $\hat{S}_i = \mathrm{diag}\{\hat{S}_{i,1}, \hat{S}_{i,2}, \ldots, \hat{S}_{i,M}\}$. Thus, it can be verified that $\hat{S}_i \to \tilde{S}$ ($i \in \mathcal{T}_F$) in a finite time T_s.

Step 2: Boundedness of $\tilde{\xi}_i(t)$, $\hat{\alpha}_{ij}(t)$ and $\hat{\beta}_{ik}(t)$ in $[0, T_s)$.

Consider the following Lyapunov functional candidate:

$$V_2 = \sum_{i=M+1}^{M+N} \tilde{\xi}_i^T \bar{Q}^{-1}\tilde{\xi}_i + \sum_{i=M+1}^{M+N}\sum_{j=1}^M (\hat{\alpha}_{ij} - \alpha)^2$$

$$+ \frac{1}{2}\lambda_{\min}(\bar{Q}^{-1})\sum_{i=M+1}^{M+N}\sum_{k=M+1}^{M+N} (\hat{\beta}_{ik} - \beta)^2, \tag{6.49}$$

where $\bar{Q}^{-1} = \mathrm{diag}\left\{Q_1^{-1}, Q_2^{-1}, \ldots, Q_M^{-1}\right\}$, and α and β are two positive constants to be determined.

Taking the time derivative of V_2 along (6.44) gives

$$
\dot{V}_2 = \sum_{i=M+1}^{M+N} \tilde{\xi}_i^T \left(\bar{Q}^{-1}\bar{S} + \bar{S}^T \bar{Q}^{-1} \right) \tilde{\xi}_i - 2 \sum_{i=M+1}^{M+N} \tilde{\xi}_i^T \bar{Q}^{-1} \hat{W}_i \tilde{\xi}_i
$$

$$
- 2 \sum_{i=M+1}^{M+N} \tilde{\xi}_i^T \bar{Q}^{-1} \sum_{k=M+1}^{M+N} \hat{\beta}_{ik} w_{ik} (\tilde{\xi}_i - \tilde{\xi}_k) + 2 \sum_{i=M+1}^{M+N} \tilde{\xi}_i^T \bar{Q}^{-1} \tilde{S}_i (\tilde{\xi}_i + \bar{z})
$$

$$
+ 2 \sum_{i=M+1}^{M+N} \sum_{j=1}^{M} (\hat{\alpha}_{ij} - \alpha) \, w_{ij} \tilde{\xi}_{i,j}^T F_j^T F_j \tilde{\xi}_{i,j}
$$

$$
+ \lambda_{\min}(\bar{Q}^{-1}) \sum_{i=M+1}^{M+N} \sum_{k=M+1}^{M+N} \left(\hat{\beta}_{ik} - \beta \right) w_{ik} \left\| \tilde{\xi}_i - \tilde{\xi}_k \right\|^2. \tag{6.50}
$$

Since $-2 \sum_{i=M+1}^{M+N} \tilde{\xi}_i^T \bar{Q}^{-1} \hat{W}_i \tilde{\xi}_i = -2 \sum_{i=M+1}^{M+N} \sum_{j=1}^{M} \hat{\alpha}_{ij} w_{ij} \tilde{\xi}_{i,j}^T F_j^T F_j \tilde{\xi}_{i,j}$, one has

$$
- 2 \sum_{i=M+1}^{M+N} \tilde{\xi}_i^T \bar{Q}^{-1} \hat{W}_i \tilde{\xi}_i + 2 \sum_{i=M+1}^{M+N} \sum_{j=1}^{M} (\hat{\alpha}_{ij} - \alpha) \, w_{ij} \tilde{\xi}_{i,j}^T F_j^T F_j \tilde{\xi}_{i,j}
$$

$$
= -2\alpha \sum_{i=M+1}^{M+N} \sum_{j=1}^{M} w_{ij} \tilde{\xi}_{i,j}^T F_j^T F_j \tilde{\xi}_{i,j}. \tag{6.51}
$$

Note that $w_{ik} = w_{ki}$ and $\hat{\beta}_{ki} = \hat{\beta}_{ik} \geqslant 0$. One can obtain

$$
- 2 \sum_{i=M+1}^{M+N} \tilde{\xi}_i^T \bar{Q}^{-1} \sum_{k=M+1}^{M+N} \hat{\beta}_{ik} w_{ik} \left(\tilde{\xi}_i - \tilde{\xi}_k \right)
$$

$$
= - \sum_{i=M+1}^{M+N} \sum_{k=M+1}^{M+N} \hat{\beta}_{ik} w_{ik} \left(\tilde{\xi}_i^T - \tilde{\xi}_k^T \right) \bar{Q}^{-1} \left(\tilde{\xi}_i - \tilde{\xi}_k \right)
$$

$$
\leqslant -\lambda_{\min}(\bar{Q}^{-1}) \sum_{i=M+1}^{M+N} \sum_{k=M+1}^{M+N} \hat{\beta}_{ik} w_{ik} \left(\tilde{\xi}_i^T - \tilde{\xi}_k^T \right) \left(\tilde{\xi}_i - \tilde{\xi}_k \right). \tag{6.52}
$$

Substituting (6.51) and (6.52) into (6.50) gives

$$
\dot{V}_2 \leqslant \sum_{i=M+1}^{M+N} \tilde{\xi}_i^T (\bar{Q}^{-1}\bar{S} + \bar{S}^T \bar{Q}^{-1}) \tilde{\xi}_i - 2\alpha \sum_{i=M+1}^{M+N} \sum_{j=1}^{M} w_{ij} \tilde{\xi}_{i,j}^T F_j^T F_j \tilde{\xi}_{i,j}
$$

$$
- 2\beta \lambda_{\min}(\bar{Q}^{-1}) \sum_{i=M+1}^{M+N} \tilde{\xi}_i^T \sum_{k=M+1}^{M+N} w_{ik} \left(\tilde{\xi}_i - \tilde{\xi}_k \right) + 2 \sum_{i=M+1}^{M+N} \tilde{\xi}_i^T \bar{Q}^{-1} \tilde{S}_i \left(\tilde{\xi}_i + \bar{z} \right). \tag{6.53}
$$

Let $\tilde{S} = \text{diag}\{\tilde{S}_{M+1}, \tilde{S}_{M+2}, \ldots, \tilde{S}_{M+N}\}$ and $\widehat{W} = \text{diag}\{\widehat{W}_{M+1}, \widehat{W}_{M+2}, \ldots, \widehat{W}_{M+N}\}$, where $\widehat{W}_i = \text{diag}_{j=1,2,\ldots,M}\{w_{ij}F_j^T F_j\}$ $(i = M+1, M+2, \ldots, M+N)$. Then it holds from (6.53) that

$$\dot{V}_2 \leqslant \tilde{\xi}^T \left(I_N \otimes (\bar{Q}^{-1}\bar{S} + \bar{S}^T \bar{Q}^{-1}) \right) \tilde{\xi} - 2\alpha\tilde{\xi}^T \widehat{W}\tilde{\xi} - 2\beta\lambda_{\min}(\bar{Q}^{-1})\tilde{\xi}^T \left(L_F \otimes I_{\bar{n}} \right) \tilde{\xi}$$
$$+ \tilde{\xi}^T \left(\left(I_N \otimes \bar{Q}^{-1} \right) \tilde{S} + \tilde{S}^T \left(I_N \otimes \bar{Q}^{-1} \right) \right) \tilde{\xi} + 2\tilde{\xi}^T \left(I_N \otimes \bar{Q}^{-1} \right) \tilde{S} \left(\mathbf{1}_N \otimes \bar{z} \right). \tag{6.54}$$

Note that $2\tilde{\xi}^T(I_N \otimes \bar{Q}^{-1})\tilde{S}(\mathbf{1}_N \otimes \bar{z}) \leqslant \tilde{\xi}^T\tilde{\xi} + \|(I_N \otimes \bar{Q}^{-1})\tilde{S}(\mathbf{1}_N \otimes \bar{z})\|^2$. Since $\tilde{S}$ and $\bar{z}$ are bounded in $[0, T_s)$, there exist positive constants μ and ε such that

$$\dot{V}_2 \leqslant \mu\tilde{\xi}^T\tilde{\xi} + \varepsilon \leqslant \frac{\mu}{\lambda_{\min}(\bar{Q}^{-1})}V_2 + \varepsilon \tag{6.55}$$

holds for $t \in [0, T_s)$. Thus, V_2 is bounded in $[0, T_s)$. So are $\tilde{\xi}_i$, $\hat{\alpha}_{ij}$, and $\hat{\beta}_{ik}$.

Step 3: Convergence of $\tilde{\xi}_i(t)$.

When $t \geqslant T_s$, one has that $\tilde{S} = 0$. Then it follows from (6.54) that

$$\dot{V}_2 \leqslant \tilde{\xi}^T \left(I_N \otimes (\bar{Q}^{-1}\bar{S} + \bar{S}^T \bar{Q}^{-1}) \right) \tilde{\xi} - 2\alpha\tilde{\xi}^T \widehat{W}\tilde{\xi} - 2\beta\lambda_{\min}(\bar{Q}^{-1})\tilde{\xi}^T \left(L_F \otimes I_{\bar{n}} \right) \tilde{\xi}. \tag{6.56}$$

Let $\widehat{\xi} = (I_N \otimes \bar{Q}^{-1})\tilde{\xi}$. It follows from (6.56) that

$$\dot{V}_2 \leqslant \widehat{\xi}^T \left(I_N \otimes (\bar{S}\bar{Q} + \bar{Q}\bar{S}^T) - 2\alpha\bar{W}_Q - 2\beta\lambda_{\min}(\bar{Q}^{-1})(L_F \otimes \bar{Q}^2) \right) \widehat{\xi}, \tag{6.57}$$

where $\bar{W}_Q = (I_N \otimes \bar{Q})\widehat{W}(I_N \otimes \bar{Q}) = \text{diag}\{\bar{W}_{M+1}, \bar{W}_{M+2}, \ldots, \bar{W}_{M+N}\}$ with $\bar{W}_i = \text{diag}_{j=1,2,\ldots,M}\{w_{ij}Q_j F_j^T F_j Q_j\}$ $(i = M+1, M+2, \ldots, M+N)$. Since the graph $\mathcal{G}_F$ is connected, there exists an orthogonal matrix $U = [\mathbf{1}_N/\sqrt{N}, H_1]$ such that $U^T L_F U = \Lambda = \text{diag}\{0, \lambda_2, \lambda_3, \ldots, \lambda_N\}$. Let $\varsigma = (U^T \otimes I_{\bar{n}})\widehat{\xi}$. Then one can obtain from (6.57) that

$$\dot{V}_2 \leqslant \varsigma^T \left(I_N \otimes (\bar{S}\bar{Q} + \bar{Q}\bar{S}^T) - 2\alpha\bar{W}_U - 2\beta\lambda_{\min}(\bar{Q}^{-1})(\Lambda \otimes \bar{Q}^2) \right) \varsigma, \tag{6.58}$$

where $\bar{W}_U = (U^T \otimes I_{\bar{n}})\bar{W}_Q(U \otimes I_{\bar{n}})$.

Let $\bar{E} = I_N \otimes (\bar{S}\bar{Q} + \bar{Q}\bar{S}^T) - 2\alpha\bar{W}_U - 2\beta\lambda_{\min}(\bar{Q}^{-1})(\Lambda \otimes \bar{Q}^2)$. In the following, we will show that there exist sufficiently large α and β such that $\bar{E} < 0$. Using the definitions of U and $\bar{W}_Q$, it can be verified that

$$\bar{W}_U = \left(U^T \otimes I_{\bar{n}} \right) \bar{W}_Q \left(U \otimes I_{\bar{n}} \right)$$
$$= \left(\begin{bmatrix} \mathbf{1}_N^T/\sqrt{N} \\ H_1^T \end{bmatrix} \otimes I_{\bar{n}} \right) \text{diag}\{\bar{W}_{M+1}, \ldots, \bar{W}_{M+N}\} \left(\begin{bmatrix} \frac{\mathbf{1}_N}{\sqrt{N}}, H_1 \end{bmatrix} \otimes I_{\bar{n}} \right)$$
$$= \begin{bmatrix} \frac{1}{N}\sum_{i=M+1}^{M+N} \bar{W}_i & Y_{12} \\ Y_{12}^T & Y_{22} \end{bmatrix}, \tag{6.59}$$

where $Y_{12} = (1_N^T/\sqrt{N} \otimes I_{\bar{n}})\bar{W}_Q(H_1 \otimes I_{\bar{n}})$ and $Y_{22} = (H_1^T \otimes I_{\bar{n}})\bar{W}_Q(H_1 \otimes I_{\bar{n}})$. Since there exist no isolated leaders under Assumption 6.3, for leader j ($j = 1, 2, \ldots, M$), one can obtain that $w_{M+1,j} + w_{M+2,j} + \cdots + w_{M+N,j} > 0$. Note that $\bar{W}_i = \text{diag}_{j=1,2,\ldots,M}\{w_{ij}Q_jF_j^TF_jQ_j\}$. One has that $\frac{1}{N}\sum_{i=M+1}^{M+N}\bar{W}_i = \frac{1}{N}\text{diag}_{j=1,2,\ldots,M}\{\sum_{i=M+1}^{M+N}w_{ij}Q_jF_j^TF_jQ_j\}$.

Let $\Lambda_{22} = \text{diag}\{\lambda_2, \lambda_3, \ldots, \lambda_N\}$. Then it follows from (6.59) that

$$
\bar{E} = \begin{bmatrix} 1 & \\ & I_{N-1} \end{bmatrix} \otimes (\bar{S}\bar{Q} + \bar{Q}\bar{S}^T) - 2\alpha \begin{bmatrix} \frac{1}{N}\sum_{i=M+1}^{M+N}\bar{W}_i & Y_{12} \\ Y_{12}^T & Y_{22} \end{bmatrix}
$$

$$
- 2\beta\lambda_{\min}(\bar{Q}^{-1})\left(\begin{bmatrix} 0 & \\ & \Lambda_{22} \end{bmatrix} \otimes \bar{Q}^2\right)
$$

$$
= \begin{bmatrix} \bar{E}_{11} & \bar{E}_{12} \\ \bar{E}_{12}^T & \bar{E}_{22} \end{bmatrix},
$$

where $\bar{E}_{11} = \bar{S}\bar{Q} + \bar{Q}\bar{S}^T - \frac{2\alpha}{N}\sum_{i=M+1}^{M+N}\bar{W}_i$, $\bar{E}_{12} = -2\alpha Y_{12}$, and $\bar{E}_{22} = I_{N-1} \otimes (\bar{S}\bar{Q} + \bar{Q}\bar{S}^T) - 2\alpha Y_{22} - 2\beta\lambda_{\min}(\bar{Q}^{-1})(\Lambda_{22} \otimes \bar{Q}^2)$. Note that $S_jQ_j + Q_jS_j^T - Q_jF_j^TF_jQ_j + I_{n_j} = 0$ and $\sum_{i=M+1}^{M+N}w_{ij} > 0$ ($j = 1, 2, \ldots, M$). Choose sufficiently large α such that $\alpha > \frac{N}{2\min\limits_{j=1,\ldots,M}\sum_{i=M+1}^{M+N}w_{ij}}$. Then one has

$$
\bar{E}_{11} = \text{diag}_{j=1,2,\ldots,M}\left\{S_jQ_j + Q_jS_j^T - \frac{2\alpha}{N}\sum_{i=M+1}^{M+N}w_{ij}Q_jF_j^TF_jQ_j\right\}
$$

$$
< 0. \tag{6.60}
$$

Select sufficiently large constant β such that

$$
\beta > \max\left\{0, \frac{\lambda_{\max}(\Omega)}{2\min\limits_{k=2,\ldots,N}\{\lambda_k\}\lambda_{\min}(\bar{Q}^2)\lambda_{\min}(\bar{Q}^{-1})}\right\},
$$

where $\Omega = I_{N-1} \otimes (\bar{S}\bar{Q} + \bar{Q}\bar{S}^T) - 2\alpha Y_{22} - \bar{E}_{12}^T\bar{E}_{11}^{-1}\bar{E}_{12}$. Then it holds that

$$
\bar{E}_{22} - \bar{E}_{12}^T\bar{E}_{11}^{-1}\bar{E}_{12}
$$

$$
= \Omega - 2\beta\lambda_{\min}(\bar{Q}^{-1})(\Lambda_{22} \otimes \bar{Q}^2)
$$

$$
< 0. \tag{6.61}
$$

Based on the Schur complement property in Lemma 2.6, it follows from (6.60) and (6.61) that $\bar{E} < 0$.

Therefore, it can be verified from (6.58) that $\dot{V}_2 \leqslant 0$ when $t \geqslant T_s$, which means that V_2 is bounded, and so are $\tilde{\xi}_i$, $\hat{\alpha}_{ij}$, and $\hat{\beta}_{ik}$. Since the updating laws (6.39) and (6.40) are monotonically increasing, one can obtain that $\hat{\alpha}_{ij}$ and $\hat{\beta}_{ik}$ converge to some positive constants. Moreover, note that $\dot{V}_2 \equiv 0$ means that $\tilde{\xi} = 0$. Taking advantage of LaSalle's Invariance principle, it can be verified that $\lim_{t\to\infty}\tilde{\xi}(t) = 0$, i.e., $\lim_{t\to\infty}(\hat{\xi}_i(t) - \bar{z}(t)) = 0$ ($i \in \mathcal{T}_F$). This completes the proof of Theorem 6.3. $\qquad\square$

Remark 6.11. *The well-informed follower assumption used in [9] is removed in this section. Although each informed follower can only receive from a subset of leaders, by combining the local estimation and the interaction with neighbouring followers, a novel adaptive distributed observer is proposed for each follower to estimate the matrices and the states of the multiple leaders. Note that the distributed observer (6.36) can estimate the leaders' matrices in finite time, while the existing results in [87, 157] only guarantee the exponential convergence. Together with the adaptive gains $\hat{\alpha}_{ij}$ and $\hat{\beta}_{ik}$, the proposed observer (6.36) and (6.37) can be designed in a fully distributed form by each follower with no need for the eigenvalue information of both the leaders' system matrices and the Laplacian matrix of the graph. Moreover, the assumption that the leader's system matrix has no eigenvalues with positive real parts used in [159] is removed in this section owing to the finite-time convergence of (6.36).*

6.3.3   Finite-time Solution of Regulator Equations

In this subsection, an adaptive algorithm is proposed to solve the regulator equations in finite time based on the estimation of the leaders' matrices. The following assumption is standard in cooperative control of heterogeneous swarm systems (see, e.g. [83–87]).

Assumption 6.4. *There exist solution pairs $(X_{i,j}, U_{i,j})$ $(i \in \mathcal{T}_F, j \in \mathcal{T}_L)$ such that the following regulation equations hold:*

$$\begin{cases} X_{i,j} S_j = A_i X_{i,j} + B_i U_{i,j}, \\ 0 = C_i X_{i,j} - F_j. \end{cases} \tag{6.62}$$

It should be pointed out that the matrices S_j and F_j $(j \in \mathcal{T}_L)$ are only available to the followers which contain the leader j as neighbour. For the other followers that do not know S_j and F_j, the regulation equations (6.62) cannot be checked directly. Motivated by the adaptive approach in [87], the estimated matrices $\hat{S}_{i,j}$ and $\hat{F}_{i,j}$ will be applied to solve the equations (6.62) adaptively.

The following lemma extends the exponential convergence in Lemma 3 of [87] to the finite-time convergence.

Lemma 6.4. *Consider the equation $Ax = b$, where $A \in \mathbb{R}^{m \times n}$ and $b \in \mathbb{R}^{m \times 1}$ satisfy $rank(A) = rank(A, b) = r \geqslant 1$. Suppose that there exist time-varying matrix $\hat{A}(t) \in \mathbb{R}^{m \times n}$ and vector $\hat{b}(t) \in \mathbb{R}^{m \times 1}$, which are bounded and piecewise continuous, such that $\hat{A}(t) \to A$ and $\hat{b}(t) \to b$ in a finite time T_1. For any bounded initial state, and positive constants $k > 0$ and $0 < \varphi < 1$, the following system*

$$\dot{x}(t) = -k \hat{A}^T(t) sig^\varphi \left(\hat{A}(t) x(t) - \hat{b}(t) \right) \tag{6.63}$$

has a unique bounded solution $x(t)$ such that $x(t) \to x^*$ in a finite time T_2, where x^* satisfies $Ax^* = b$.

Proof. As shown in [87], there exists an orthogonal matrix $P \in \mathbb{R}^{n \times n}$ such that $AP = \begin{bmatrix} \bar{A} & 0_{m \times (n-r)} \end{bmatrix}$, where $\bar{A} \in \mathbb{R}^{m \times r}$ is of full column rank. So there is a unique vector $\bar{x}_1^* \in \mathbb{R}^r$ such that $\bar{A}\bar{x}_1^* = b$. Let $x^* = P\bar{x}^* = P \begin{bmatrix} \bar{x}_1^* \\ \bar{x}_2^* \end{bmatrix}$, where $\bar{x}_2^* \in \mathbb{R}^{n-r}$ denotes any column vector. Then one can obtain that $Ax^* = AP \begin{bmatrix} \bar{x}_1^* \\ \bar{x}_2^* \end{bmatrix} = \bar{A}\bar{x}_1^* = b$. Let $\bar{x} = P^T x$. It follows from (6.63) that

$$\dot{\bar{x}} = -kP^T \hat{A}^T \mathrm{sig}^\varphi(\hat{A}P\bar{x} - \hat{b}). \tag{6.64}$$

Note that $\hat{A}(t) \to A$ and $\hat{b}(t) \to b$ in a finite time T_1. The following proof is divided into two steps. Firstly, we will show that $\bar{x}$ is bounded in $[0, T_1)$. Then, it will be proved that $\bar{x} \to \bar{x}^*$ in finite time when $t \geqslant T_1$.

Step 1: Boundedness of $\bar{x}$ in $[0, T_1)$.

Let $V_3 = \bar{x}^T \bar{x}$. Then, the time derivative of V_3 along (6.64) is $\dot{V}_3 = -2k\bar{x}^T P^T \hat{A}^T \mathrm{sig}^\varphi(\hat{A}P\bar{x} - \hat{b})$. Let $\hat{z} = \hat{A}P\bar{x} - \hat{b} = [\hat{z}_1, \hat{z}_2, \ldots, \hat{z}_m]^T \in \mathbb{R}^m$. It follows that

$$\dot{V}_3 = -2k \sum_{i=1}^{m} |\hat{z}_i|^{1+\varphi} - 2k\hat{b}^T \mathrm{sig}^\varphi(\hat{z})$$

$$\leqslant -2k\hat{b}^T \mathrm{sig}^\varphi(\hat{z})$$

$$\leqslant 2k \sum_{i=1}^{m} \left|\hat{b}_i\right| |\hat{z}_i|^\varphi. \tag{6.65}$$

From Lemma 2.5, one gets that $\left|\hat{b}_i\right||\hat{z}_i|^\varphi \leqslant \frac{1}{1+\varphi}\left|\hat{b}_i\right|^{1+\varphi} + \frac{\varphi}{1+\varphi}|\hat{z}_i|^{1+\varphi}$ and $|\hat{z}_i|^{1+\varphi} \leqslant \frac{1+\varphi}{2}|\hat{z}_i|^2 + \frac{1-\varphi}{2}$. So $\left|\hat{b}_i\right||\hat{z}_i|^\varphi \leqslant \frac{\varphi}{2}|\hat{z}_i|^2 + \frac{1}{1+\varphi}\left|\hat{b}_i\right|^{1+\varphi} + \frac{\varphi(1-\varphi)}{2(1+\varphi)}$. It holds from (6.65) that

$$\dot{V}_3 \leqslant \varphi k \sum_{i=1}^{m} |\hat{z}_i|^2 + \hat{\vartheta}, \tag{6.66}$$

where $\hat{\vartheta} = k \sum_{i=1}^{m} \left(\frac{2}{1+\varphi}\left|\hat{b}_i\right|^{1+\varphi} + \frac{\varphi(1-\varphi)}{(1+\varphi)} \right)$. Note that $\sum_{i=1}^{m} |\hat{z}_i|^2 = \|\hat{z}\|^2 = \|\hat{A}P\bar{x} - \hat{b}\|^2$, and $\hat{A}$ and $\hat{b}$ are bounded in $[0, T_1)$. There exist positive constants σ and ϖ such that $\|\hat{A}P\bar{x} - \hat{b}\|^2 \leqslant \sigma\bar{x}^T\bar{x} + \varpi$. Since $\hat{b}$ is bounded in $[0, T_1)$, one can obtain that $|\hat{\vartheta}| \leqslant \vartheta$ in $[0, T_1)$, where ϑ is a positive constant. Then it can be verified from (6.66) that $\dot{V}_3 \leqslant \varphi k \sigma V_3 + \varphi k \varpi + \vartheta$, which implies that V_3 is bounded in $[0, T_1)$, and so is $\bar{x}$.

Step 2: Finite-time convergence of $\bar{x}$.

When $t \geqslant T_1$, one has that $\hat{A}(t) = A$ and $\hat{b}(t) = b$. Then, it follows from (6.64) that

$$\dot{\bar{x}} = -kP^T A^T \mathrm{sig}^\varphi(AP\bar{x} - b). \tag{6.67}$$

Let $\bar{x} = \begin{bmatrix} \bar{x}_1 \\ \bar{x}_2 \end{bmatrix}$, where $\bar{x}_1 \in \mathbb{R}^r$ and $\bar{x}_2 \in \mathbb{R}^{n-r}$. Note that $AP = [\,\bar{A}\ \ 0_{m \times (n-r)}\,]$. Then (6.67) can be divided into

$$\begin{cases} \dot{\bar{x}}_1 = -k\bar{A}^T \mathrm{sig}^\varphi\left(\bar{A}\bar{x}_1 - b\right), \\ \dot{\bar{x}}_2 = 0. \end{cases} \tag{6.68}$$

Thus, there is a constant vector $\bar{x}_2^* \in \mathbb{R}^{n-r}$ such that $\bar{x}_2(t) \to \bar{x}_2^*$ in a finite time T_1.

Let $\tilde{x}_1 = \bar{x}_1 - \bar{x}_1^*$. Note that $\bar{A}\bar{x}_1^* = b$. It holds from (6.68) that

$$\dot{\tilde{x}}_1 = -k\bar{A}^T \mathrm{sig}^\varphi\left(\bar{A}\tilde{x}_1\right). \tag{6.69}$$

Consider the Lyapunov functional candidate $V_4 = \tilde{x}_1^T \tilde{x}_1$. Then the time derivative of V_4 along (6.69) is described by $\dot{V}_4 = -2k\tilde{x}_1^T \bar{A}^T \mathrm{sig}^\varphi(\bar{A}\tilde{x}_1)$. Let $\tilde{y} = \bar{A}\tilde{x}_1 \in \mathbb{R}^m$. Then it follows that $\dot{V}_4 = -2k\tilde{y}^T \mathrm{sig}^\varphi(\tilde{y}) = -2k\sum_{i=1}^{m} |\tilde{y}_i|^{1+\varphi} \leqslant -2k(\|\tilde{y}\|^2)^{(1+\varphi)/2}$. Since $\bar{A}^T \bar{A}$ is a positive definite matrix, one can obtain that $\tilde{y}^T \tilde{y} = \tilde{x}_1^T \bar{A}^T \bar{A}\tilde{x}_1 \geqslant \lambda_{\min}(\bar{A}^T \bar{A})V_4$. Thus it holds that

$$\dot{V}_4 \leqslant -2k\left[\lambda_{\min}(\bar{A}^T \bar{A})\right]^{(1+\varphi)/2} V_4^{(1+\varphi)/2}. \tag{6.70}$$

According to Lemma 2.27, it can be verified that $\tilde{x}_1(t) = \bar{x}_1(t) - \bar{x}_1^* \to 0$ in a finite time T_2, where $T_2 = T_1 + T_x$ and T_x can be estimated by $T_x \leqslant \frac{2V_4^{(1-\varphi)/2}(\tilde{x}_1(T_1))}{c_x(1-\varphi)}$, $c_x = 2k[\lambda_{\min}(\bar{A}^T \bar{A})]^{(1+\varphi)/2}$.

Therefore, one can get that $x(t) = P\bar{x}(t) \to P\begin{bmatrix} \bar{x}_1^* \\ \bar{x}_2^* \end{bmatrix} = x^*$ in a finite time $T_2 = T_1 + T_x$, where x^* satisfies $Ax^* = b$. This completes the proof of Lemma 6.4. □

Let $\vartheta_{i,j} = \mathrm{vec}\left(\begin{bmatrix} X_{i,j} \\ U_{i,j} \end{bmatrix}\right)$, $\Psi_{i,j} = S_j^T \otimes \begin{bmatrix} I_{n_i} & 0 \\ 0 & 0 \end{bmatrix} - I_{n_j} \otimes \begin{bmatrix} A_i & B_i \\ C_i & 0 \end{bmatrix}$,

and $b_{i,j} = \mathrm{vec}\left(\begin{bmatrix} 0 \\ -F_j \end{bmatrix}\right)$ $(i \in \mathcal{T}_F, j \in \mathcal{T}_L)$. According to [149], the regulation equation (6.62) can be transformed to

$$\Psi_{i,j} \vartheta_{i,j} = b_{i,j}. \tag{6.71}$$

Let $\hat{\Psi}_{i,j}(t) = \hat{S}_{i,j}^T(t) \otimes \begin{bmatrix} I_{n_i} & 0 \\ 0 & 0 \end{bmatrix} - I_{n_j} \otimes \begin{bmatrix} A_i & B_i \\ C_i & 0 \end{bmatrix}$ and $\hat{b}_{i,j}(t) = \mathrm{vec}\left(\begin{bmatrix} 0 \\ -\hat{F}_{i,j}(t) \end{bmatrix}\right)$. Note that $\hat{S}_{i,j}(t) \to S_j$ in a finite time T_s and $\hat{F}_{i,j}(t) \to F_j$ in a finite time T_f. So it follows that $\hat{\Psi}_{i,j}(t) \to \Psi_{i,j}$ and $\hat{b}_{i,j}(t) \to b_{i,j}$ in a finite time $T_m = \max\{T_s, T_f\}$. Based on Lemma 6.4, one can obtain the following results directly. So the proof is omitted.

Lemma 6.5. *Suppose that Assumption 6.4 holds. For $i \in \mathcal{T}_F$ and $j \in \mathcal{T}_L$, consider the following system:*

$$\dot{\hat{\vartheta}}_{i,j}(t) = -\kappa \hat{\Psi}_{i,j}^T(t) sig^\phi \left(\hat{\Psi}_{i,j}(t) \hat{\vartheta}_{i,j}(t) - \hat{b}_{i,j}(t) \right), \qquad (6.72)$$

where $\kappa > 0$ and $0 < \phi < 1$. For any bounded initial state, the system (6.72) has a unique bounded solution $\hat{\vartheta}_{i,j}(t)$ such that $\hat{\vartheta}_{i,j}(t) \to \vartheta_{i,j}^*$ in a finite time T_r, where $\vartheta_{i,j}^* = vec \left(\begin{bmatrix} X_{i,j}^* \\ U_{i,j}^* \end{bmatrix} \right)$ satisfies the equation (6.71). Moreover, let $\begin{bmatrix} \hat{X}_{i,j}(t) \\ \hat{U}_{i,j}(t) \end{bmatrix} = mtx_{n_j}^{n_i+m_i} \left(\hat{\vartheta}_{i,j}(t) \right)$, where $\hat{X}_{i,j}(t) \in \mathbb{R}^{n_i \times n_j}$ and $\hat{U}_{i,j}(t) \in \mathbb{R}^{m_i \times n_j}$. Then, it holds that $\hat{X}_{i,j}(t) \to X_{i,j}^*$ and $\hat{U}_{i,j}(t) \to U_{i,j}^*$ in a finite time T_r.

Remark 6.12. *In [87], the adaptive algorithm to solve the regulator equations can achieve exponential convergence. In order to guarantee the stability of the algorithm, the control gain μ_3 in [87] is required to be sufficiently large (see Remark 4 in [87] for more details). However, the lower bound of μ_3 is related to the spectrum of the leader's system matrix, which is actually global information. Note that Lemma 6.5 enhances the exponential convergence in [87] to the finite-time convergence. Then, the gain κ in (6.72) can be freely chosen without requiring the eigenvalue information of the leader's system matrix, which is advantageous to design a fully distributed output formation tracking protocol.*

6.3.4 Formation Tracking Protocol Design and Analysis

In this subsection, local exosystem is applied to generate the desired time-varying output formation of each follower. Then, a time-varying output formation tracking protocol and an algorithm to ascertain the control parameters are presented. It is proved that the outputs of the followers can not only realize the expected formation shape but also track the predefined convex combination of multiple leaders.

For follower i ($i \in \mathcal{T}_F$), the desired time-varying output formation $h_{y,i}(t)$ is generated by the following local exosystem:

$$\begin{cases} \dot{h}_i(t) = H_i h_i(t) + R_i r_i(t), \\ h_{y,i}(t) = Y_i h_i(t), \end{cases} \qquad (6.73)$$

where $h_i(t) \in \mathbb{R}^{n_{hi}}$, $r_i(t) \in \mathbb{R}^{m_{ri}}$, and $h_{y,i}(t) \in \mathbb{R}^p$. The local exosystem (6.73) is only known to the follower i itself. The bounded external input $r_i(t)$ is applied to generate more time-varying formation types.

Remark 6.13. *Many typical formation shapes can be generated by (6.73). For example, consider a heterogeneous swarm system moving in the XY plane.*

Let $H_i = 0$, $R_i = 0$, and $Y_i = 1$ along X-axis and Y-axis. Then, the local exosystem (6.73) can specify any constant formation configuration studied in [103]. Moreover, choose $H_i = \begin{bmatrix} 0 & 1 \\ -\omega^2 & 0 \end{bmatrix}$, $R_i = 0_{2\times1}$, and $Y_i = \begin{bmatrix} 1 & 0 \end{bmatrix}$ along X-axis and Y-axis, where ω denotes a positive constant. In this case, rotating circular formation considered in [48] can be generated by (6.73).

Assumption 6.5. *The following local regulation equations*

$$\begin{cases} X_{hi}H_i = A_i X_{hi} + B_i U_{hi} \\ 0 = C_i X_{hi} - Y_i \end{cases} \tag{6.74}$$

have solution pairs (X_{hi}, U_{hi}) $(i \in \mathcal{T}_F)$.

Similar to Assumption 6.4, Assumption 6.5 requires that the local regulation equations (6.74) have solutions. Since H_i and Y_i $(i \in \mathcal{T}_F)$ are known to the i-th follower itself, one can calculate (X_{hi}, U_{hi}) from (6.74) directly.

For the i-th follower $(i \in \mathcal{T}_F)$, consider the following time-varying output formation tracking protocol:

$$u_i(t) = K_i^{(1)}\hat{x}_i(t) + K_{hi}^{(2)}h_i(t) + \sum_{j=1}^{M}\rho_j\hat{K}_{i,j}^{(2)}(t)\hat{\xi}_{i,j}(t) + \tau_i(t), \tag{6.75}$$

where $\hat{x}_i(t) \in \mathbb{R}^{n_i}$ is the estimated state of $x_i(t)$, $\tau_i(t) \in \mathbb{R}^{m_i}$ denotes the formation compensation input, and $K_i^{(1)}$, $K_{hi}^{(2)}$, and $\hat{K}_{i,j}^{(2)}(t)$ stand for gain matrices to be ascertained by the following algorithm. In (6.75), the first and third terms are used to drive the followers to track the multiple leaders based on the output regulation strategy. The role of the second term is to introduce the desired formation information for each follower. The fourth term in (6.75) is applied to expand the feasible time-varying output formation tracking set.

Algorithm 6.2. *For the i-th follower $(i \in \mathcal{T}_F)$, the output formation tracking protocol (6.75) can be designed by the following steps.*
   Step 1. *Since $\operatorname{rank}(B_i) = m_i$, there exists a non-singular matrix $\Pi_i = [\widehat{B}_i^T, \tilde{B}_i^T]^T$ with $\widehat{B}_i \in \mathbb{R}^{m_i \times n_i}$ and $\tilde{B}_i \in \mathbb{R}^{(n_i-m_i)\times n_i}$ such that $\widehat{B}_i B_i = I_{mi}$ and $\tilde{B}_i B_i = 0$. For the desired output formation vector $h_{y,i}(t)$ generated by the local exosystem (6.73), choose X_{hi} and U_{hi} satisfying the local regulation equation (6.74) such that the following time-varying output formation tracking feasible condition holds:*

$$\lim_{t\to\infty}\left(\tilde{B}_i X_{hi}R_i r_i(t)\right) = 0. \tag{6.76}$$

If the condition (6.76) is satisfied for each follower, then let the formation compensation input $\tau_i(t) = \widehat{B}_i X_{hi}R_i r_i(t)$, otherwise the given formation $h_{y,i}(t)$ is not feasible under the proposed protocol and the algorithm stops.

Step 2. *Based on Theorem 6.3, construct the distributed observer to get* $\hat{S}_{i,j}(t)$, $\hat{F}_{i,j}(t)$, *and* $\hat{\xi}_{i,j}(t)$. *Then, calculate the adaptive solution* $\hat{X}_{i,j}(t)$ *and* $\hat{U}_{i,j}(t)$ *of (6.62) using the approach in Lemma 6.5.*

Step 3. *Select* $K_i^{(1)}$ *to make* $A_i + B_i K_i^{(1)}$ *Hurwitz. Let* $K_{hi}^{(2)} = U_{hi} - K_i^{(1)} X_{hi}$ *and* $\hat{K}_{i,j}^{(2)}(t) = \hat{U}_{i,j}(t) - K_i^{(1)} \hat{X}_{i,j}(t)$.

Step 4. *Design the following Luenberger observer for* $\hat{x}_i(t)$:

$$\dot{\hat{x}}_i(t) = A_i \hat{x}_i(t) + B_i u_i(t) + L_{oi}\left(C_i \hat{x}_i(t) - y_i(t)\right), \tag{6.77}$$

where $L_{oi} \in \mathbb{R}^{n_i \times p}$ *denotes a gain matrix such that* $A_i + L_{oi} C_i$ *is Hurwitz.*

Remark 6.14. *As shown in [9], not all time-varying formation vectors can be realized even for homogeneous swarm systems (see Illustrative Example 2 in [9] for more details). The feasible condition (6.76) means that the predefined formations should be compatible with the dynamic constraints of the heterogeneous swarm systems. The external input* $r_i(t)$ *and the formation compensation input* $\tau_i(t)$ *are applied to expand the feasible formation set such that more time-varying formation types can be realized by heterogeneous swarm systems. In [103, 104, 119], the expected formations were generated by autonomous exosystem, which can be viewed as a special case of (6.73) by letting* $r_i(t) \equiv 0$. *In this case, the feasible condition (6.76) always holds.*

The following theorem guarantees that the heterogeneous swarm system can realize the expected output formation tracking under the proposed protocol.

Theorem 6.4. *Suppose that Assumptions 6.3–6.5 hold. If the given formation vector* $h_y(t)$ *generated by (6.73) satisfies the feasible condition (6.76), then the desired time-varying output formation tracking is realized by heterogeneous swarm system (6.32) and (6.33) with multiple leaders under the distributed control protocol (6.75) designed by Algorithm 6.2.*

Proof. Let $\tilde{X}_{i,j}(t) = \hat{X}_{i,j}(t) - X_{i,j}^*$, $\tilde{U}_{i,j}(t) = \hat{U}_{i,j}(t) - U_{i,j}^*$, and $\tilde{\xi}_{i,j}(t) = \hat{\xi}_{i,j}(t) - z_j(t)$ ($i \in \mathcal{T}_F$, $j \in \mathcal{T}_L$). It can be verified from Theorem 6.3 and Lemma 6.5 that $\tilde{X}_{i,j}(t) \to 0$ and $\tilde{U}_{i,j}(t) \to 0$ in a finite time T_r, and $\lim_{t \to \infty} \tilde{\xi}_{i,j}(t) = 0$. Let $\tilde{x}_{oi}(t) = \hat{x}_i(t) - x_i(t)$. Then it follows from (6.77) that $\dot{\tilde{x}}_{oi}(t) = (A_i + L_{oi} C_i) \tilde{x}_{oi}(t)$. Since $A_i + L_{oi} C_i$ is Hurwitz, one has that $\lim_{t \to \infty} \tilde{x}_{oi}(t) = 0$.

Substituting (6.75) into (6.33) leads to

$$\dot{x}_i = A_i x_i + B_i K_i^{(1)} \hat{x}_i + B_i K_{hi}^{(2)} h_i + \sum_{j=1}^{M} \rho_j B_i \hat{K}_{i,j}^{(2)} \hat{\xi}_{i,j} + B_i \tau_i. \tag{6.78}$$

Let $\tilde{x}_i = x_i - X_{hi}h_i - \sum_{j=1}^{M} \rho_j X_{i,j}^* z_j$. Then, it holds that

$$\dot{\tilde{x}}_i = \left(A_i + B_i K_i^{(1)}\right) x_i + B_i K_i^{(1)} \tilde{x}_{oi} + B_i K_{hi}^{(2)} h_i + B_i \tau_i - X_{hi} H_i h_i$$
$$- X_{hi} R_i r_i + \sum_{j=1}^{M} \rho_j B_i \hat{K}_{i,j}^{(2)} \hat{\xi}_{i,j} - \sum_{j=1}^{M} \rho_j X_{i,j}^* S_j z_j. \tag{6.79}$$

Note that $K_{hi}^{(2)} = U_{hi} - K_i^{(1)} X_{hi}$ and the local regulation equation (6.74) holds. One can obtain from (6.79) that

$$\dot{\tilde{x}}_i = \left(A_i + B_i K_i^{(1)}\right)(x_i - X_{hi}h_i) + B_i K_i^{(1)} \tilde{x}_{oi} + B_i \tau_i - X_{hi} R_i r_i$$
$$+ \sum_{j=1}^{M} \rho_j B_i \hat{K}_{i,j}^{(2)} \hat{\xi}_{i,j} - \sum_{j=1}^{M} \rho_j X_{i,j}^* S_j z_j. \tag{6.80}$$

Since $\hat{K}_{i,j}^{(2)} = \hat{U}_{i,j} - K_i^{(1)} \hat{X}_{i,j}$ and $X_{i,j}^* S_j = A_i X_{i,j}^* + B_i U_{i,j}^*$, it follows from (6.80) that

$$\dot{\tilde{x}}_i = \left(A_i + B_i K_i^{(1)}\right) \tilde{x}_i + B_i K_i^{(1)} \tilde{x}_{oi} + B_i \tau_i - X_{hi} R_i r_i$$
$$+ \sum_{j=1}^{M} \rho_j B_i \left(\tilde{K}_{i,j}^{(2)} z_j + K_{i,j}^{(2)*} \tilde{\xi}_{i,j} + \tilde{K}_{i,j}^{(2)} \tilde{\xi}_{i,j}\right), \tag{6.81}$$

where $\tilde{K}_{i,j}^{(2)} = \tilde{U}_{i,j} - K_i^{(1)} \tilde{X}_{i,j}$ and $K_{i,j}^{(2)*} = U_{i,j}^* - K_i^{(1)} X_{i,j}^*$. Note that $\tilde{X}_{i,j} \to 0$ and $\tilde{U}_{i,j} \to 0$ in a finite time T_r. So $\tilde{K}_{i,j}^{(2)} \to 0$ in a finite time T_r.

For $t \in [0, T_r)$, one can obtain that $\tilde{x}_{oi}$, z_j, $\tilde{\xi}_{i,j}$, and $\tilde{K}_{i,j}^{(2)}$ are bounded, and so is $\tilde{x}_i$. When $t \geq T_r$, one gets that $\tilde{K}_{i,j}^{(2)} = 0$. Then (6.81) can be transformed to

$$\dot{\tilde{x}}_i = (A_i + B_i K_i^{(1)}) \tilde{x}_i + B_i K_i^{(1)} \tilde{x}_{oi} + B_i \tau_i - X_{hi} R_i r_i + \sum_{j=1}^{M} \rho_j B_i K_{i,j}^{(2)*} \tilde{\xi}_{i,j}. \tag{6.82}$$

Note that the output formation tracking feasible condition (6.76) holds and $\tau_i = \widehat{B}_i X_{hi} R_i r_i$. Since $\Pi_i = [\widehat{B}_i^T, \tilde{B}_i^T]^T$ is nonsingular, it can be verified that $\lim_{t \to \infty} (B_i \tau_i - X_{hi} R_i r_i) = 0$. Furthermore, recall that $\lim_{t \to \infty} \tilde{x}_{oi}(t) = 0$, $\lim_{t \to \infty} \tilde{\xi}_{i,j}(t) = 0$, and $A_i + B_i K_i^{(1)}$ is Hurwitz. Based on the input-to-state stability theory, one can obtain that $\lim_{t \to \infty} \tilde{x}_i(t) = 0$.

Let $e_i(t) = y_i(t) - h_{y,i}(t) - \sum_{j=1}^{M} \rho_j q_j(t)$ denote the output formation tracking error for follower i ($i \in \mathcal{T}_F$). Then, it holds that $e_i(t) = C_i \tilde{x}_i(t)$. So one has that $\lim_{t \to \infty} e_i(t) = 0$, which implies that the desired time-varying output formation tracking is realized by heterogeneous swarm system (6.32) and (6.33) with multiple leaders. This completes the proof of Theorem 6.4. $\square$

Remark 6.15. *With the aid of the modified adaptive distributed observer in Theorem 6.3 and the improved adaptive solution algorithm in Lemma 6.5, the proposed output formation tracking approach can be performed by each follower in a fully distributed form. Under the containment-based strategy in [9, 54, 108], the formation reference is restricted to the convex combination of leaders. However, we can choose any values for ρ_j in this section. In addition, the convex combinations in [9, 54, 108] depend on the interaction topology and cannot be predefined. Thus, compared with [9, 54, 108], a more flexible formation reference can be selected in this section.*

Remark 6.16. *Different from the containment control, where the followers need converge to the convex hull spanned by the multiple leaders, the control object of formation tracking in this chapter is to drive the followers to not only realize a given time-varying formation but also track the same convex combination of the leaders. There are time-varying formation and its derivative in the design of output formation tracking protocol (6.75), and all the followers need reach an agreement on the common formation reference (i.e., $\sum_{j=1}^{M} \rho_j q_j(t)$), which is quite different from containment control problems. Although the containment-based control strategy was applied to solve formation tracking problems with multiple leaders in [9, 54, 108], the existing results all depend on the well-informed follower assumption. This assumption is removed in this section, and then the containment-based strategy is abandoned. Instead, a novel output formation tracking approach is presented using a distributed observer scheme, where the followers can estimate the states of all the leaders.*

6.3.5 Numerical Simulation

Consider a heterogeneous swarm system with ten agents, where $\mathcal{T}_L = \{1, 2, 3\}$ and $\mathcal{T}_F = \{4, 5, \ldots, 10\}$. The interaction topology is given in Fig. 6.13, where one can see that Assumption 6.3 holds. The heterogeneous swarm system is assumed to move in the three dimensional space (i.e., the XYZ space). The dynamics of each agent is described by (6.32) and (6.33) with $z_j = [(z_j^X)^T, (z_j^Y)^T, (z_j^Z)^T]^T$, $q_j = [(q_j^X)^T, (q_j^Y)^T, (q_j^Z)^T]^T$ $(j \in \mathcal{T}_L)$, $x_i = [(x_i^X)^T, (x_i^Y)^T, (x_i^Z)^T]^T$, and $y_i = [(y_i^X)^T, (y_i^Y)^T, (y_i^Z)^T]^T$ $(i \in \mathcal{T}_F)$. The leaders are modeled as $S_1 = S_2 = I_3 \otimes \begin{bmatrix} 0 & 1 \\ 0 & 0 \end{bmatrix}$, $F_1 = F_2 = I_3 \otimes [\, 1 \quad 0\,]$,

$$S_3 = I_3 \otimes \begin{bmatrix} 0 & 1 & 0 \\ 0 & 0 & 1 \\ 0 & 0 & 0 \end{bmatrix}, \text{ and } F_3 = I_3 \otimes [\, 1 \quad 0 \quad 0\,].$$ Consider the following dynamics of the followers:

Follower 4: $A_4 = 0_{3 \times 3}$, $B_4 = I_3$, $C_4 = I_3$.

Followers 5-9: $A_i = I_3 \otimes \begin{bmatrix} 0 & 1 \\ i-5 & i-5 \end{bmatrix}$, $B_i = I_3 \otimes \begin{bmatrix} 0 \\ i-4 \end{bmatrix}$, $C_i = I_3 \otimes [\, 1 \quad 0\,]$, $i = 5, 6, \ldots, 9$.

Follower 10: $A_{10} = I_3 \otimes \begin{bmatrix} 0 & 1 & 0 \\ 0 & 0 & 1 \\ 1 & 2 & 1 \end{bmatrix}$, $B_{10} = I_3 \otimes \begin{bmatrix} 0 \\ 0 \\ 1 \end{bmatrix}$, $C_{10} = I_3 \otimes$

$[\, 1 \quad 0 \quad 0 \,]$.

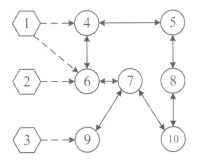

FIGURE 6.13: Interaction topology.

The outputs of the followers need to realize a rotating circular formation tracking, where the desired formation is specified by $h_{y,i}(t) = \begin{bmatrix} 0 \\ 3\cos(t + 2(i-1)\pi/7) \\ 3\sin(t + 2(i-1)\pi/7) \end{bmatrix}$. When $h_{y,i}(t)$ is realized, the outputs of followers along the X-axis will achieve consensus with the formation reference, and the outputs of followers in the YZ plane will form a circular formation rotating around the three leaders. To generate $h_{y,i}(t)$ ($i \in \mathcal{T}_F$), the local exosystem (6.73) is designed as $H_i = I_3 \otimes \begin{bmatrix} 0 & 1 \\ -1 & 0 \end{bmatrix}$, $R_i = 0_{6\times3}$, and $Y_i = I_3 \otimes [\, 1 \quad 0 \,]$. The initial state of (6.73) is set as $h_i^X(0) = [0,0]^T$, $h_i^Y(0) = [3\cos(2(i-4)\pi/7), -3\sin(2(i-4)\pi/7)]^T$, and $h_i^Z(0) = [3\sin(2(i-4)\pi/7), 3\cos(2(i-4)\pi/7)]^T$ ($i \in \mathcal{T}_F$). Choose $\rho_1 = \frac{1}{3}$, $\rho_2 = \frac{1}{3}$, and $\rho_3 = \frac{1}{3}$. Then, the combination of the leaders' outputs (i.e., the formation reference) is defined as $\frac{1}{3}(q_1(t) + q_2(t) + q_3(t))$.

Follow Algorithm 6.2 to ascertain the output formation tracking protocol (6.75). For the desired formation $h_{y,i}(t)$ ($i \in \mathcal{T}_F$) generated by the local exosystem (6.73), choose $X_{h4} = I_3 \otimes [\, 1 \quad 0 \,]$, $U_{h4} = I_3 \otimes [\, 0 \quad 1 \,]$, $X_{hk} = I_6$,

$U_{hk} = [\, -1 \quad -(k-5)/(k-4) \,]$, $k = 5, 6, \ldots, 9$, $X_{h10} = I_3 \otimes \begin{bmatrix} 1 & 0 \\ 0 & 1 \\ -1 & 0 \end{bmatrix}$,

and $U_{h10} = I_3 \otimes [\, 0 \quad -3 \,]$ such that the local regulation equations (6.74) hold. Note that $R_i = 0_{6\times3}$. It can be verified that each follower satisfies the output formation tracking feasible condition (6.76) and the compensation input $\tau_i(t) = 0$. The distributed observer is designed as $\eta = 2$ and $\gamma = \frac{1}{2}$ to get $\hat{S}_{i,j}(t)$, $\hat{F}_{i,j}(t)$, and $\hat{\xi}_{i,j}(t)$. Select $\kappa = 2$ and $\phi = \frac{1}{2}$ in the online regulator equation solver (6.72). The gain matrix $K_i^{(1)}$ of each follower is chosen as $K_4^{(1)} =$

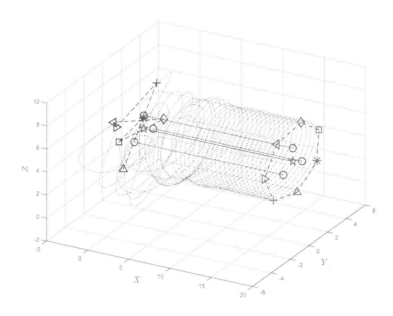

FIGURE 6.14: Output trajectories of the heterogeneous swarm system within $t = 30$s.

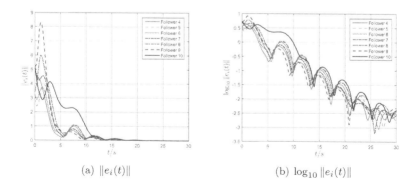

(a) $\|e_i(t)\|$ (b) $\log_{10} \|e_i(t)\|$

FIGURE 6.15: Time-varying output formation tracking errors of the followers.

$-2I_3$, $K_i^{(1)} = I_3 \otimes [\ -(i-3)/(i-4) \quad -(i-3)/(i-4) \ ]$ $(i = 5,6,\ldots,9)$, and $K_{10}^{(1)} = I_3 \otimes [\ -5 \quad -8 \quad -5 \ ]$. Design the observer gain L_{oi} in (6.77) as $L_{o4} = -6I_3$, $L_{oi} = I_3 \otimes [\ -(i+4) \quad -(i^2-5 \ )]^T$ $(i = 5,6,\ldots,9)$, and $L_{o10} = I_3 \otimes [\ -16 \quad -92 \quad -245 \ ]^T$. The initial states of the leaders are set to be $z_1(0) = [0,0.6,-1,0,5-\sqrt{3}/3,0]^T$, $z_2(0) = [0,0.6,1,0,5-\sqrt{3}/3,0]^T$, and $z_3(0) = [0,0.6,0,0,0,0,5+2\sqrt{3}/3,0,0]^T$, and the initial states $x_i(0)$, $\hat{\xi}_i(0)$, and $\hat{x}_i(0)$ $(i \in \mathcal{T}_F)$ are generated by random numbers. The other initial values are selected to be $\alpha_{ij}(0) = 5$, $\beta_{ik}(0) = 5$, $\hat{s}_i(0) = 0$, $\hat{f}_i(0) = 0$, and $\hat{\vartheta}_{i,j}(0) = 0$.

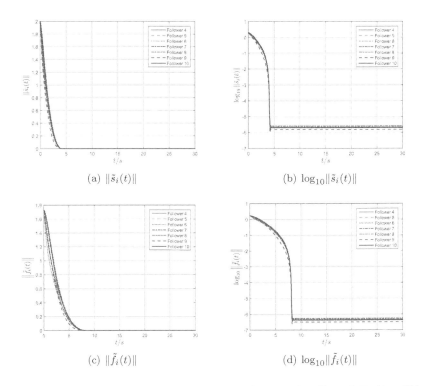

(a) $\|\tilde{s}_i(t)\|$

(b) $\log_{10}\|\tilde{s}_i(t)\|$

(c) $\|\tilde{f}_i(t)\|$

(d) $\log_{10}\|\tilde{f}_i(t)\|$

FIGURE 6.16: Estimation errors of each follower for S_j and F_j $(j \in \mathcal{T}_L)$.

Fig. 6.14 shows the out trajectories of the heterogeneous swarm system within $t = 30$s, where the leaders are marked by circles, the followers are depicted by upward-pointing triangle, asterisk, square, diamond, left-pointing triangle, right-pointing triangle, and plus sign respectively, and the convex combination of the three leaders is represented by pentagram. In Fig. 6.15, the time-varying output formation tracking errors of the followers are given. Figs. 6.16 and 6.17 depict the estimation errors of each follower for the leader's matrices S_j and F_j and the leader's states z_j $(j \in \mathcal{T}_L)$, respectively. To show the behaviours near zero more clearly, the error signals are also plotted on a logarithmic scale $\log_{10}$ in Figs. 6.15-6.17. The adaptive gains $\hat{\alpha}_{ij}(t)$ and $\hat{\beta}_{ik}(t)$

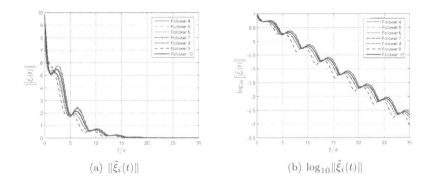

(a) $\|\tilde{\xi}_i(t)\|$ (b) $\log_{10}\|\tilde{\xi}_i(t)\|$

FIGURE 6.17: Estimation error of each follower for z_j $(j \in \mathcal{T}_L)$.

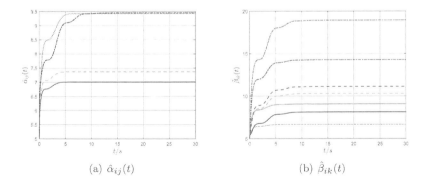

(a) $\hat{\alpha}_{ij}(t)$ (b) $\hat{\beta}_{ik}(t)$

FIGURE 6.18: Adaptive updating gains.

are shown in Fig. 6.18, where one can see that all the adaptive parameters converge to some positive constants finally. As shown in Figs. 6.14–6.17, the outputs of the followers form a rotating circular formation, whose centre is the given convex combination of the three leaders. Therefore, the expected time-varying output formation tracking is realized by the heterogeneous swarm system with multiple leaders under the proposed adaptive protocol.

6.4   Conclusions

Time-varying formation tracking problems for heterogeneous swarm systems with multiple leaders were studied in this chapter. Firstly, for the case with directed switching topologies, based on the well-informed follower assumption, a distributed time-varying output formation tracking protocol was designed. Sufficient conditions to achieve formation tracking with multiple leaders were given by using the piecewise Lyapunov stability theory. Furthermore, the well-informed follower assumption was removed, and the formation tracking problems with incomplete information of multiple leaders were discussed. A distributed observer was designed for each follower to estimate the dynamical matrices and the states of multiple leaders, and an adaptive algorithm was proposed to solve the regulator equations in finite time. A fully distributed time-varying output formation tracking protocol and a design algorithm were proposed. It was proved that the desired formation tracking with multiple leaders can be achieved by heterogeneous swarm systems without requiring the well-informed follower assumption. The results in this chapter are mainly based on [108] and [109].

Chapter 7

Formation-containment Tracking Control for Heterogeneous Swarm Systems

7.1   Introduction

In the formation tracking problems with multiple leaders in Chapter 6 and the containment control problems in [110–117], it is usually assumed that there is no interaction and collaboration among multiple leaders. However, in practical application scenarios, the leaders also need to coordinate to maintain a desired time-varying formation and track the reference trajectory or a specific target such that the task requirements, e.g. enclosing and surveillance, can be satisfied. In this case, the leader layer and the follower layer have different collaborative goals, and the leader layer also has a coupling effect on the cooperative goals of the follower layer. How to model, analyze, and design the above-mentioned formation tracking control problem with multiple coordination layers is a complex problem that needs to be further solved.

For high-order heterogeneous swarm systems with different intra-layer cooperative control objectives and inter-layer coordination couplings, the definition and framework for formation-containment tracking control are given in this chapter. The agents are classified into three types, i.e., tracking-leader, formation-leader, and follower. The tracking-leader is applied to generate the macroscopic trajectory of the whole swarm system. Under the formation-containment tracking control, the multiple formation-leaders can form the desired time-varying formation and track the reference trajectory simultaneously, while the followers need to move into the formation formed by the multiple formation-leaders. The main contents of this chapter are summarized as follows.

Since containment control is an important part for formation-containment tracking, predefined containment control problems for heterogeneous swarm systems with switching topologies are studied firstly in this chapter. A distributed observer is designed for each follower to estimate the whole states of all the leaders. Based on the estimated states, a predefined containment tracking controller is constructed for the followers, where the desired convex combinations of the multiple leaders are specified by several given constant weights. Then, formation-containment tracking control problems for heterogeneous swarm systems with leaders' unknown inputs on switching graphs are further investigated. A distributed formation-containment tracking controller is proposed by using output regulation control. Under the influences of hierarchical coordination couplings, an algorithm to design the proposed controller and sufficient conditions to achieve formation-containment tracking are presented.

7.2   Predefined Containment Control

In this section, containment control problems for fully heterogeneous swarm systems with switching topologies are studied, where both the leaders and the followers can have different dynamics. Firstly, a distributed observer is constructed for each follower to estimate the states of all the nonidentical leaders using the neighbouring interaction. Then, based on the estimated states of the multiple leaders, an output containment controller is proposed for followers using the output regulation strategy, where several predefined weights are applied for the followers to specify the desired convex combinations of the leaders. Thus, the given containment format is independent of the interaction topology. In light of the common Lyapunov stability and output regulation theory, it is proved that the desired output containment can be realized by fully heterogeneous swarm systems under the influences of switching topologies. Finally, a simulation example is provided to verify the effectiveness of theoretical results.

7.2.1   Problem Description

A heterogeneous swarm system composed of M leaders and N followers is considered in this section. Let $\mathcal{O}_L = \{1, 2, \ldots, M\}$ denote the leader set and $\mathcal{O}_F = \{M + 1, M + 2, \ldots, M + N\}$ represent the follower set. A leader has no neighbour, and a follower has at least one neighbour.

It is assumed that the graph may be switching in this section. Let $H = \{1, 2, \ldots, h\}$ denote the index set of all the possible graphs. Let $[t_k, t_{k+1})$ ($k = 0, 1, 2 \cdots$) represent an infinite sequence of time intervals, where $t_0 = 0$, $t_{k+1} - t_k \geqslant \tau_d > 0$, and each interval is uniformly bounded and non-overlapping. The graph changes at the switching sequence t_{k+1}. Define the switching signal as $\sigma(t): [0, \infty) \to \{1, 2, \ldots, h\}$. Let $\mathcal{G}_{\sigma(t)}$ and $L_{\sigma(t)}$ denote the graph and the Laplacian matrix at t, respectively.

Consider the following dynamics for the j-th leader ($j \in \mathcal{O}_L$):

$$\begin{cases} \dot{z}_j(t) = S_j z_j(t), \\ q_j(t) = F_j z_j(t), \end{cases} \tag{7.1}$$

where $z_j(t) \in \mathbb{R}^{n_j}$ and $q_j(t) \in \mathbb{R}^p$ are the state and the output. For the i-th follower ($i \in \mathcal{O}_F$), its dynamics is modeled as

$$\begin{cases} \dot{x}_i(t) = A_i x_i(t) + B_i u_i(t), \\ y_i(t) = C_i x_i(t), \end{cases} \tag{7.2}$$

where $x_i(t) \in \mathbb{R}^{n_i}$, $u_i(t) \in \mathbb{R}^{m_i}$, and $y_i(t) \in \mathbb{R}^p$ are the state, the control input, and the output, respectively. The pair (A_i, B_i) is stabilizable, and the pair (C_i, A_i) is detectable.

Remark 7.1. *In [110–117], although the followers are heterogeneous, it is required that the multiple leaders must have identical dynamics. Note that both the leaders and the followers have heterogeneous dynamics in (7.1) and (7.2). Thus, this section focuses on the containment control problems of fully heterogeneous swarm systems with non-identical leaders.*

In the existing containment control results such as [110–117], the convex combinations of multiple leaders for each follower to track are determined explicitly by the topologies. For example, assume that there is a homogeneous swarm system with $\dot{x}_i = A x_i + B u_i$, $i = 1, 2, \ldots, M + N$. Let $x_L = [x_1^T, x_2^T, \ldots, x_M^T]^T$ and $x_F = [x_{M+1}^T, x_{M+2}^T, \ldots, x_{M+N}^T]^T$ indicate respectively the whole states of leaders and followers. The Laplacian matrix of the graphs can be divided as $L_{\sigma(t)} = \begin{bmatrix} 0 & 0 \\ L_1^{\sigma(t)} & L_2^{\sigma(t)} \end{bmatrix}$. Using the control strategy in [10], when the containment is achieved, we can get

$$\lim_{t \to \infty} \left[x_F - \left(-(L_2^{\sigma(t)})^{-1} L_1^{\sigma(t)} \otimes I_n \right) x_L \right] = 0,$$

where the convex combinations of multiple leaders (i.e., the desired tracking targets for followers) are determined by the topologies $-(L_2^{\sigma(t)})^{-1} L_1^{\sigma(t)}$ directly. Thus, for swarm systems on switching graphs, the convex combinations of the leaders will vary along with the possible topology, and then the followers' transient outputs or states may go beyond the convex hull constituted by the leaders under the influence of switching graphs.

To explain this point more intuitively, the following simulation example is given.

Illustrative example 7.1. *Consider a homogeneous swarm system described by $\dot{x}_i = A x_i + B u_i$ ($i = 1, 2, \ldots, 6$), where $x_i \in \mathbb{R}^3$, $A = \begin{bmatrix} 0 & 1 & 0 \\ 0 & 0 & 1 \\ 0 & -1 & 0 \end{bmatrix}$ and*

$B = \begin{bmatrix} 0 \\ 0 \\ 1 \end{bmatrix}$. *There are 3 leaders and 3 followers with* $\mathcal{O}_L = \{1,2,3\}$ *and* $\mathcal{O}_F =$
$\{4,5,6\}$. *The graph is assumed to switch between* $\mathcal{G}_1$ *and* $\mathcal{G}_2$ *each 15s, and the possible graphs are shown in Fig. 7.1. Let* $x_L = [x_1^T, x_2^T, x_3^T]^T$ *and* $x_F = [x_4^T, x_5^T, x_6^T]^T$. *Then, based on the existing containment approaches, we can get the containment error* $\tilde{x}_C = x_F - \left(-(L_2^{\sigma(t)})^{-1} L_1^{\sigma(t)} \otimes I_n \right) x_L$, *and the convex combinations of multiple leaders are decided by* $-(L_2^{\sigma(t)})^{-1} L_1^{\sigma(t)}$. *Under the containment protocol (7) in [10], the state snapshots near the first switching time* $t = 15.0s$ *of the swarm system are provided in Fig. 7.2, where the leaders are marked by diamond, circle, and cross respectively, and the followers are symbolized as squares. Besides, Fig. 7.3 shows the curve of containment error* $\|\tilde{x}_C\|$. *From Figs. 7.2 and 7.3, we can see that the transient states of the three followers obviously go beyond the convex hull constituted by the leaders under the influence of switching graphs, and the desired containment cannot be realized. The reason is that the convex combinations of multiple leaders described by* $-(L_2^{\sigma(t)})^{-1} L_1^{\sigma(t)}$ *are affected by the switching graphs. Thus, the existing containment control approaches in [110–117] are not applicable to switching graphs in general.*

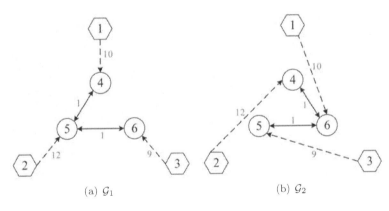

(a) $\mathcal{G}_1$ (b) $\mathcal{G}_2$

FIGURE 7.1: Possible switching graphs in Illustration example 7.1.

Motivated by this phenomenon, a novel predefined containment control framwork is proposed for heterogeneous swarm systems on switching graphs in this section.

Definition 7.1. *For any* $i \in \mathcal{O}_F$, *if there are given non-negative constants* $\rho_{i,j}$ $(j \in \mathcal{O}_L)$ *satisfying* $\sum_{j=1}^M \rho_{i,j} = 1$ *such that*

$$\lim_{t \to \infty} \left(y_i(t) - \sum_{j=1}^M \rho_{i,j} q_j(t) \right) = 0, \tag{7.3}$$

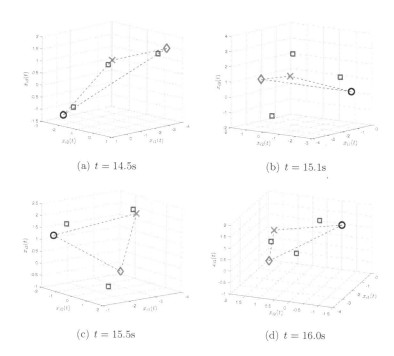

(a) $t = 14.5$s

(b) $t = 15.1$s

(c) $t = 15.5$s

(d) $t = 16.0$s

FIGURE 7.2: State snapshots near the switching time $t = 15.0$s in Illustration example 7.1.

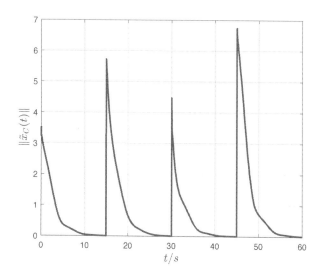

FIGURE 7.3: Containment control error in Illustration example 7.1.

then the heterogeneous swarm system (7.1) and (7.2) realizes predefined output containment.

The constants $\rho_{i,j}$ in Definition 7.1 are applied to specify the desired convex combinations of leaders for each follower to track. Under the proposed predefined containment control approach, the weights $\rho_{i,j}$ are independent on the topology, and we can design specific values of $\rho_{i,j}$ in advance based on practical mission requirements.

7.2.2 Predefined Containment Controller Design and Stability Analysis

Without loss of generality, it is required that there exist no isolated leaders, i.e., each leader can be acquired by at least one follower. If there is an isolated leader, then change the number of leaders to be $M-1$ and all the results still hold. Let the graph among the followers at t be $\mathcal{G}_F^{\sigma(t)}$, whose associated Laplacian matrix is described by $L_F^{\sigma(t)}$.

Assumption 7.1. *Each possible graph $\mathcal{G}_F^{\sigma(t)}$ among the followers is undirected and connected.*

Assumption 7.2. *The following regulation equations have solution pairs $(X_{i,j}, U_{i,j})$ ($i \in \mathcal{O}_F$, $j \in \mathcal{O}_L$):*

$$\begin{cases} X_{i,j}S_j = A_i X_{i,j} + B_i U_{i,j}, \\ 0 = C_i X_{i,j} - F_j. \end{cases} \tag{7.4}$$

For follower i ($i \in \mathcal{O}_F$), consider the following output containment protocol:

$$\dot{\hat{\xi}}_i = \bar{S}\hat{\xi}_i - \begin{bmatrix} \hat{\alpha}_{i1}w_{i1}^\sigma(\hat{\xi}_{i,1}-z_1) \\ \hat{\alpha}_{i2}w_{i2}^\sigma(\hat{\xi}_{i,2}-z_2) \\ \vdots \\ \hat{\alpha}_{iM}w_{iM}^\sigma(\hat{\xi}_{i,M}-z_M) \end{bmatrix} - \sum_{k=M+1}^{M+N} \hat{\beta}_{ik}w_{ik}^\sigma(\hat{\xi}_i-\hat{\xi}_k),$$

$$\dot{\hat{\alpha}}_{ij} = w_{ij}^\sigma \left\| \hat{\xi}_{i,j} - z_j \right\|^2, \ j \in \mathcal{O}_L,$$

$$\dot{\hat{\beta}}_{ik} = w_{ik}^\sigma \left\| \hat{\xi}_i - \hat{\xi}_k \right\|^2, \ k \in \mathcal{O}_F,$$

$$\dot{\hat{x}}_i = A_i\hat{x}_i + B_iu_i + L_{oi}(C_i\hat{x}_i - y_i),$$

$$u_i = K_i^{(1)}\hat{x}_i + \sum_{j=1}^{M} \rho_{i,j}K_{i,j}^{(2)}\hat{\xi}_{i,j}, \tag{7.5}$$

where $\hat{\xi}_i(t) = [\hat{\xi}_{i,1}^T(t), \hat{\xi}_{i,2}^T(t), \dots, \hat{\xi}_{i,M}^T(t)]^T$ with $\hat{\xi}_{i,j}(t)$ being the i-th follower's estimated state for $z_j(t)$ ($j \in \mathcal{O}_L$), $\bar{S} = \text{diag}\{S_1, S_2, \dots, S_M\}$, $\hat{\alpha}_{ij}(t)$ ($j \in \mathcal{O}_L$)

and $\hat{\beta}_{ik}(t)$ $(k \in \mathcal{O}_F)$ are two adaptive gains, $\hat{x}_i(t)$ denotes the estimated state of $x_i(t)$ in the Luenberger observer, and $K_i^{(1)}$, $K_{i,j}^{(2)}$, and L_{oi} are constant gain matrices to be determined.

The following theorem shows that the given output containment can be realized under the proposed control protocol.

Theorem 7.1. *Suppose that Assumptions 7.1 and 7.2 hold. Choose $K_i^{(1)}$ such that $A_i + B_i K_i^{(1)}$ is Hurwitz, and let $K_{i,j}^{(2)} = U_{i,j} - K_i^{(1)} X_{i,j}$ $(i \in \mathcal{O}_F, j \in \mathcal{O}_L)$. The gain matrix L_{oi} $(i \in \mathcal{O}_F)$ is selected to make $A_i + L_{oi} C_i$ Hurwitz. Then, the heterogeneous swarm system (7.1) and (7.2) with switching topologies can achieve the expected output containment under the distributed control protocol (7.5).*

Proof. Let $\tilde{\xi}_i = [\tilde{\xi}_{i,1}^T, \tilde{\xi}_{i,2}^T, \ldots, \tilde{\xi}_{i,M}^T]^T$ $(i \in \mathcal{O}_F)$, where $\tilde{\xi}_{i,j} = \hat{\xi}_{i,j} - z_j$ $(j = 1, 2, \ldots, M)$. In the following, we will show that $\lim_{t \to \infty} \tilde{\xi}_{i,j} = 0$ firstly. Let $\bar{z} = [z_1^T, z_2^T, \ldots, z_M^T]^T$. Note that $\tilde{\xi}_i = \hat{\xi}_i - \bar{z}$ and $\dot{\bar{z}} = \bar{S}\bar{z}$. We can obtain from (7.5) that

$$\dot{\tilde{\xi}}_i = \bar{S}\tilde{\xi}_i - \hat{W}_i^\sigma \tilde{\xi}_i - \sum_{k=M+1}^{M+N} \hat{\beta}_{ik} w_{ik}^\sigma \left(\tilde{\xi}_i - \tilde{\xi}_k \right), \tag{7.6}$$

where $\hat{W}_i^\sigma = \mathrm{diag}\{\hat{\alpha}_{i1} w_{i1}^\sigma I_{n_1}, \hat{\alpha}_{i2} w_{i2}^\sigma I_{n_2}, \ldots, \hat{\alpha}_{iM} w_{iM}^\sigma I_{n_M}\}$. Consider the following common Lyapunov functional candidate:

$$V = \sum_{i=M+1}^{M+N} \tilde{\xi}_i^T \tilde{\xi}_i + \sum_{i=M+1}^{M+N} \sum_{j=1}^{M} (\hat{\alpha}_{ij} - \alpha)^2 + \frac{1}{2} \sum_{i=M+1}^{M+N} \sum_{k=M+1}^{M+N} \left(\hat{\beta}_{ik} - \beta \right)^2. \tag{7.7}$$

For each interval $[t_k, t_{k+1})$, follow the similar steps in the proof of Theorem 6.3, and we can get $\dot{V} \leqslant 0$. Note that V in (7.7) is a common Lyapunov function for the switched system (7.6). Then, if sufficiently large α and β are chosen, we can obtain that $\lim_{t \to \infty} \tilde{\xi}(t) = 0$, i.e., $\lim_{t \to \infty} (\hat{\xi}_{i,j}(t) - z_j(t)) = 0$ $(i \in \mathcal{O}_F, j \in \mathcal{O}_L)$.

Let $\tilde{x}_{oi}(t) = \hat{x}_i(t) - x_i(t)$ $(i \in \mathcal{O}_F)$. We can obtain from the Luenberger observer in (7.5) that $\lim_{t \to \infty} \tilde{x}_{oi}(t) = 0$. Substituting u_i into (7.2) gives

$$\dot{x}_i = A_i x_i + B_i K_i^{(1)} \hat{x}_i + \sum_{j=1}^{M} \rho_{i,j} B_i K_{i,j}^{(2)} \hat{\xi}_{i,j}. \tag{7.8}$$

Let $\tilde{x}_i = x_i - \sum_{j=1}^{M} \rho_{i,j} X_{i,j} z_j$. Then, we get from (7.8) that

$$\dot{\tilde{x}}_i = (A_i + B_i K_i^{(1)}) x_i + B_i K_i^{(1)} \tilde{x}_{oi} + \sum_{j=1}^{M} \rho_{i,j} B_i K_{i,j}^{(2)} \hat{\xi}_{i,j}$$

$$- \sum_{j=1}^{M} \rho_{i,j} X_{i,j} S_j z_j. \tag{7.9}$$

Note that $K_{i,j}^{(2)} = U_{i,j} - K_i^{(1)} X_{i,j}$ and $X_{i,j} S_j = A_i X_{i,j} + B_i U_{i,j}$. It follows that

$$\dot{\tilde{x}}_i = (A_i + B_i K_i^{(1)}) \tilde{x}_i + B_i K_i^{(1)} \tilde{x}_{oi} + \sum_{j=1}^{M} \rho_{i,j} B_i K_{i,j}^{(2)} \tilde{\xi}_{i,j}. \qquad (7.10)$$

Since $A_i + B_i K_i^{(1)}$ is Hurwitz, $\lim_{t\to\infty} \tilde{x}_{oi}(t) = 0$, and $\lim_{t\to\infty} \tilde{\xi}_{i,j}(t) = 0$, based on the input-to-state stability theory, it can be verified from (7.10) that $\lim_{t\to\infty} \tilde{x}_i(t) = 0$. Let $\tilde{y}_{Ci} = y_i(t) - \sum_{j=1}^{M} \rho_{i,j} q_j(t)$ ($i \in \mathcal{O}_F$). Note that $C_i X_{i,j} - F_j = 0$. Then, we have that $\tilde{y}_{Ci} = C_i \tilde{x}_i$. Thus, it follows that $\lim_{t\to\infty} \tilde{y}_{Ci}(t) = 0$, which means that the predefined output containment is realized by the heterogeneous swarm system (7.1) and (7.2). This completes the proof of Theorem 7.1. ∎

Remark 7.2. *Based on the estimated states of the multiple non-identical leaders, an output containment protocol is proposed for heterogeneous swarm system (7.1) and (7.2), where the predefined weights $\rho_{i,j}$ ($i \in \mathcal{O}_F$, $j = 1, 2, \ldots, M$) are used to specify the desired convex combinations of the leaders. Therefore, the follower's expected tracking values are independent of the graph, and the given containment can be achieved by heterogeneous swarm systems with switching topologies. In [110–117], the convex combinations of the leaders are determined by the interaction topology, so the existing approaches cannot be applied to the switching topologies in general.*

7.2.3   Simulation Example

Suppose that there is a heterogeneous swarm system with 3 leaders and 6 followers, where $\mathcal{O}_L = \{1, 2, 3\}$ and $\mathcal{O}_F = \{4, 5, \ldots, 9\}$. The heterogeneous swarm system moves in the XY plane. All possible graphs with 0-1 weights are shown in Fig. 7.4, and the switching signal is given in Fig. 7.5.

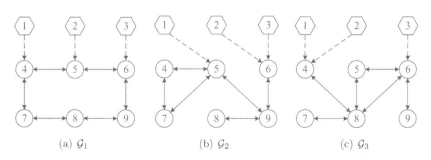

(a) $\mathcal{G}_1$ (b) $\mathcal{G}_2$ (c) $\mathcal{G}_3$

FIGURE 7.4: Possible graphs.

Consider the following heterogeneous dynamics of the leaders: $S_1 = I_2 \otimes$
$\begin{bmatrix} 0 & 1 \\ 0 & 0 \end{bmatrix}$, $S_2 = I_2 \otimes \begin{bmatrix} 0 & 1 \\ -1 & 0 \end{bmatrix}$, $S_3 = I_2 \otimes \begin{bmatrix} 0 & 1 \\ -4 & 0 \end{bmatrix}$, $F_1 = F_2 = F_3 =$

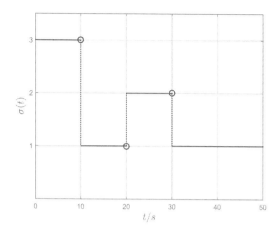

FIGURE 7.5: Switching signal.

$I_2 \otimes \begin{bmatrix} 1 & 0 \end{bmatrix}$. The dynamics of the followers are described by:

Followers 4,5,6: $A_i = I_2 \otimes \begin{bmatrix} 0 & 1 \\ 1 & i \end{bmatrix}$, $B_i = I_2 \otimes \begin{bmatrix} 0 \\ 1 \end{bmatrix}$, $C_i = I_2 \otimes \begin{bmatrix} 1 & 0 \end{bmatrix}$, $i = 4, 5, 6$.

Followers 7,8,9: $A_k = I_2 \otimes \begin{bmatrix} 0 & 1 & 0 \\ 0 & 0 & 1 \\ 1 & k-7 & k-6 \end{bmatrix}$, $B_k = I_2 \otimes \begin{bmatrix} 0 \\ 0 \\ 1 \end{bmatrix}$, $C_k = I_2 \otimes \begin{bmatrix} 1 & 0 & 0 \end{bmatrix}$, $k = 7, 8, 9$.

To specify the expected output containment, choose $\rho_{4,1} = \frac{1}{6}$, $\rho_{4,2} = \frac{1}{3}$, $\rho_{4,3} = \frac{1}{2}$, $\rho_{5,1} = \frac{1}{3}$, $\rho_{5,2} = \frac{1}{2}$, $\rho_{5,3} = \frac{1}{6}$, $\rho_{6,1} = \frac{1}{2}$, $\rho_{6,2} = \frac{1}{6}$, $\rho_{6,3} = \frac{1}{3}$, $\rho_{7,1} = \frac{1}{3}$, $\rho_{7,2} = \frac{1}{3}$, $\rho_{7,3} = \frac{1}{3}$, $\rho_{8,1} = \frac{1}{5}$, $\rho_{8,2} = \frac{2}{5}$, $\rho_{8,3} = \frac{2}{5}$, $\rho_{9,1} = \frac{2}{3}$, $\rho_{9,2} = \frac{1}{4}$, $\rho_{9,3} = \frac{1}{12}$. For each follower, the gain matrices $K_i^{(1)}$ and L_{oi} ($i = 4, 5, \ldots, 9$) are designed as follows: $K_4^{(1)} = I_2 \otimes [-3, -6]$, $L_{o4} = I_2 \otimes [-13, -73]^T$, $K_5^{(1)} = I_2 \otimes [-3, -7]$, $L_{o5} = I_2 \otimes [-14, -91]^T$, $K_6^{(1)} = I_2 \otimes [-3, -8]$, $L_{o6} = I_2 \otimes [-15, -111]^T$, $K_7^{(1)} = I_2 \otimes [-5, -6, -5]$, $L_{o7} = I_2 \otimes [-16, -90, -211]^T$, $K_8^{(1)} = I_2 \otimes [-5, -7, -6]$, $L_{o8} = I_2 \otimes [-17, -109, -356]^T$, $K_9^{(1)} = I_2 \otimes [-5, -8, -7]$, $L_{o9} = I_2 \otimes [-18, -130, -547]^T$. Select the following (X_{ij}, U_{ij}) such that the regulation equations (7.4) hold:

Followers 4,5,6: $X_{ij} = I_4$ ($j = 1, 2, 3$), $U_{i1} = I_2 \otimes [-1, -i]$, $U_{i2} = I_2 \otimes [-2, -i]$, $U_{i3} = I_2 \otimes [-5, -i]$, $i = 4, 5, 6$.

Followers 7,8,9: $X_{k1} = I_2 \otimes \begin{bmatrix} 1 & 0 \\ 0 & 1 \\ 0 & 0 \end{bmatrix}$, $X_{k2} = I_2 \otimes \begin{bmatrix} 1 & 0 \\ 0 & 1 \\ -1 & 0 \end{bmatrix}$, $X_{k3} = I_2 \otimes \begin{bmatrix} 1 & 0 \\ 0 & 1 \\ -4 & 0 \end{bmatrix}$, $k = 7, 8, 9$. $U_{71} = [-1, 0]$, $U_{72} = [0, -1]$, $U_{73} = [3, -4]$,

$U_{81} = [-1, -1]$, $U_{82} = [1, -2]$, $U_{83} = [7, -5]$, $U_{91} = [-1, -2]$, $U_{92} = [2, -3]$, $U_{93} = [11, -6]$.

The initial states of leaders are set as $z_1(0) = [1, 0, 1, 0]^T$, $z_2(0) = [4, 0, 0, 4]^T$, and $z_3(0) = [0, -6, 3, 0]^T$. The followers' initial states $x_i(0)$ and their estimations $\hat{x}_i(0)$ $(i = 4, 5, \ldots, 9)$ are generated by random numbers between -2 and 2. Let the initial values $\hat{\xi}_i(0) = 0$, $\hat{\alpha}_{ij}(0) = 2$, and $\hat{\beta}_{ik}(0) = 2$.

The output snapshots of the heterogeneous swarm system at $t = 0, 6, 35, 50$s are given in Fig. 7.6, where the followers are marked by squares, and the leaders are denoted by diamond, triangle, and circle. Fig. 7.7 shows the output containment errors, and Fig. 7.8 gives the estimation errors of $\bar{z}(t) = [z_1^T(t), z_2^T(t), z_3^T(t)]^T$ in the distributed observer. The adaptive gains $\hat{\alpha}_{ij}(t)$ and $\hat{\beta}_{ik}(t)$ are shown in Fig. 7.9. From Figs. 7.8 and 7.9, we can see that each follower can estimate the states of multiple leaders asymptotically, and all the adaptive gains converge to some constants. As shown in Figs. 7.6 and 7.7, the outputs of followers converge to the convex hull spanned by the three heterogeneous leaders, which implies that the predefined output containment is realized by the heterogeneous swarm system with switching topologies.

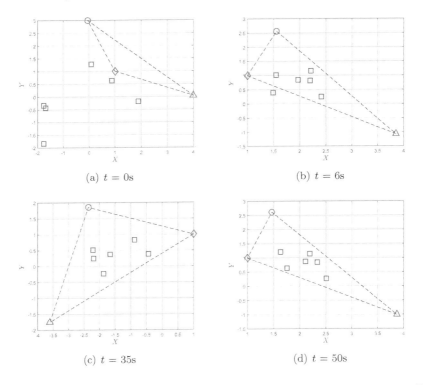

FIGURE 7.6: Output snapshots of the heterogeneous swarm system at different time instants.

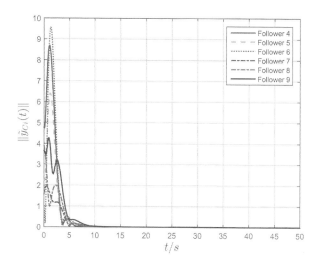

FIGURE 7.7: Output containment errors.

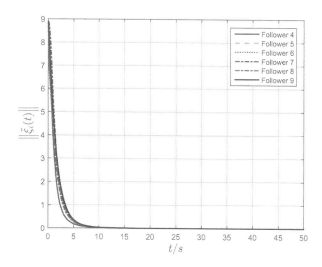

FIGURE 7.8: Estimation errors of $\bar{z}(t)$ for each follower.

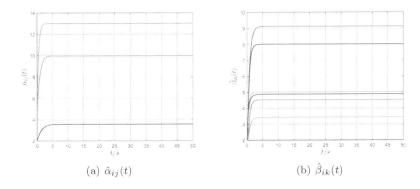

(a) $\hat{\alpha}_{ij}(t)$ (b) $\hat{\beta}_{ik}(t)$

FIGURE 7.9: Adaptive gains.

7.3   Formation-containment Tracking Control

In Section 7.2, only containment control problems are considered, where there exists no interaction and cooperation between multiple leaders. This section will further studies the formation-containment tracking control problems for heterogeneous swarm systems with leaders' unknown inputs on switching graphs. A tracking-leader is introduced to generate the macroscopic trajectory of the whole swarm system. Then, multiple formation-leaders are required to form the desired time-varying formation and track the reference trajectory simultaneously, while the followers need to move into the formation formed by the multiple formation-leaders. A distributed formation-containment tracking controller is proposed by using output regulation control and sliding mode control. Under the influences of hierarchical coordination couplings, an algorithm to design the proposed controller and sufficient conditions for heterogeneous swarm systems to achieve formation-containment tracking are presented.

7.3.1   Problem Description

Consider a heterogeneous swarm system with $N + M + 1$ agents, where the tracking-leader is denoted by $i = 0$, the formation-leaders are represented by $i = 1, 2, \ldots, N$, and the followers are labeled by $i = N + 1, N + 2, \ldots, N + M$. Consider the following definition for tracking-leader, formation-leader, and follower.

Definition 7.2. *According to the cooperative mission requirements, the agents in a swarm system are classified into three types, i.e., the tracking-leader, the formation-leader, and the follower. The tracking-leader is applied to generate the macroscopic trajectory of the whole swarm system. The multiple formation-leaders are required to form the desired time-varying formation and* track the reference trajectory generated by the tracking-leader simultaneously, while the followers need to move into the convex hull formed by the multiple formation-leaders. The tracking-leader has no neighbour, and the neighbours of a formation-leader only include leaders. A follower can have formation-leaders or other followers as its neighbours.

As shown in Fig. 7.10, a group of UAVs and UGVs move across the hazardous areas cooperatively in a formation-containment tracking form. In this example, the tracking-leader is the reference trajectory from the starting point to the destination. The formation-leaders are the UAVs with powerful detection and navigation sensors, and the followers are the UGVs with weaker perception. By formation-containment tracking control, the leader UAVs can achieve a desired formation tracking, and the follower UGVs will enter inside the formation formed by UAVs using neighbouring relative interaction. In this way, it can be ensured that the follower UGVs are in the safe area formed by the leader UAVs during the movement, and the cooperative crossing is realized at a relatively small cost.

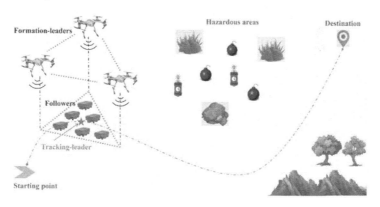

FIGURE 7.10: Example for UAVs and UGVs cooperative transportation.

The model of tracking leader is described by

$$\dot{v}_0(t) = Sv_0(t) + Er_0(t),$$
$$y_0(t) = Fv_0(t), \tag{7.11}$$

where $v_0(t) \in \mathbb{R}^q$, $r_0(t) \in \mathbb{R}^l$, and $y_0(t) \in \mathbb{R}^p$ denote the state, the control input, and the output of the tracking leader, respectively. It is required that $v_0(t)$ is bounded. The dynamics of formation-leader and follower are denoted by

$$\dot{x}_i(t) = A_i x_i(t) + B_i u_i(t),$$
$$y_i(t) = C_i x_i(t), \tag{7.12}$$

where $x_i(t) \in \mathbb{R}^{n_i}$, $u_i(t) \in \mathbb{R}^{m_i}$, and $y_i(t) \in \mathbb{R}^p$ are the state, the input, and the output of agent i $(i = 1, 2, \ldots, N + M)$.

Considering the case where the tracking-leader is a non-cooperative target, it is assumed that $r_0(t)$ is unknown to all the formation-leaders and followers for the controller design and satisfies the following bounded condition.

Assumption 7.3. *The input $r_0(t)$ of the tracking-leader is bounded, and there exists a positive constant η such that $\|r_0(t)\|_\infty \leqslant \eta$.*

For formation-leader i $(i = 1, 2, \ldots, N)$, the desired time-varying output formation is specified by a vector $h_y(t) = [h_{y1}^T(t), h_{y2}^T(t), \ldots, h_{yN}^T(t)]^T$.

Definition 7.3. *For any given bounded initial states, if*

$$\lim_{t \to \infty} (y_i(t) - h_{yi}(t) - y_0(t)) = 0, \; i = 1, 2, \ldots, N, \tag{7.13}$$

then the formation-leaders are said to achieve the desired time-varying output formation tracking.

Definition 7.4. *For each follower k $(k \in \{N + 1, N + 2, \ldots, N + M\})$, if there are non-negative constants $\rho_{k,j}$ $(j = 1, 2, \ldots, N)$ satisfying $\sum_{j=1}^{N} \rho_{k,j} = 1$ such that*

$$\lim_{t \to \infty} \left(y_k(t) - \sum_{j=1}^{N} \rho_{k,j} y_j(t) \right) = 0 \tag{7.14}$$

holds, then the swarm system (7.12) is said to realize containment.

Based on Definitions 7.3 and 7.4, the definition for formation-containment tracking is given in the following.

Definition 7.5. *For any formation-leader i $(i \in \{1, 2, \ldots, N\})$ and follower k $(k \in \{N + 1, N + 2, \ldots, N + M\})$, if (7.13) and (7.14) hold simultaneously, then the heterogeneous swarm system (7.11) and (7.12) achieves the desired formation-containment tracking.*

In [67–69], only formation-containment stabilization problems were considered , where the macroscopic trajectory of the whole swarm system cannot be controlled effectively. To overcome this defect, formation-containment tracking problems with a tracking-leader are further studied in this section. From Definitions 7.3-7.5, we can see that time-varying formation tracking, containment, and consensus problems can be unified to the proposed general framework for formation-containment tracking control.

7.3.2 Formation-containment Tracking Controller Design and Stability Analysis

Similar to Section 7.2, the topologies are assumed to be switching in this section, where the subscript set of all the possible graphs is denoted by $\mathcal{Z} =$

$\{1, 2, \ldots, z\}$. The topology and the Laplacian matrix at t are denoted by $\bar{\mathcal{G}}_{\sigma(t)}$ and $\bar{L}_{\sigma(t)}$, respectively. To guarantee that all the formation-leaders could play a role in containment control, it is required that the union of each follower's neighbouring set includes all the formation-leaders.

Assumption 7.4. *For each possible graph $\bar{\mathcal{G}}_{\sigma(t)}$, the topology $\mathcal{G}^L_{\sigma(t)}$ among leaders has a spanning tree rooted by the tracking-leader, and the topology between formation-leaders is undirected. Moreover, the undirected graph $\mathcal{G}^F_{\sigma(t)}$ associated with followers is connected.*

The Laplacian matrices for the graphs $\mathcal{G}^L_{\sigma(t)}$ and $\mathcal{G}^F_{\sigma(t)}$ are denoted by $L^{\sigma(t)}_L$ and $L^{\sigma(t)}_F$, respectively. Under Assumption 7.4, $L^{\sigma(t)}_L$ can be divided into $L^{\sigma(t)}_L = \begin{bmatrix} 0 & 0_{1 \times N} \\ L^{\sigma(t)}_{12} & L^{\sigma(t)}_{11} \end{bmatrix}$. It follows from Lemma 2.1 and Lemma 2.2 that $L^{\sigma(t)}_{11}$ is a positive definite matrix.

Similar to Subsection 6.3.4, for the formation-leader i $(i = 1, 2, \ldots, N)$, the desired time-varying output formation vector $h_{yi}(t)$ is generated by the following local exosystem:

$$
\begin{aligned}
\dot{h}_i(t) &= H_i h_i(t) + R_i r_i(t), \\
h_{yi}(t) &= Y_i h_i(t),
\end{aligned}
\tag{7.15}
$$

where $h_i(t) \in \mathbb{R}^{n_{hi}}$, $r_i(t) \in \mathbb{R}^{n_{ri}}$, and $h_{yi}(t) \in \mathbb{R}^p$. The purpose of the bounded external input $r_i(t)$ in (7.15) is to generate more general time-varying formation types. Besides, it is required that $h_i(t)$ is bounded.

Assumption 7.5. *The following regulator equations*

$$
\begin{aligned}
X_i S &= A_i X_i + B_i U_i \\
0 &= C_i X_i - F
\end{aligned}
$$

have solution pairs (X_i, U_i), $i = 1, 2, \ldots, N$.

Assumption 7.6. *The following local regulator equations*

$$
\begin{aligned}
X_{hi} H_i &= A_i X_{hi} + B_i U_{hi} \\
0 &= C_i X_{hi} - Y_i
\end{aligned}
$$

have solution pairs (X_{hi}, U_{hi}), $i = 1, 2, \ldots, N$.

Assumption 7.7. *The following regulator equations*

$$
\begin{aligned}
X_{i,j} A_j &= A_i X_{i,j} + B_i U_{i,j} \\
0 &= C_i X_{i,j} - C_j
\end{aligned}
$$

have solution pairs $(X_{i,j}, U_{i,j})$, $i = N+1, N+2, \ldots, N+M$, $j = 1, 2, \ldots, N$.

Consider the following formation-containment tracking controller:

$$\begin{cases} \dot{\hat{v}}_i = S\hat{v}_i - \sum_{j=0}^{N} \hat{\gamma}_{ij} w_{ij}^{\sigma} \left((\hat{v}_i - \hat{v}_j) + E \operatorname{sgn} \left(E^T (\hat{v}_i - \hat{v}_j) \right) \right), \\ u_i = K_{1i} x_i + K_{hi} h_i + K_{2i} \hat{v}_i - \mu \Upsilon_i \operatorname{sgn} \left(\Gamma_i^T \hat{\varsigma}_i \right) + \tau_i, \quad i = 1, 2, \ldots, N. \end{cases}$$

$$(7.16)$$

$$\begin{cases} \dot{\hat{\xi}}_i = \bar{A}\hat{\xi}_i - \hat{W}_i^{\sigma} \left(\hat{\xi}_i - \bar{x}_L + \bar{B} \operatorname{sgn}(\bar{B}^T(\hat{\xi}_i - \bar{x}_L)) \right) - \sum_{k=N+1}^{N+M} \hat{\beta}_{ik} w_{ik}^{\sigma}(\hat{\xi}_i - \hat{\xi}_k) \\ \qquad - \sum_{k=N+1}^{N+M} \hat{\beta}_{ik} w_{ik}^{\sigma} \bar{B} \operatorname{sgn} \left(\bar{B}^T(\hat{\xi}_i - \hat{\xi}_k) \right), \\ u_i = K_{3i} x_i + \sum_{j=1}^{N} \rho_{i,j} K_{i,j}^{(4)} \hat{\xi}_{i,j} \\ \qquad - \kappa \sum_{j=1}^{N} \rho_{i,j} \Upsilon_{i,j} \operatorname{sgn} \left(Q_{i,j}^T \hat{\delta}_i \right), \quad i = N+1, N+2, \ldots, N+M. \end{cases}$$

$$(7.17)$$

In the controller (7.16) for formation-leader i ($i = 1, 2, \ldots, N$), $\hat{v}_i$ denotes the distributed estimations for v_0 with $\hat{v}_0 = v_0$, and $\hat{\gamma}_{ij}$ are adaptive control gains to be constructed. $\hat{\varsigma}_i$ is defined as $\hat{\varsigma}_i = x_i - X_{hi} h_i - X_i \hat{v}_i$, τ_i represents the time-varying formation tracking compensation input, μ is a positive constant to be determined, and K_{1i}, K_{hi}, K_{2i}, Υ_i, and Γ_i are gain matrices to be designed.

In the controller (7.17) for follower i ($i = N + 1, N + 2, \ldots, N + M$), $\bar{A} = \operatorname{diag}\{A_1, A_2, \ldots, A_N\}$, $\bar{B} = \operatorname{diag}\{B_1, B_2, \ldots, B_N\}$, $\hat{W}_i^{\sigma} = \operatorname{diag}\{\hat{\alpha}_{i1} w_{i1}^{\sigma} I_{n_1}, \hat{\alpha}_{i2} w_{i2}^{\sigma} I_{n_2}, \ldots, \hat{\alpha}_{iN} w_{iN}^{\sigma} I_{n_N}\}$, $\bar{x}_L = [x_1^T, x_2^T, \ldots, x_N^T]^T$, $\hat{\xi}_i = [\hat{\xi}_{i,1}^T, \hat{\xi}_{i,2}^T, \ldots, \hat{\xi}_{i,N}^T]^T$, where $\hat{\xi}_{i,j}$ denotes the estimation of i-th follower for x_j ($j \in \{1, 2, \ldots, N\}$), and $\hat{\alpha}_{ij}$ ($j \in \{1, 2, \ldots, N\}$) and $\hat{\beta}_{ik}$ ($k \in \{N+1, N+2, \ldots, N+M\}$) are adaptive gains. $\hat{\delta}_i$ is defined as $\hat{\delta}_i = x_i - \sum_{j=1}^{N} \rho_{i,j} X_{i,j} \hat{\xi}_{i,j}$, where non-negative constants $\rho_{i,j}$ satisfying $\sum_{j=1}^{N} \rho_{i,j} = 1$ are several predefined weights to specify the desired convex combinations of multiple formation-leaders. κ is a positive constant to be designed, and K_{3i}, $K_{i,j}^{(4)}$, $\Upsilon_{i,j}$, and $Q_{i,j}$ are gain matrices to be determined.

An algorithm to design the controllers (7.16) and (7.17) is given in the following.

Algorithm 7.1. *For each formation-leader and follower, the formation-containment tracking controller (7.16) and (7.17) can be designed by the following steps.*

Step 1. *Choose matrices* (X_i, U_i) $(i = 1, 2, \ldots, N)$, (X_{hi}, U_{hi}) $(i = 1, 2, \ldots, N)$, *and* $(X_{i,j}, U_{i,j})$ $(i = N+1, N+2, \ldots, N+M, \ j = 1, 2, \ldots, N)$ *such that the regulator equations in Assumptions 7.5-7.7 hold.*

Step 2. *Design the controller (7.16) for the formation-leader* i $(i = 1, 2, \ldots, N)$. *For the desired time-varying output formation vector* $h_{yi}(t)$ *generated by the local exosystem (7.15), check whether there exists compensation input* $\tau_i(t)$ *such that the following formation-containment tracking feasibility condition holds:*

$$\lim_{t \to \infty} \left(B_i \tau_i(t) - X_{hi} R_i r_i(t) \right) = 0, \ i = 1, 2, \ldots, N. \tag{7.18}$$

If condition (7.18) holds for each formation-leader, then the algorithm continues. Otherwise, the desired output formation $h_y(t)$ *is not feasible under the proposed controller, and the algorithm stops. The updating law for the adaptive gain* $\hat{\gamma}_{ij}$ *is*

$$\dot{\hat{\gamma}}_{ij} = w_{ij}^\sigma \left(\|\hat{v}_i - \hat{v}_j\|^2 + \|E^T (\hat{v}_i - \hat{v}_j)\|_1 \right), \ j \in \{0, 1, \ldots, N\}, \tag{7.19}$$

where the initial values $\hat{\gamma}_{ij}(0) \geqslant 0$ *and* $\hat{\gamma}_{ik}(0) = \hat{\gamma}_{ki}(0)$, $k = 1, 2, \ldots, N$. *Choose sufficiently large* μ *such that* $\mu \geqslant \eta$. *Design the gain matrix* K_{1i} *to make* $A_i + B_i K_{1i}$ *Hurwitz. Let* $K_{hi} = U_{hi} - K_{1i} X_{hi}$ *and* $K_{2i} = U_i - K_{1i} X_i$. *Select* Υ_i *such that* $B_i \Upsilon_i - X_i E = 0$. *Let* $\Gamma_i = \Phi_i B_i \Upsilon_i$, *where* Φ_i *is a positive definite matrix satisfying the following Lyapunov equation:* $\Phi_i (A_i + B_i K_{1i}) + (A_i + B_i K_{1i})^T \Phi_i = -I_{n_i}$.

Step 3. *Design the controller (7.17) for the follower* i $(i = N+1, N+2, \ldots, N+M)$. *The adaptive gains* $\hat{\alpha}_{ij}$ $(j \in \{1, 2, \ldots, N\})$ *and* $\hat{\beta}_{ik}$ $(k \in \{N+1, N+2, \ldots, N+M\})$ *are updated by*

$$\dot{\hat{\alpha}}_{ij} = w_{ij}^\sigma \left(\|\hat{\xi}_{i,j} - x_j\|^2 + \|B_j^T (\hat{\xi}_{i,j} - x_j)\|_1 \right), \ j \in \{1, 2, \ldots, N\}, \tag{7.20}$$

$$\dot{\hat{\beta}}_{ik} = w_{ik}^\sigma \left(\|\hat{\xi}_i - \hat{\xi}_k\|^2 + \|\bar{B}^T(\hat{\xi}_i - \hat{\xi}_k)\|_1 \right), \ k \in \{N+1, N+2, \ldots, N+M\}, \tag{7.21}$$

where the initial values $\hat{\alpha}_{ij}(0) \geqslant 0$, $\hat{\beta}_{ik}(0) \geqslant 0$, *and* $\hat{\beta}_{ik}(0) = \hat{\beta}_{ki}(0)$. *Select sufficiently large positive constant* κ *such that* $\kappa \geqslant \max_{j=1,\ldots,N} \{\vartheta_j\}$, *where* ϑ_j $(j \in \{1, 2, \ldots, N\})$ *denotes the upper bound of control input for the* j-*th formation-leader, i.e.,* $\|u_j\|_\infty \leqslant \vartheta_j$. *Similarly, design gain matrix* K_{3i} *such that* $A_i + B_i K_{3i}$ *is Hurwitz, let* $K_{i,j}^{(4)} = U_{i,j} - K_{3i} X_{i,j}$ $(j \in \{1, 2, \ldots, N\})$, *and choose* $\Upsilon_{i,j}$ $(j \in \{1, 2, \ldots, N\})$ *to satisfy* $B_i \Upsilon_{i,j} - X_{i,j} B_j = 0$. *Besides, let* $Q_{i,j} = P_i B_i \Upsilon_{i,j}$, *where* P_i *is a positive definite matrix solved by the following equation:* $P_i (A_i + B_i K_{3i}) + (A_i + B_i K_{3i})^T P_i = -I_{n_i}$.

The following theorem gives sufficient conditions to achieve formation-containment tracking.

Theorem 7.2. *Suppose that Assumptions 7.3-7.7 hold. If the desired formation vector $h_y(t)$ generated by (7.15) satisfies the feasibility condition (7.18),
then heterogeneous swarm system (7.11) and (7.12) with switching topologies
can realize the desired formation-containment tracking under the controller
(7.16) and (7.17) designed by Algorithm 7.1.*

Proof. In the following, we will prove that the formation-leaders can achieve
the desired output formation tracking firstly. Then, the outputs of followers are
shown to converge into the convex hull formed by multiple formation-leaders.

For formation-leader i ($i \in \{1, 2, \ldots, N\}$), let $\tilde{v}_i = \hat{v}_i - v_0$ denote the
estimation error for the state of the tracking-leader. It follows from (7.11) and
(7.16) that

$$
\dot{\tilde{v}}_i = S\tilde{v}_i - \sum_{j=1}^{N} \hat{\gamma}_{ij} w_{ij}^{\sigma} \left((\tilde{v}_i - \tilde{v}_j) + E \operatorname{sgn} \left(E^T (\tilde{v}_i - \tilde{v}_j) \right) \right)
$$
$$
- \hat{\gamma}_{i0} w_{i0}^{\sigma} \left(\tilde{v}_i + E \operatorname{sgn} \left(E^T \tilde{v}_i \right) \right) - E r_0. \tag{7.22}
$$

Consider the following common Lyapunov function:

$$
V_1 = \sum_{i=1}^{N} \tilde{v}_i^T \tilde{v}_i + \frac{1}{2} \sum_{i=1}^{N} \sum_{j=1, j \neq i}^{N} (\hat{\gamma}_{ij} - \gamma)^2 + \sum_{i=1}^{N} (\hat{\gamma}_{i0} - \gamma)^2, \tag{7.23}
$$

where γ is a positive constant to be determined. Taking the derivative of V_1
along the trajectory (7.22) gives

$$
\dot{V}_1 = \sum_{i=1}^{N} \tilde{v}_i^T \left(S + S^T \right) \tilde{v}_i - 2 \sum_{i=1}^{N} \hat{\gamma}_{i0} w_{i0}^{\sigma} \tilde{v}_i^T \left(\tilde{v}_i + E \operatorname{sgn} \left(E^T \tilde{v}_i \right) \right)
$$
$$
- 2 \sum_{i=1}^{N} \tilde{v}_i^T \sum_{j=1}^{N} \hat{\gamma}_{ij} w_{ij}^{\sigma} \left((\tilde{v}_i - \tilde{v}_j) + E \operatorname{sgn} \left(E^T (\tilde{v}_i - \tilde{v}_j) \right) \right) - 2 \sum_{i=1}^{N} \tilde{v}_i^T E r_0
$$
$$
+ \sum_{i=1}^{N} \sum_{j=1}^{N} (\hat{\gamma}_{ij} - \gamma) w_{ij}^{\sigma} \left(\| \tilde{v}_i - \tilde{v}_j \|^2 + \left\| E^T (\tilde{v}_i - \tilde{v}_j) \right\|_1 \right)
$$
$$
+ 2 \sum_{i=1}^{N} (\hat{\gamma}_{i0} - \gamma) w_{i0}^{\sigma} \left(\| \tilde{v}_i \|^2 + \left\| E^T \tilde{v}_i \right\|_1 \right). \tag{7.24}
$$

Since $w_{ij}^{\sigma} = w_{ji}^{\sigma}$, $\hat{\gamma}_{ij} = \hat{\gamma}_{ji}$, and $\operatorname{sgn} \left(E^T (\tilde{v}_i - \tilde{v}_j) \right) = -\operatorname{sgn} \left(E^T (\tilde{v}_j - \tilde{v}_i) \right)$,
we get

$$
- 2 \sum_{i=1}^{N} \tilde{v}_i^T \sum_{j=1}^{N} \hat{\gamma}_{ij} w_{ij}^{\sigma} \left((\tilde{v}_i - \tilde{v}_j) + E \operatorname{sgn} \left(E^T (\tilde{v}_i - \tilde{v}_j) \right) \right)
$$
$$
= - \sum_{i=1}^{N} \sum_{j=1}^{N} \hat{\gamma}_{ij} w_{ij}^{\sigma} \left(\| \tilde{v}_i - \tilde{v}_j \|^2 + \left\| E^T (\tilde{v}_i - \tilde{v}_j) \right\|_1 \right). \tag{7.25}
$$

Substituting (7.25) into (7.24) gives

$$\dot{V}_1 = \sum_{i=1}^{N} \tilde{v}_i^T \left(S + S^T\right) \tilde{v}_i - 2\gamma \sum_{i=1}^{N} \tilde{v}_i^T \sum_{j=1}^{N} w_{ij}^{\sigma} \left(\tilde{v}_i - \tilde{v}_j\right) - 2\gamma \sum_{i=1}^{N} w_{i0}^{\sigma} \tilde{v}_i^T \tilde{v}_i + \Delta,$$

(7.26)

where $\Delta = -\gamma \sum_{i=1}^{N} \sum_{j=1}^{N} w_{ij}^{\sigma} \left\|E^T \left(\tilde{v}_i - \tilde{v}_j\right)\right\|_1 - 2\gamma \sum_{i=1}^{N} w_{i0}^{\sigma} \left\|E^T \tilde{v}_i\right\|_1 - 2\sum_{i=1}^{N} \tilde{v}_i^T E r_0$.

Under Assumption 7.3, it holds that

$$-2\sum_{i=1}^{N} \tilde{v}_i^T E r_0 \leqslant 2\eta \sum_{i=1}^{N} \left\|E^T \tilde{v}_i\right\|_1$$

$$\leqslant 2\eta N \max_{i=1,\ldots,N} \left\|E^T \tilde{v}_i\right\|_1.$$

(7.27)

At time instant t, assume that $\left\|E^T \tilde{v}_i\right\|_1$ reaches the maximum for the formation-leader $\bar{k}$ ($\bar{k} \in \{1, 2, \ldots, N\}$), i.e., $\left\|E^T \tilde{v}_{\bar{k}}\right\|_1 = \max_{i=1,\ldots,N} \left\|E^T \tilde{v}_i\right\|_1$. Consider the following two cases.

(i) If the formation-leader $\bar{k}$ can receive from the tracking-leader directly, i.e., $w_{\bar{k}0} > 0$, then we can choose sufficiently large γ such that $\gamma \geqslant \frac{\eta N}{w_{\bar{k}0}}$. It follows that

$$\Delta \leqslant -2\gamma w_{\bar{k}0}^{\sigma} \left\|E^T \tilde{v}_{\bar{k}}\right\|_1 + 2\eta N \left\|E^T \tilde{v}_{\bar{k}}\right\|_1$$

$$\leqslant 0.$$

(7.28)

(ii) If the formation-leader $\bar{k}$ cannot communicate with the tracking-leader directly, i.e., $w_{\bar{k}0} = 0$, then under Assumption 7.4, there exists at least one informed formation-leader having a path from itself to the formation-leader $\bar{k}$. This informed formation-leader is assumed to be $\bar{k}_m$, and the corresponding undirected path is denoted by $\left(\bar{k}, \bar{k}_1\right), \left(\bar{k}_1, \bar{k}_2\right), \ldots, \left(\bar{k}_{m-1}, \bar{k}_m\right)$ with m representing a positive integer. We can obtain that $w_{\bar{k}_1 \bar{k}} = w_{\bar{k} \bar{k}_1} > 0$, $\ldots$, $w_{\bar{k}_m \bar{k}_{m-1}} = w_{\bar{k}_{m-1} \bar{k}_m} > 0$, $w_{\bar{k}_m 0} > 0$. According to triangle inequality, it holds that

$$\left\|E^T \tilde{v}_{\bar{k}}\right\|_1 = \left\|E^T \left(\tilde{v}_{\bar{k}} - \tilde{v}_{\bar{k}_1} + \tilde{v}_{\bar{k}_1} - \tilde{v}_{\bar{k}_2} + \ldots + \tilde{v}_{\bar{k}_{m-1}} - \tilde{v}_{\bar{k}_m} + \tilde{v}_{\bar{k}_m}\right)\right\|_1$$

$$\leqslant \left\|E^T \left(\tilde{v}_{\bar{k}} - \tilde{v}_{\bar{k}_1}\right)\right\| + \left\|E^T \left(\tilde{v}_{\bar{k}_1} - \tilde{v}_{\bar{k}_2}\right)\right\| + \cdots$$

$$+ \left\|E^T \left(\tilde{v}_{\bar{k}_{m-1}} - \tilde{v}_{\bar{k}_m}\right)\right\| + \left\|E^T \tilde{v}_{\bar{k}_m}\right\|.$$

Let $\gamma \geqslant \max \left\{\frac{\eta N}{w_{\bar{k}\bar{k}_1}}, \frac{\eta N}{w_{\bar{k}_1 \bar{k}_2}}, \ldots, \frac{\eta N}{w_{\bar{k}_{m-1} \bar{k}_m}}, \frac{\eta N}{w_{\bar{k}_m 0}}\right\}$. Then, we have

$$\Delta \leqslant 2\eta N \left\|E^T \tilde{v}_{\bar{k}}\right\|_1 - 2\gamma w_{\bar{k}\bar{k}_1} \left\|E^T \left(\tilde{v}_{\bar{k}} - \tilde{v}_{\bar{k}_1}\right)\right\| - 2\gamma w_{\bar{k}_1 \bar{k}_2} \left\|E^T \left(\tilde{v}_{\bar{k}_1} - \tilde{v}_{\bar{k}_2}\right)\right\|$$

$$- \cdots - 2\gamma w_{\bar{k}_{m-1} \bar{k}_m} \left\|E^T \left(\tilde{v}_{\bar{k}_{m-1}} - \tilde{v}_{\bar{k}_m}\right)\right\| - 2\gamma w_{\bar{k}_m 0} \left\|E^T \tilde{v}_{\bar{k}_m}\right\|$$

$$\leqslant 0.$$

(7.29)

Based on the above two cases, for any $w_{ij}^\sigma > 0$ and $w_{i0}^\sigma > 0$ ($i = 1, 2, \ldots, N$, $j = 1, 2, \ldots, N$) in all possible graphs, choose sufficiently large γ such that $\gamma \geqslant \max\left\{\frac{\eta N}{w_{ij}^\sigma}, \frac{\eta N}{w_{i0}^\sigma}\right\}$. Then, we can obtain $\Delta \leqslant 0$ ($\forall \sigma(t) \in \mathcal{Z}$). It follows from (7.26) that

$$\dot{V}_1 \leqslant \sum_{i=1}^N \tilde{v}_i^T \left(S + S^T\right) \tilde{v}_i - 2\gamma \sum_{i=1}^N \tilde{v}_i^T \sum_{j=1}^N w_{ij}^\sigma \left(\tilde{v}_i - \tilde{v}_j\right) - 2\gamma \sum_{i=1}^N w_{i0}^\sigma \tilde{v}_i^T \tilde{v}_i. \tag{7.30}$$

Let $\tilde{v} = [\tilde{v}_1^T, \tilde{v}_2^T, \ldots, \tilde{v}_N^T]^T$, and we get

$$\dot{V}_1 \leqslant \tilde{v}^T \left(I_N \otimes \left(S + S^T\right)\right) \tilde{v} - 2\gamma \tilde{v}^T \left(L_{11}^\sigma \otimes I_q\right) \tilde{v}. \tag{7.31}$$

Since L_{11}^σ is a positive definite matrix under Assumption 7.4, there exists an orthogonal matrix $U_{\sigma(t)} \in \mathbb{R}^{N \times N}$ such that $U_\sigma^T L_{11}^\sigma U_\sigma = \Xi_\sigma = \text{diag}\{\lambda_1^\sigma, \lambda_2^\sigma, \ldots, \lambda_N^\sigma\}$. Let $\zeta = [\zeta_1^T, \zeta_2^T, \ldots, \zeta_N^T]^T = \left(U_\sigma^T \otimes I_n\right) \tilde{v}$. The equation (7.31) can be transformed to

$$\dot{V}_1 \leqslant \zeta^T \left(I_N \otimes \left(S + S^T\right)\right) \zeta - 2\gamma \zeta^T \left(\Xi_\sigma \otimes I_q\right) \zeta$$

$$= \sum_{i=1}^N \zeta_i^T \left(S + S^T - 2\gamma \lambda_i^\sigma I_q\right) \zeta_i. \tag{7.32}$$

Choose $\gamma > \max\limits_{i=1,\ldots,N} \frac{\lambda_{\max}\left(S + S^T\right)}{2\lambda_i^\sigma}$ ($\forall \sigma(t) \in \mathcal{Z}$). It can verified that $\dot{V}_1 \leqslant 0$. Then, following the similar analysis in the proof of Theorem 5.2, we can obtain that $\hat{\gamma}_{ij}$ converge to some positive constants and $\lim\limits_{t \to \infty} \tilde{v}_i(t) = 0$, which implies that $\lim\limits_{t \to \infty} \left(\hat{v}_i(t) - v_0(t)\right) = 0$, $i = 1, 2, \ldots, N$.

For the formation-leader i ($i = 1, 2, \ldots, N$), substitute the controller (7.16) into the system (7.12). Let $\varsigma_i = x_i - X_{hi}h_i - X_i v_0$, then it follows that

$$\dot{\varsigma}_i = (A_i + B_i K_{1i}) x_i + B_i K_{hi} h_i - X_{hi} H_i h_i + B_i K_{2i} v_0 - X_i S v_0$$
$$- \mu B_i \Upsilon_i \, \text{sgn} \left(\Gamma_i \hat{\varsigma}_i\right) - X_i E r_0 + B_i \tau_i - X_{hi} R_i r_i + B_i K_{2i} \tilde{v}_i. \tag{7.33}$$

Taking $K_{hi} = U_{hi} - K_{1i} X_{hi}$, $K_{2i} = U_i - K_{1i} X_i$, and $B_i \Upsilon_i - X_i E = 0$ into (7.33) gives

$$\dot{\varsigma}_i = (A_i + B_i K_{1i}) \varsigma_i - \mu B_i \Upsilon_i \, \text{sgn} \left(\Gamma_i \hat{\varsigma}_i\right) - B_i \Upsilon_i r_0 + B_i \tau_i - X_{hi} R_i r_i + B_i K_{2i} \tilde{v}_i. \tag{7.34}$$

Since the desired formation satisfies the feasibility condition (7.18), we have $B_i \tau_i - X_{hi} R_i r_i \to 0$. Moreover, based on $\lim\limits_{t \to \infty} \tilde{v}_i(t) = 0$, it holds that $\vartheta_i = B_i \tau_i - X_{hi} R_i r_i + B_i K_{2i} \tilde{v}_i \to 0$. Consider the Lyapunov function $V_{2i} = \varsigma_i^T \Phi_i \varsigma_i$. Taking the derivative of V_{2i} along trajectory (7.34) gives

$$\dot{V}_{2i} = \varsigma_i^T \left(\Phi_i (A_i + B_i K_{1i}) + (A_i + B_i K_{1i})^T \Phi_i\right) \varsigma_i$$
$$- 2\mu \varsigma_i^T \Gamma_i \, \text{sgn} \left(\Gamma_i^T \hat{\varsigma}_i\right) - 2\varsigma_i^T \Gamma_i r_0 + 2\varsigma_i^T \Phi_i \vartheta_i. \tag{7.35}$$

Let $\tilde{\varsigma}_i = \hat{\varsigma}_i - \varsigma_i$, then we have $\dot{\tilde{\varsigma}}_i = -X_i \tilde{v}_i \to 0$. Note that $\mu \geqslant \eta$. It follows that

$$
\begin{aligned}
& - 2\mu \varsigma_i^T \Gamma_i \operatorname{sgn}\left(\Gamma_i^T \hat{\varsigma}_i\right) - 2\varsigma_i^T \Gamma_i r_0 \\
&= -2\mu \hat{\varsigma}_i^T \Gamma_i \operatorname{sgn}\left(\Gamma_i^T \hat{\varsigma}_i\right) + 2\mu \tilde{\varsigma}_i^T \Gamma_i \operatorname{sgn}\left(\Gamma_i^T \hat{\varsigma}_i\right) - 2\hat{\varsigma}_i^T \Gamma_i r_0 + 2\tilde{\varsigma}_i^T \Gamma_i r_0 \\
&\leqslant -2\left(\mu - \eta\right)\left\|\Gamma_i^T \hat{\varsigma}_i\right\|_1 + 2\left(\mu + \eta\right)\left\|\Gamma_i^T \tilde{\varsigma}_i\right\|_1 \\
&\leqslant 2\left(\mu + \eta\right)\left\|\Gamma_i^T \tilde{\varsigma}_i\right\|_1 .
\end{aligned}
\tag{7.36}
$$

According to Lemma 2.3, we can obtain

$$
2\varsigma_i^T \Phi_i \tilde{\vartheta}_i \leqslant \frac{1}{2}\varsigma_i^T \varsigma_i + 2\left\|\Phi_i \tilde{\vartheta}_i\right\|^2 .
\tag{7.37}
$$

Substituting these inequalities into (7.35) gives

$$
\dot{V}_{2i} \leqslant -\frac{1}{2\lambda_{\max}\left(\Phi_i\right)} V_{2i} + 2\left(\mu + \eta\right)\left\|\Gamma_i^T \tilde{\varsigma}_i\right\|_1 + 2\left\|\Phi_i \tilde{\vartheta}_i\right\|^2 .
\tag{7.38}
$$

Since $\tilde{\varsigma}_i$ and $\tilde{\vartheta}_i$ are bounded, $\lim_{t\to\infty} \tilde{\varsigma}_i(t) = 0$, and $\lim_{t\to\infty} \tilde{\vartheta}_i(t) = 0$, according to Lemma 2.19 in [150], it follows that $\lim_{t\to\infty} V_{2i}(t) = 0$, which means that $\lim_{t\to\infty} \varsigma_i(t) = 0$. Let $\tilde{y}_i = y_i - h_{yi} - y_0$ represent time-varying output formation tracking error. Since $0 = C_i X_i - F$ and $0 = C_i X_{hi} - Y_i$, it follows that $\tilde{y}_i = C_i \varsigma_i$. Thus, it can be verified that $\lim_{t\to\infty} \tilde{y}_i(t) = 0$, i.e., the formation-leaders achieve the desired time-varying output formation tracking.

In the following, we will prove that the outputs of followers can enter into the convex hull formed by multiple formation-leaders under the proposed controller (7.17). Let $\tilde{\xi}_i = \hat{\xi}_i - \bar{x}_L = [\tilde{\xi}_{i,1}^T, \tilde{\xi}_{i,2}^T, \dots, \tilde{\xi}_{i,N}^T]^T$ ($i \in \{N+1, N+2, \dots, N+M\}$), where $\tilde{\xi}_{i,j} = \hat{\xi}_{i,j} - x_j$ ($j = 1, 2, \dots, N$). It follows from (7.12) and (7.17) that

$$
\begin{aligned}
\dot{\tilde{\xi}}_i = \bar{A}\tilde{\xi}_i - \hat{W}_i^\sigma \left(\tilde{\xi}_i + \bar{B}\operatorname{sgn}\left(\bar{B}^T \tilde{\xi}_i\right)\right) - \sum_{k=N+1}^{N+M} \hat{\beta}_{ik} w_{ik}^\sigma \left(\tilde{\xi}_i - \tilde{\xi}_k\right) \\
- \sum_{k=N+1}^{N+M} \hat{\beta}_{ik} w_{ik}^\sigma \bar{B} \operatorname{sgn}\left(\bar{B}^T \left(\tilde{\xi}_i - \tilde{\xi}_k\right)\right) - \bar{B}\bar{u}_L ,
\end{aligned}
\tag{7.39}
$$

where $\bar{u}_L = [u_1^T, u_2^T, \dots, u_N^T]^T$.

Consider the following common Lyapunov function:

$$
V_3 = \sum_{i=N+1}^{N+M} \tilde{\xi}_i^T \tilde{\xi}_i + \sum_{i=N+1}^{N+M} \sum_{j=1}^{N} \left(\hat{\alpha}_{ij} - \alpha\right)^2 + \frac{1}{2} \sum_{i=N+1}^{N+M} \sum_{k=N+1, k\neq i}^{N+M} \left(\hat{\beta}_{ik} - \beta\right)^2 ,
\tag{7.40}
$$

where α and β are two positive constants to be determined. Taking the derivative of V_3 along (7.39) gives

$$
\dot{V}_3 = \sum_{i=N+1}^{N+M} \tilde{\xi}_i^T (\bar{A} + \bar{A}^T) \tilde{\xi}_i - 2 \sum_{i=N+1}^{N+M} \tilde{\xi}_i^T \hat{W}_i^\sigma \tilde{\xi}_i
$$
$$
- 2 \sum_{i=N+1}^{N+M} \tilde{\xi}_i^T \sum_{k=N+1}^{N+M} \hat{\beta}_{ik} w_{ik}^\sigma (\tilde{\xi}_i - \tilde{\xi}_k) + 2 \sum_{i=N+1}^{N+M} \sum_{j=1}^{N} (\hat{\alpha}_{ij} - \alpha) \, w_{ij}^\sigma \tilde{\xi}_{i,j}^T \tilde{\xi}_{i,j}
$$
$$
+ \sum_{i=N+1}^{N+M} \sum_{k=N+1}^{N+M} (\hat{\beta}_{ik} - \beta) w_{ik}^\sigma \left\| \tilde{\xi}_i - \tilde{\xi}_k \right\|^2 + \Pi, \tag{7.41}
$$

where

$$
\Pi = -2 \sum_{i=N+1}^{N+M} \tilde{\xi}_i^T \hat{W}_i^\sigma \bar{B} \, \mathrm{sgn}(\bar{B}^T \tilde{\xi}_i) - 2 \sum_{i=N+1}^{N+M} \tilde{\xi}_i^T \sum_{k=N+1}^{N+M} \hat{\beta}_{ik} w_{ik}^\sigma \bar{B} \, \mathrm{sgn}(\bar{B}^T (\tilde{\xi}_i - \tilde{\xi}_k))
$$
$$
- 2 \sum_{i=N+1}^{N+M} \tilde{\xi}_i^T \bar{B} \bar{u}_L + 2 \sum_{i=N+1}^{N+M} \sum_{j=1}^{N} (\hat{\alpha}_{ij} - \alpha) \, w_{ij}^\sigma \left\| B_j^T \tilde{\xi}_{i,j} \right\|_1
$$
$$
+ \sum_{i=N+1}^{N+M} \sum_{k=N+1}^{N+M} \left(\hat{\beta}_{ik} - \beta \right) w_{ik}^\sigma \left\| \bar{B}^T \left(\tilde{\xi}_i - \tilde{\xi}_k \right) \right\|_1. \tag{7.42}
$$

Since $\hat{W}_i^\sigma = \mathrm{diag}\, \{\hat{\alpha}_{i1} w_{i1}^\sigma I_{n_1}, \hat{\alpha}_{i2} w_{i2}^\sigma I_{n_2}, \ldots, \hat{\alpha}_{iN} w_{iN}^\sigma I_{n_N} \}$, we can obtain

$$
-2 \sum_{i=N+1}^{N+M} \tilde{\xi}_i^T \hat{W}_i^\sigma \bar{B} \, \mathrm{sgn} \left(\bar{B}^T \tilde{\xi}_i \right) + 2 \sum_{i=N+1}^{N+M} \sum_{j=1}^{N} (\hat{\alpha}_{ij} - \alpha) \, w_{ij}^\sigma \left\| B_j^T \tilde{\xi}_{i,j} \right\|_1
$$
$$
= -2\alpha \sum_{i=N+1}^{N+M} \sum_{j=1}^{N} w_{ij}^\sigma \left\| B_j^T \tilde{\xi}_{i,j} \right\|_1.
$$

Note that $\mathrm{sgn} \left(\bar{B}^T \left(\tilde{\xi}_k - \tilde{\xi}_i \right) \right) = -\mathrm{sgn} \left(\bar{B}^T \left(\tilde{\xi}_i - \tilde{\xi}_k \right) \right)$, $w_{ik}^\sigma = w_{ki}^\sigma$, and $\hat{\beta}_{ik} = \hat{\beta}_{ki}$. It follows that

$$
-2 \sum_{i=N+1}^{N+M} \tilde{\xi}_i^T \sum_{k=N+1}^{N+M} \hat{\beta}_{ik} w_{ik}^\sigma \bar{B} \, \mathrm{sgn} \left(\bar{B}^T \left(\tilde{\xi}_i - \tilde{\xi}_k \right) \right)
$$
$$
+ \sum_{i=N+1}^{N+M} \sum_{k=N+1}^{N+M} \left(\hat{\beta}_{ik} - \beta \right) w_{ik}^\sigma \left\| \bar{B}^T \left(\tilde{\xi}_i - \tilde{\xi}_k \right) \right\|_1
$$
$$
= -\beta \sum_{i=N+1}^{N+M} \sum_{k=N+1}^{N+M} w_{ik}^\sigma \sum_{j=1}^{M} \left\| B_j^T \left(\tilde{\xi}_{i,j} - \tilde{\xi}_{k,j} \right) \right\|_1.
$$

Let $\mathcal{O}_L = \{1, 2, \ldots, N\}$ and $\mathcal{O}_F = \{N+1, N+2, \ldots, N+M\}$. Because the control inputs $u_j(t)$ of the formation-leaders are bounded, it can be verified that

$$-2 \sum_{i=N+1}^{N+M} \tilde{\xi}_i^T \bar{B} \bar{u}_L \leqslant 2\vartheta \sum_{i=N+1}^{N+M} \sum_{j=1}^{N} \left\| B_j^T \tilde{\xi}_{i,j} \right\|_1$$

$$\leqslant 2\vartheta MN \max_{i \in \mathcal{O}_F, j \in \mathcal{O}_L} \left\| B_j^T \tilde{\xi}_{i,j} \right\|_1,$$

where $\vartheta = \max_{j=1,\ldots,N} \{\vartheta_j\}$. At the time instant t, assume that $\left\| B_j^T \tilde{\xi}_{i,j} \right\|_1$ gets its maximum for follower $\bar{i}$ ($\bar{i} \in \mathcal{O}_F$) and formation-leader $\bar{j}$ ($\bar{j} \in \mathcal{O}_L$), i.e., $\left\| B_{\bar{j}}^T \tilde{\xi}_{\bar{i},\bar{j}} \right\|_1 = \max_{i \in \mathcal{O}_F, j \in \mathcal{O}_L} \left\| B_j^T \tilde{\xi}_{i,j} \right\|_1$. It holds from (7.42) that

$$\Pi \leqslant -\beta \sum_{i=N+1}^{N+M} \sum_{k=N+1}^{N+M} w_{ik}^\sigma \sum_{j=1}^{N} \left\| B_j^T (\tilde{\xi}_{i,j} - \tilde{\xi}_{k,j}) \right\|_1 - 2\alpha \sum_{i=N+1}^{N+M} \sum_{j=1}^{N} w_{ij}^\sigma \left\| B_j^T \tilde{\xi}_{i,j} \right\|_1$$

$$+ 2\vartheta MN \left\| B_{\bar{j}}^T \tilde{\xi}_{\bar{i},\bar{j}} \right\|_1. \tag{7.43}$$

Similar to the analysis in (7.28) and (7.29), considering the two cases $\bar{j} \in \mathcal{N}_{\bar{i}}$ and $\bar{j} \notin \mathcal{N}_{\bar{i}}$ respectively, choose sufficiently large α and β such that $\alpha \geqslant \max_{i \in \mathcal{O}_F, j \in \mathcal{O}_L} \left\{ \frac{\vartheta MN}{w_{ij}^\sigma} \right\}$ and $\beta \geqslant \max_{i,k \in \mathcal{O}_F} \left\{ \frac{\vartheta MN}{w_{ik}^\sigma} \right\}$ hold for all $w_{ij}^\sigma > 0$ and $w_{ik}^\sigma > 0$ in each possible graph. Then, it follows that $\Pi \leqslant 0$ ($\forall \sigma(t) \in \mathcal{Z}$).

Furthermore, it can be verified that

$$-2 \sum_{i=N+1}^{N+M} \tilde{\xi}_i^T \hat{W}_i^\sigma \tilde{\xi}_i + 2 \sum_{i=N+1}^{N+M} \sum_{j=1}^{N} (\hat{\alpha}_{ij} - \alpha) w_{ij}^\sigma \tilde{\xi}_{i,j}^T \tilde{\xi}_{i,j}$$

$$= -2\alpha \sum_{i=N+1}^{N+M} \sum_{j=1}^{N} w_{ij}^\sigma \tilde{\xi}_{i,j}^T \tilde{\xi}_{i,j},$$

$$-2 \sum_{i=N+1}^{N+M} \tilde{\xi}_i^T \sum_{k=N+1}^{N+M} \hat{\beta}_{ik} w_{ik}^\sigma \left(\tilde{\xi}_i - \tilde{\xi}_k \right) + \sum_{i=N+1}^{N+M} \sum_{k=N+1}^{N+M} \left(\hat{\beta}_{ik} - \beta \right) w_{ik}^\sigma \left\| \tilde{\xi}_i - \tilde{\xi}_k \right\|^2$$

$$= -2\beta \sum_{i=N+1}^{N+M} \tilde{\xi}_i^T \sum_{k=N+1}^{N+M} w_{ik}^\sigma \left(\tilde{\xi}_i - \tilde{\xi}_k \right).$$

Substituting $\Pi \leqslant 0$ into (7.41) gives

$$\dot{V}_3 \leqslant \sum_{i=N+1}^{N+M} \tilde{\xi}_i^T \left(\bar{A} + \bar{A}^T \right) \tilde{\xi}_i - 2\alpha \sum_{i=N+1}^{N+M} \sum_{j=1}^{N} w_{ij}^\sigma \tilde{\xi}_{i,j}^T \tilde{\xi}_{i,j}$$

$$- 2\beta \sum_{i=N+1}^{N+M} \tilde{\xi}_i^T \sum_{k=N+1}^{N+M} w_{ik}^\sigma \left(\tilde{\xi}_i - \tilde{\xi}_k \right). \tag{7.44}$$

Let $\tilde{\xi} = [\tilde{\xi}_{N+1}^T, \tilde{\xi}_{N+2}^T, \ldots, \tilde{\xi}_{N+M}^T]^T$ and $\bar{W}_\sigma = \mathrm{diag}\left\{\bar{W}_{N+1}^\sigma, \bar{W}_{N+2}^\sigma, \ldots, \bar{W}_{N+M}^\sigma\right\}$, where $\bar{W}_i^\sigma = \mathrm{diag}\{w_{i1}^\sigma I_{n_1}, w_{i2}^\sigma I_{n_2}, \ldots, w_{iN}^\sigma I_{n_N}\}$ $(i = N+1, N+2, \ldots, N+M)$. We can transform (7.44) to

$$\dot{V}_3 \leqslant \tilde{\xi}^T \left(I_M \otimes (\bar{A} + \bar{A}^T)\right) \tilde{\xi} - 2\alpha \tilde{\xi}^T \bar{W}_\sigma \tilde{\xi} - 2\beta \tilde{\xi}^T (L_F^\sigma \otimes I_{\bar{n}})\tilde{\xi}, \qquad (7.45)$$

where $\bar{n} = n_1 + n_2 + \cdots + n_N$. Under Assumption 7.4, L_F^σ is a positive semidefinite matrix, and there exists an orthogonal matrix $U_\sigma = [\mathbf{1}_M/\sqrt{M}, H_1^\sigma]$ such that $U_\sigma^T L_F^\sigma U_\sigma = \Lambda_\sigma = \mathrm{diag}\{0, \lambda_2^\sigma, \lambda_3^\sigma, \ldots, \lambda_M^\sigma\}$. Let $\varsigma = \left(U_\sigma^T \otimes I_{\bar{n}}\right) \tilde{\xi}$. Then, it holds from (7.45) that

$$\dot{V}_3 \leqslant \varsigma^T \left(I_M \otimes (\bar{A} + \bar{A}^T)\right) \varsigma - 2\alpha \varsigma^T \left(U_\sigma^T \otimes I_{\bar{n}}\right) \bar{W}_\sigma \left(U_\sigma \otimes I_{\bar{n}}\right) \varsigma$$
$$- 2\beta \varsigma^T \left(\Lambda_\sigma \otimes I_{\bar{n}}\right) \varsigma. \qquad (7.46)$$

Let $\bar{\Omega}_\sigma = I_M \otimes (\bar{A} + \bar{A}^T) - 2\alpha \left(U_\sigma^T \otimes I_{\bar{n}}\right) \bar{W}_\sigma \left(U_\sigma \otimes I_{\bar{n}}\right) - 2\beta \left(\Lambda_\sigma \otimes I_{\bar{n}}\right)$. We will prove that there are sufficiently large α and β such that $\bar{\Omega}_\sigma < 0$ $(\forall \sigma(t) \in \mathcal{Z})$ in the following. Let

$$\Xi_\sigma = \left(U_\sigma^T \otimes I_{\bar{n}}\right) \bar{W}_\sigma \left(U_\sigma \otimes I_{\bar{n}}\right)$$
$$= \begin{bmatrix} \Xi_{11}^\sigma & \Xi_{12}^\sigma \\ (\Xi_{12}^\sigma)^T & \Xi_{22}^\sigma \end{bmatrix},$$

where $\Xi_{11}^\sigma = \frac{1}{M} \sum_{i=N+1}^{N+M} \bar{W}_i^\sigma$, $\Xi_{12}^\sigma = (\mathbf{1}_N^T/\sqrt{M} \otimes I_{\bar{n}})\bar{W}_\sigma (H_1^\sigma \otimes I_{\bar{n}})$, and $\Xi_{22}^\sigma = \left((H_1^\sigma)^T \otimes I_{\bar{n}}\right) \bar{W}_\sigma (H_1^\sigma \otimes I_{\bar{n}})$. Since the union of the followers' neighbouring sets includes all the formation-leaders, we have $\sum_{i=N+1}^{N+M} w_{ij}^\sigma > 0$, $j = 1, 2, \ldots, N$. Because $\bar{W}_i^\sigma = \mathrm{diag}\{w_{i1}^\sigma I_{n_1}, w_{i2}^\sigma I_{n_2}, \ldots, w_{iN}^\sigma I_{n_N}\}$, it holds that $\Xi_{11}^\sigma = \frac{1}{M} \sum_{i=N+1}^{N+M} \bar{W}_i^\sigma > 0$. Let $\Lambda_{22}^\sigma = \mathrm{diag}\{\lambda_2^\sigma, \lambda_3^\sigma, \ldots, \lambda_M^\sigma\}$. Then, we can obtain

$$\bar{\Omega}_\sigma = \begin{bmatrix} 1 & \\ & I_{M-1} \end{bmatrix} \otimes (\bar{A} + \bar{A}^T) - 2\alpha \begin{bmatrix} \Xi_{11}^\sigma & \Xi_{12}^\sigma \\ (\Xi_{12}^\sigma)^T & \Xi_{22}^\sigma \end{bmatrix} - 2\beta \left(\begin{bmatrix} 0 & \\ & \Lambda_{22}^\sigma \end{bmatrix} \otimes I_{\bar{n}}\right)$$
$$= \begin{bmatrix} \bar{\Omega}_{11}^\sigma & \bar{\Omega}_{12}^\sigma \\ (\bar{\Omega}_{12}^\sigma)^T & \bar{\Omega}_{22}^\sigma \end{bmatrix}, \qquad (7.47)$$

where $\bar{\Omega}_{11}^\sigma = \bar{A} + \bar{A}^T - 2\alpha \Xi_{11}^\sigma$, $\bar{\Omega}_{12}^\sigma = -2\alpha \Xi_{12}^\sigma$, and $\bar{\Omega}_{22}^\sigma = I_{M-1} \otimes (\bar{A} + \bar{A}^T) - 2\alpha \Xi_{22}^\sigma - 2\beta (\Lambda_{22}^\sigma \otimes I_{\bar{n}})$.

Choose sufficiently large α and β such that $\alpha > \dfrac{M \lambda_{\max}(\bar{A} + \bar{A}^T)}{2 \min\limits_{j=1,\ldots,N} \sum_{i=N+1}^{N+M} w_{ij}^\sigma}$ and $\beta > \dfrac{\lambda_{\max}(\Theta_\sigma)}{2 \min\limits_{k=2,\ldots,M}\{\lambda_k^\sigma\}}$ $(\forall \sigma(t) \in \mathcal{Z})$, where $\Theta_\sigma = I_{M-1} \otimes (\bar{A} + \bar{A}^T) - 2\alpha \Xi_{22}^\sigma - (\bar{\Omega}_{12}^\sigma)^T (\bar{\Omega}_{11}^\sigma)^{-1} \bar{\Omega}_{12}^\sigma$. Then, we get $\bar{\Omega}_{11}^\sigma = \bar{A} + \bar{A}^T - 2\alpha \Xi_{11}^\sigma < 0$ and $\bar{\Omega}_{22}^\sigma - (\bar{\Omega}_{12}^\sigma)^T (\bar{\Omega}_{11}^\sigma)^{-1} \bar{\Omega}_{12}^\sigma < 0$. According to Lemma 2.6, it holds that $\bar{\Omega}_\sigma < 0$

($\forall \sigma(t) \in \mathcal{Z}$). We can obtain from (7.46) that $\dot{V}_3 \leqslant 0$. Thus, it can be verified that adaptive gains $\hat{a}_{ij}$ and $\hat{\beta}_{ik}$ converge to some positive constants and $\lim_{t \to \infty} \tilde{\xi}(t) = 0$, i.e., $\lim_{t \to \infty} \left(\hat{\xi}_i(t) - \bar{x}_L(t) \right) = 0$ ($i \in \{N+1, N+2, \ldots, N+M\}$).

Let $\delta_i = x_i - \sum_{j=1}^{N} \rho_{i,j} X_{i,j} x_j$ ($i \in \{N+1, N+2, \ldots, N+M\}$). It follows that

$$\dot{\delta}_i = A_i x_i + B_i K_{3i} x_i + \sum_{j=1}^{N} \rho_{i,j} B_i K_{i,j}^{(4)} \hat{\xi}_{i,j} - \sum_{j=1}^{N} \rho_{i,j} X_{i,j} A_j x_j$$

$$- \kappa \sum_{j=1}^{N} \rho_{i,j} B_i \Upsilon_{i,j} \operatorname{sgn} \left(Q_{i,j}^T \hat{\delta}_i \right) - \sum_{j=1}^{N} \rho_{i,j} X_{i,j} B_j u_j. \tag{7.48}$$

Substituting $\hat{\xi}_{i,j} = x_j + \tilde{\xi}_{i,j}$, $K_{i,j}^{(4)} = U_{i,j} - K_{3i} X_{i,j}$, $X_{i,j} A_j = A_i X_{i,j} + B_i U_{i,j}$, and $B_i \Upsilon_{i,j} - X_{i,j} B_j = 0$ into (7.48) gives

$$\dot{\delta}_i = (A_i + B_i K_{3i}) \delta_i + \sum_{j=1}^{N} \rho_{i,j} B_i K_{i,j}^{(4)} \tilde{\xi}_{i,j} - \kappa \sum_{j=1}^{N} \rho_{i,j} B_i \Upsilon_{i,j} \operatorname{sgn} \left(Q_{i,j}^T \hat{\delta}_i \right)$$

$$- \sum_{j=1}^{N} \rho_{i,j} B_i \Upsilon_{i,j} u_j. \tag{7.49}$$

Let $\tilde{\psi}_i = \sum_{j=1}^{N} \rho_{i,j} B_i K_{i,j}^{(4)} \tilde{\xi}_{i,j}$. Since $\lim_{t \to \infty} \tilde{\xi}_{i,j}(t) = 0$ ($j \in \{1, 2, \ldots, N\}$), we can obtain that $\lim_{t \to \infty} \tilde{\psi}_i(t) = 0$.

Consider the Lyapunov function $V_{4i} = \delta_i^T P_i \delta_i$. Taking the derivative of V_{4i} along (7.49) gives

$$\dot{V}_{4i} = -\delta_i^T \delta_i + 2\delta_i^T P_i \tilde{\psi}_i - 2\kappa \sum_{j=1}^{N} \rho_{i,j} \delta_i^T Q_{i,j} \operatorname{sgn} \left(Q_{i,j}^T \hat{\delta}_i \right) - 2 \sum_{j=1}^{N} \rho_{i,j} \delta_i^T Q_{i,j} u_j. \tag{7.50}$$

Let $\tilde{\delta}_i = \hat{\delta}_i - \delta_i = -\sum_{j=1}^{N} \rho_{i,j} X_{i,j} \tilde{\xi}_{i,j}$. It holds that $\lim_{t \to \infty} \tilde{\delta}_i(t) = 0$. Since $\kappa \geqslant \max_{j=1,\ldots,N} \{\vartheta_j\}$, following the similar steps in (7.35)-(7.38) leads to

$$\dot{V}_{4i} \leqslant -\frac{1}{2\lambda_{\max}(P_i)} V_{4i} + 2\left\| P_i \tilde{\psi}_i \right\|^2 + 2(\kappa + \vartheta) \sum_{j=1}^{N} \rho_{i,j} \left\| Q_{i,j}^T \tilde{\delta}_i \right\|_1. \tag{7.51}$$

Since $\lim_{t \to \infty} \tilde{\psi}_i(t) = 0$ and $\lim_{t \to \infty} \tilde{\delta}_i(t) = 0$, it follows from (7.51) that $\lim_{t \to \infty} V_{4i}(t) = 0$, which means that $\lim_{t \to \infty} \delta_i(t) = 0$. Define the output containment error as $\tilde{y}_{Ci} = y_i - \sum_{j=1}^{N} \rho_{i,j} y_j$ ($i \in \{N+1, N+2, \ldots, N+M\}$). Since $C_i X_{i,j} - C_j = 0$, we can obtain that $\tilde{y}_{Ci} = C_i \delta_i$, and then

$\lim_{t\to\infty} \tilde{y}_{Ci}(t) = 0$. Thus, the outputs of followers can converge into the convex hull formed by multiple formation-leaders. To sum up, the heterogeneous swarm system (7.11) and (7.12) on switching graphs can realize the desired output formation-containment tracking under the proposed controller (7.16) and (7.17). This completes the proof of Theorem 7.2. $\square$

In the containment control [110–117], there exists no interaction and cooperation between multiple leaders. In [67–69], only formation-containment stabilization control problems were considered, where the macroscopic trajectory of the whole swarm system cannot be controlled effectively. In this section, formation-containment tracking problems for heterogeneous swarm systems with a tracking-leader having unknown time-varying input are further studied. From the proof of Theorem 7.2, we can see that both the states and the control inputs of formation-leaders will have coupling influences on the containment control of followers. Thus, the formation-containment tracking problem considered in this section cannot be decoupled into separative formation tracking and containment directly.

In Section 5.3, node-based adaptive control approaches were presented to deal with the unknown input of a leader, where both the adaptive gains and the non-linear functions to compensate for the leader's input were assigned to each node, and the compensation terms depended on the neighbouring error aggregated at each node. To analyze the stability of the node-based controllers in Section 5.3, Lyapunov functions were constructed by using the topology information, which implies that these analysis approaches are not applied to the switching graphs directly. Different from Section 5.3, adaptive gains and non-linear compensation functions are assigned to each edge of the possible graphs. Under the edge-based distributed observer, common Lyapunov functions are constructed as V_1 and V_3, and the stability of the switched system is analyzed based on V_1 and V_3. Thus, the proposed distributed observers in (7.16) and (7.17) can dispose of the influences of both leaders' unknown inputs and switching graphs.

In the existing results on containment or formation-containment control [67–69, 110–116, 118–120], the desired convex combinations of multiple leaders for the followers to track are determined by the topologies directly. In comparison with these previous results, the proposed predefined containment control framework in this chapter has two main advantages. Firstly, containment control can be achieved for totally heterogeneous swarm systems with non-identical formation-leaders having unknown inputs. Secondly, the convex combinations of multiple formation-leaders can be specified by several predefined weights $\rho_{i,j}$ ($i \in \mathcal{O}_F$, $j \in \mathcal{O}_L$). Under this containment framework, the desired tracking values of the followers do not rely on the topology, which makes this approach applicable to switching graphs. Besides, using the predefined weights, we can specify the relative output relations between formation-leaders and followers in advance, while the existing containment approaches can only guarantee the followers to converge into the convex hull. These weights $\rho_{i,j}$ can be chosen arbitrarily according to the mission requirements in this section, which is more feasible and reasonable for practical applications.

7.3.3 Simulation Example

The proposed formation-containment tracking control approach is applied to cooperative transportation of a heterogeneous UAV-UGV swarm system, where the tracking-leader $i = 0$ denotes the macroscopic reference trajectory of the whole swarm system, the formation-leaders $i = 1, 2, 3, 4$ are UAVs, and the followers $i = 5, 6, \ldots, 10$ are UGVs. These UAVs are required to achieve the desired formation tracking at a given height, and all the UGVs need to converge to the projection of convex hull formed by multiple UAVs on the ground. Since the height channel of multi-rotor UAVs can be controlled separately, only the movement in the XY plane is considered for the heterogeneous robot swarm system in this example. As shown in Fig. 7.11, the interaction graphs among the UAV and UGV robots are assumed to be switching each 5s with the initial topology being $\bar{\mathcal{G}}_1$.

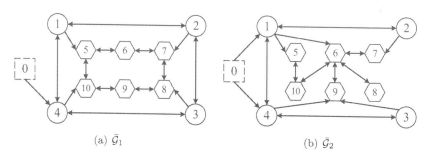

(a) $\bar{\mathcal{G}}_1$ (b) $\bar{\mathcal{G}}_2$

FIGURE 7.11: Possible graphs.

Consider the following model for the tracking-leader: $S = I_2 \otimes \begin{bmatrix} 0 & 1 \\ 0 & 0 \end{bmatrix}$, $E = I_2 \otimes \begin{bmatrix} 0 \\ 1 \end{bmatrix}$, $F = I_2 \otimes [\, 1 \quad 0 \,]$, where its time-varying control input is described by $r_0(t) = [0.1 \cos(0.1t), 0.1 \sin(0.1t)]^T$. Based on the inner and outer loop control framework in [8], the UAV model in the position and velocity loop can be described by (7.12) approximately, where $A_i = I_2 \otimes \begin{bmatrix} 0 & 1 \\ \varpi_{xi} & \varpi_{vi} \end{bmatrix}$, $B_i = I_2 \otimes \begin{bmatrix} 0 \\ 1 \end{bmatrix}$, $C_i = I_2 \otimes [\, 1 \quad 0 \,]$ $(i = 1, 2, 3, 4)$. Let $\varpi_{x1} = \varpi_{x2} = 0$, $\varpi_{v1} = \varpi_{v2} = 0$, $\varpi_{x3} = \varpi_{x4} = -1$, and $\varpi_{v3} = \varpi_{v4} = -1$. Using the feedback linearization approach in [45], from the view of formation control, we can also use (7.12) to describe the UGV kinematics approximately, where $A_i = 0_{2 \times 2}$, $B_i = I_2$, and $C_i = I_2$ $(i = 5, 6, \ldots, 10)$. In this example, the state of a UAV is composed of position and velocity, and the output is position. For a UGV, the position denotes both state and output of it.

Choose four UAVs as formation-leaders, which fly at a given height $h_Z = 5$m. These UAVs are required to achieve a square formation in the XY plane and the desired formation is specified by $h_y = [h_{y1}^T, h_{y2}^T, h_{y3}^T, h_{y4}^T]^T$, where $h_{y1} = [-1, 1]^T$, $h_{y2} = [1, 1]^T$, $h_{y3} = [1, -1]^T$, and $h_{y4} = [-1, -1]^T$. To generate h_y, the matrices in exosystem (7.15) can be designed as $H_i = 0_{2 \times 2}$, $R_i = 0_{2 \times 2}$, and $Y_i = I_2$ ($i = 1, 2, 3, 4$). Six UGVs need to converge to the projection of convex hull on the ground formed by multiple UAVs. To specify the desired tracking target for each UGV, define the weight vectors $\rho_i = [\rho_{i,1}, \rho_{i,2}, \rho_{i,3}, \rho_{i,4}]$ ($i = 5, 6, \ldots, 10$) and choose $\rho_5 = [\frac{1}{16}, \frac{3}{16}, \frac{5}{16}, \frac{7}{16}]$, $\rho_6 = [\frac{1}{16}, \frac{7}{16}, \frac{3}{16}, \frac{5}{16}]$, $\rho_7 = [\frac{3}{16}, \frac{7}{16}, \frac{5}{16}, \frac{1}{16}]$, $\rho_8 = [\frac{5}{16}, \frac{1}{16}, \frac{3}{16}, \frac{7}{16}]$, $\rho_9 = [\frac{7}{16}, \frac{1}{16}, \frac{5}{16}, \frac{3}{16}]$, $\rho_{10} = [\frac{7}{16}, \frac{5}{16}, \frac{3}{16}, \frac{1}{16}]$.

Based on Algorithm 7.1, the formation-containment tracking controller (7.16) and (7.17) can be designed in the following. Firstly, choose $X_1 = X_2 = I_4$, $U_1 = U_2 = I_2 \otimes [\, 0 \quad 0 \,]$, $X_3 = X_4 = I_4$, $U_3 = U_4 = I_2 \otimes [\, 1 \quad 1 \,]$, $X_{hi} = I_2 \otimes [\, 1 \quad 0 \,]^T$, $U_{hi} = -\varpi_{xi} I_2$ ($i = 1, 2, 3, 4$), $X_{i,j} = I_2 \otimes [\, 1 \quad 0 \,]$, $U_{i,j} = I_2 \otimes [\, 0 \quad 1 \,]$ ($i = 5, 6, \ldots, 10$, $j = 1, 2, 3, 4$) such that the regulator equations in Assumptions 7.5-7.7 hold. Then, design the controller (7.16) for formation-leader UAV i ($i = 1, 2, 3, 4$). Since $R_i = 0_{2 \times 2}$, we can choose $\tau_i = 0_{2 \times 1}$ such that the feasibility condition (7.18) is satisfied for each UAV. Select the initial values of adaptive gains $\hat{\gamma}_{ij}$ as $\hat{\gamma}_{ij}(0) = 2$ ($i = 1, 2, 3, 4$, $j = 0, 1, \ldots, 4$). Let $\mu = 1$, $K_{11} = K_{12} = I_2 \otimes [\, -2 \quad -2 \,]$, $K_{13} = K_{14} = I_2 \otimes [\, -1 \quad -1 \,]$, $\Upsilon_i = I_2$ ($i = 1, 2, 3, 4$). Finally, determine the controller (7.17) for follower UGV i ($i = 5, 6, \ldots, 10$). Let the initial values of $\hat{\alpha}_{ij}$ ($j = 1, 2, 3, 4$) and $\hat{\beta}_{ik}$ ($k = 5, 6, \ldots, 10$) be $\hat{\alpha}_{ij}(0) = \hat{\beta}_{ik}(0) = 20$, and choose $K_{3i} = -I_2$ and $\Upsilon_{i,j} = 0_{2 \times 2}$ ($j = 1, 2, 3, 4$). The initial state of tracking-leader is set as $v_0(0) = [-20, 0, 0, -1]^T$ and the initial states of formation-leaders and followers are generated by random numbers.

Fig. 7.12 shows the position trajectories within $t = 30$s and position snapshots at different time instants ($t = 0, 15, 25, 30$s) for UAV-UGV heterogeneous swarm system, where the tracking-leader ($i = 0$) is denoted by pentagram, the formation-leader UAVs ($i = 1, 2, 3, 4$) are marked by diamond, upward-pointing triangle, circle, and right-pointing triangle respectively, and the follower UGVs ($i = 5, 6, \ldots, 10$) are represented by squares. Figs. 7.13 and 7.14 give the output formation tracking errors of formation-leader UAVs and the output containment errors of follower UGVs. As shown in Figs. 7.12-7.14, four UAVs realize the desired square formation and track the trajectory of tracking-leader, while six follower UGVs can converge to the projection of convex hull on the ground formed by multiple UAVs. Therefore, heterogeneous UAV-UGV swarm system with switching graphs can achieve the desired output formation-containment tracking under the proposed controller.

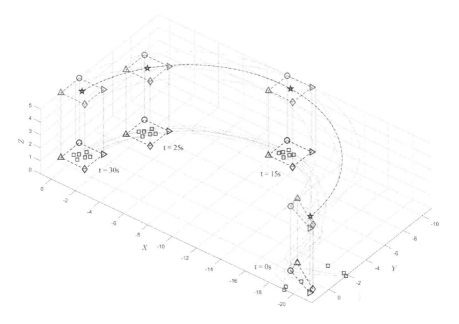

FIGURE 7.12: Position trajectories within $t = 30$s and snapshots at $t = 0, 15, 25, 30$s of UAV-UGV swarm system.

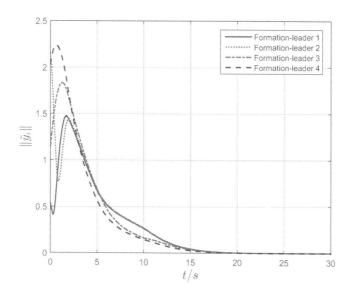

FIGURE 7.13: Output formation tracking errors for formation-leader UAVs.

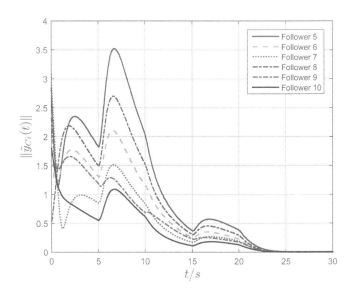

FIGURE 7.14: Output containment errors for follower UGVs.

7.4   Conclusions

In this chapter, predefined containment control problems for heterogeneous swarm systems on switching topologies were studied firstly. A distributed observer was designed for each follower to estimate the whole states of all the leaders. Based on the estimated states, a predefined containment tracking controller was constructed for the followers, where the desired convex combinations of the multiple leaders were specified by several given constant weights. Furthermore, formation-containment tracking control problems for heterogeneous swarm systems with leaders' unknown inputs and switching graphs were investigated. A distributed formation-containment tracking controller was proposed by using output regulation control and sliding-mode control. Under the influences of hierarchical coordination couplings, an algorithm to design the proposed controller was designed, and sufficient conditions to achieve formation-containment tracking were presented using common Lyapunov stability theory. The results in this chapter are mainly based on [10].

Chapter 8

Experiments on Formation Tracking for UAV and UGV Swarm Systems

8.1 Introduction

In this chapter, the proposed formation tracking approaches are applied to the practical UAV and UGV swarm systems, and several formation experiments are carried out to further verify the effectiveness of the theoretical results. The formation tracking and formation-containment control experiments for quadrotor UAV swarm systems are given firstly. Then, the UAV-UGV heterogeneous swarm cooperative platform is used to test the given time-varying formation tracking controller. The main contents of this chapter are given as follows.

Firstly, time-varying formation tracking problems for quadrotor UAV swarm systems with switching directed topologies are considered. A formation tracking control protocol is constructed utilizing local neighbouring information, and an algorithm is given to design the proposed controller. The obtained theoretical results are applied to solve the target enclosing problems of the UAV swarm systems. A flight experiment for three follower quadrotor UAVs to enclose a leader quadrotor UAV is carried out to verify the effectiveness of the proposed control approach.

Then, formation-containment control problems for a group of quadrotor UAVs are investigated. Based on the neighbouring information of UAVs, formation-containment control protocols are constructed, and sufficient conditions for UAV swarm systems to achieve formation-containment are presented. An approach to determine the gain matrices of the formation-containment protocol is proposed by solving an algebraic Riccati equation. A formation-containment platform with five quadrotor UAVs is introduced, and both simulation and experiment results are given to demonstrate the effectiveness of the formation-containment controller.

Furthermore, time-varying formation tracking problems for UAV-UGV heterogeneous swarm systems are studied. Based on the distributed observer for the virtual leader, a time-varying formation tracking controller is proposed and an algorithm to design the control parameters is presented. The proposed control approach is applied to air-ground cooperative reconnaissance

DOI: 10.1201/9781003263470-8   199 scene, where both simulation and experiment results on formation tracking for UAV-UGV heterogeneous swarm system are presented.

8.2   Time-varying Formation Tracking for UAV Swarm Systems

In this section, time-varying formation tracking control problems for a team of quadrotor UAVs with switching and directed interaction topologies are investigated, where the follower UAVs can realize a given time-varying formation while tracking the leader UAV. A formation tracking control protocol is firstly constructed utilizing local neighbouring information, and an algorithm composed of four steps is provided to design the proposed protocol. It is proved that the UAV swarm system can realize the desired formation tracking using the designed protocol if the dwell time for the switching directed topologies is larger than a fixed threshold and the time-varying formation tracking feasibility condition is satisfied. Based on the ultra-wideband (UWB) positioning technology, a UAV swarm formation control platform with four quadrotor UAVs is given. The obtained theoretical results are applied to solve the target enclosing problems of the UAV swarm systems. A flight experiment for three follower quadrotor UAVs to enclose a leader quadrotor UAV is carried out to verify the effectiveness of the presented results.

8.2.1   Problem Description

Assume that there exists a UAV swarm system with N UAVs. A UAV is called a leader UAV (leader) if it has no neighbours otherwise is called a follower UAV (follower). Without loss of generality, assume that UAV 1 is the leader UAV and the rest $N - 1$ UAVs are the follower UAVs. Let $F = \{2, 3, \ldots, N\}$ stand for the set of the follower UAVs. For each quadrotor UAV, the dynamics of attitude has a much smaller time constant than the trajectory dynamics. By applying the two-time scale separation principle, the time-varying formation tracking can be carried out in an outer/inner loop architecture [129, 130], where the attitudes are stabilized in the inner loop, and the outer loop tracks the trajectory.

Because the time-varying formation tracking control in this section only focuses on the position and the velocity, we can pay attention to the outer loop and model the leader UAV and follower UAVs by (8.1a) and (8.1b) in the outer loop, respectively. The model of the leader is described by

$$\begin{cases} \dot{x}_1(t) = v_1(t), \\ \dot{v}_1(t) = \alpha_x x_1(t) + \alpha_v v_1(t), \end{cases} \tag{8.1a}$$

where $x_1(t) \in \mathbb{R}^n$ and $v_1(t) \in \mathbb{R}^n$ represent the position and velocity of the leader UAV, α_x and α_v are known damping constants, and $n \geq 1$ represents the dimension of the space. The dynamics of the i-th follower UAV is denoted by

$$\begin{cases} \dot{x}_i(t) = v_i(t), \\ \dot{v}_i(t) = \alpha_x x_i(t) + \alpha_v v_i(t) + u_i(t), \end{cases} \tag{8.1b}$$

where $x_i(t) \in \mathbb{R}^n$, $v_i(t) \in \mathbb{R}^n$ and $u_i(t) \in \mathbb{R}^n$ represent respectively the position, velocity, and control input vectors of the follower UAV i ($i \in F$). In the following, let $n = 1$ to simplify the description. Note that all the results can be utilized to the higher dimensional case directly by applying the Kronecker product.

Let $h_F(t) = [h_2^T(t), h_3^T(t), \ldots, h_N^T(t)]^T$ represent the formation vector that specifies the time-varying formation for all follower UAVs with $h_i(t) = [h_{ix}(t), h_{iv}(t)]^T$ ($i \in F$) piecewise continuously differentiable. Because the predefined formation is realized by the physical movement of each follower UAV in practice, it is logical to have the piecewise continuously differentiability assumption on $h_i(t)$. Let $\xi_k(t) = [x_k(t), v_k(t)]^T$ ($k = 1, 2, \ldots, N$).

Definition 8.1. UAV swarm system (8.1) is said to achieve the time-varying formation tracking if for any given bounded initial states, it holds that

$$\lim_{t \to \infty} (\xi_i(t) - h_i(t) - \xi_1(t)) = 0 \ (i \in F). \tag{8.2}$$

Remark 8.1. *In (8.2), $h_i(t)$ ($i \in F$) is utilized to describe the expected time-varying formation for the followers. Note that $h_i(t)$ ($i \in F$) is not the global waypoints for the UAVs to follow. In fact, $h_i(t)$ stands for the relative offset of $\xi_i(t)$ with respect to $\xi_1(t)$. When the UAV swarm system (8.1) achieves the expected time-varying formation tracking, the leader UAV can locate inside or outside the time-varying formation formed by follower UAVs. If $\lim_{t \to \infty} \sum_{i=2}^{N} h_i(t) = 0$, it can be obtained from (8.2) that $\lim_{t \to \infty} (\sum_{i=2}^{N} \xi_i(t)/(N-1) - \xi_1(t)) = 0$, which means that $\xi_1(t)$ lies in the centre of the time-varying formation specified by $h_F(t)$. Therefore, by choosing $\lim_{t \to \infty} \sum_{i=2}^{N} h_i(t) = 0$, Definition 8.1 turns into the definitions for target pursuing or enclosing.*

For UAV swarm system (8.1) with switching directed topologies, construct the following time-varying formation tracking protocol:

$$\begin{aligned} u_i(t) = &K \sum_{j=2}^{N} w_{\sigma(t)}^{ij}(t) \left((\xi_i(t) - h_i(t)) - (\xi_j(t) - h_j(t)) \right) \\ &+ K w_{\sigma(t)}^{i1}(t) \left((\xi_i(t) - h_i(t)) - \xi_1(t) \right) - \alpha h_i(t) + \dot{h}_{iv}(t), \end{aligned} \tag{8.3}$$

where $i \in F$, $\alpha = [\alpha_x, \alpha_v]$, and $K = [k_{11}, k_{12}]$ represents a constant gain matrix to be determined.

Considering the leader-follower topology structure of the UAV swarm system (8.1), it can be obtained that the Laplacian matrix $L_{\sigma(t)}$ is written as

$$L_{\sigma(t)} = \begin{bmatrix} 0 & 0 \\ l^{lf}_{\sigma(t)} & L^{ff}_{\sigma(t)} \end{bmatrix}, \tag{8.4}$$

where $l^{lf}_{\sigma(t)} \in \mathbb{R}^{(N-1)\times 1}$ stands for the interactions from the leader UAV to the follower UAVs and $L^{ff}_{\sigma(t)} \in \mathbb{R}^{(N-1)\times(N-1)}$. Let $\xi_F(t) = [\xi_2^T(t), \xi_3^T(t), \ldots, \xi_N^T(t)]^T$, $B_1 = [1,0]^T$ and $B_2 = [0,1]^T$. Under the formation tracking protocol (8.3) with switching and directed topologies, the augmented dynamics of UAV swarm system (8.1) can be written as

$$\begin{cases} \dot{\xi}_1(t) = \left(B_1 B_2^T + B_2\alpha\right)\xi_1(t), \\ \dot{\xi}_F(t) = \left(I_{N-1} \otimes \left(B_1 B_2^T + B_2\alpha\right) + \left(L^{ff}_{\sigma(t)} \otimes B_2 K\right)\right)\xi_F(t) \\ \qquad + \left(l^{lf}_{\sigma(t)} \otimes B_2 K\right)\xi_1(t) + \left(I_{N-1} \otimes B_2 B_2^T\right)\dot{h}_F(t) \\ \qquad - \left(L^{ff}_{\sigma(t)} \otimes B_2 K + I_{N-1} \otimes B_2\alpha\right)h_F(t). \end{cases} \tag{8.5}$$

For UAV swarm system (8.5) with switching and directed graphs, this section mainly focuses on the following two problems: (i) how to design the time-varying formation tracking protocol (8.3) under the influence of switching directed topologies, and (ii) how to verify the validity of the theoretical results using a UAV experimental platform.

8.2.2   Formation Tracking Controller Design

In this subsection, an algorithm to design the time-varying formation tracking protocol (8.3) is proposed firstly. Then, the stability of the designed protocol is proved based on the piecewise Lyapunov function theory.

Assumption 8.1. *For each follower UAV i $(i \in F)$, there is at least a path from the leader UAV-1 to it in each possible interaction topology $G_{\sigma(t)}$.*

If Assumption 8.1 holds, one can get that $G_{\sigma(t)}$ has a spanning tree with the leader UAV-1 as the root.

Lemma 8.1 ([19]). *If Assumption 8.1 holds, then $L_{\sigma(t)}$ has a simple zero eigenvalue with $\mathbf{1}_N$ being its corresponding right eigenvector, and the real parts of all the other $N-1$ eigenvalues are positive and the same as those of $L^{ff}_{\sigma(t)}$.*

Let $\lambda_i(L^{ff}_{\sigma(t)})$ $(i = 1, 2, \ldots, N-1)$ stand for the eigenvalues of $L^{ff}_{\sigma(t)}$ and $\gamma^{\min}_{\sigma(t)} = \min\left\{\mathrm{Re}(\lambda_i(L^{ff}_{\sigma(t)}))\right\}$. From Lemma 8.1 and (8.4), we know that if Assumption 8.1 is satisfied, then $\gamma^{\min}_{\sigma(t)}$ is positive. From Lemma 2 in [132], we can obtain that for any $\gamma_{\sigma(t)} \in (0, \gamma_{\sigma(t)}^{\min})$, there is a positive definite matrix $\Upsilon_{\sigma(t)} \in \mathbb{R}^{(N-1) \times (N-1)}$ satisfying

$$\left(L_{\sigma(t)}^{ff}\right)^T \Upsilon_{\sigma(t)} + \Upsilon_{\sigma(t)} L_{\sigma(t)}^{ff} - 2\gamma_{\sigma(t)} \Upsilon_{\sigma(t)} > 0. \tag{8.6}$$

Let $\xi(t) = \left[\xi_1^T(t), \xi_F^T(t)\right]^T$. From (8.5), one gets that

$$\dot{\xi}(t) = \left(\begin{bmatrix} B_1 B_2^T + B_2\alpha & 0 \\ 0 & I_{N-1} \otimes \left(B_1 B_2^T + B_2\alpha\right) \end{bmatrix} + L_{\sigma(t)} \otimes B_2 K \right) \xi(t)$$
$$- \begin{bmatrix} 0 \\ L_{\sigma(t)}^{ff} \otimes B_2 K + I_{N-1} \otimes B_2\alpha \end{bmatrix} h_F(t) \tag{8.7}$$
$$+ \begin{bmatrix} 0 \\ I_{N-1} \otimes B_2 B_2^T \end{bmatrix} \dot{h}_F(t).$$

Define $\psi_i(t) = \xi_i(t) - h_i(t)$ $(i \in F)$, $\psi_F(t) = \left[\psi_2^T(t), \psi_3^T(t), ..., \psi_N^T(t)\right]^T$, $\psi(t) = \left[\xi_1^T(t), \psi_F^T(t)\right]^T$. Then, it holds that

$$\xi(t) = \psi(t) + \begin{bmatrix} 0 \\ I \end{bmatrix} h_F(t). \tag{8.8}$$

Substituting (8.8) into (8.7), it can be obtained that

$$\dot{\psi}(t) = \begin{bmatrix} B_1 B_2^T + B_2\alpha & 0 \\ 0 & I_{N-1} \otimes \left(B_1 B_2^T + B_2\alpha\right) \end{bmatrix} \psi(t)$$
$$- \begin{bmatrix} 0 \\ L_{\sigma(t)}^{ff} \otimes B_2 K + I_{N-1} \otimes B_2\alpha \end{bmatrix} h_F(t) \tag{8.9}$$
$$+ \left(L_{\sigma(t)} \otimes B_2 K\right) \psi(t)$$
$$+ \begin{bmatrix} 0 \\ I_{N-1} \otimes B_2 B_2^T \end{bmatrix} \dot{h}_F(t) - \begin{bmatrix} 0 \\ I \end{bmatrix} \dot{h}_F(t).$$

From (8.4), it can be obtained that

$$\left(L_{\sigma(t)} \otimes B_2 K\right) \begin{bmatrix} 0 \\ I \end{bmatrix} = \begin{bmatrix} 0 \\ L_{\sigma(t)}^{ff} \otimes B_2 K \end{bmatrix}. \tag{8.10}$$

Note that $B_1 = [1,0]^T$ and $B_2 = [0,1]^T$. we get

$$\begin{bmatrix} 0 \\ I_{N-1} \otimes B_2 B_2^T \end{bmatrix} - \begin{bmatrix} 0 \\ I \end{bmatrix} = \begin{bmatrix} 0 \\ I_{N-1} \otimes B_1 B_1^T \end{bmatrix}. \tag{8.11}$$

From (8.9), (8.10) and (8.11), one has

$$\dot{\psi}(t) = \begin{bmatrix} B_1 B_2^T + B_2\alpha & 0 \\ 0 & I_{N-1} \otimes \left(B_1 B_2^T + B_2\alpha\right) \end{bmatrix} \psi(t)$$
$$+ \left(L_{\sigma(t)} \otimes B_2 K\right) \psi(t) - \begin{bmatrix} 0 \\ I_{N-1} \otimes B_1 B_1^T \end{bmatrix} \dot{h}_F(t) \tag{8.12}$$
$$+ \begin{bmatrix} 0 \\ I_{N-1} \otimes B_1 B_2^T \end{bmatrix} h_F(t).$$

Let $T = \begin{bmatrix} 1 & 0 \\ \mathbf{1}_{N-1} & I_{N-1} \end{bmatrix}$. It holds that $T^{-1} = \begin{bmatrix} 1 & 0 \\ -\mathbf{1}_{N-1} & I_{N-1} \end{bmatrix}$. It can be verified from Lemma 8.1 that $L_{\sigma(t)}\mathbf{1}_N = 0$, which yields

$$l^{lf}_{\sigma(t)} + L^{ff}_{\sigma(t)}\mathbf{1}_{N-1} = 0. \tag{8.13}$$

Based on (8.13), one gets

$$T^{-1}L_{\sigma(t)}T = \begin{bmatrix} 0 & 0 \\ 0 & L^{ff}_{\sigma(t)} \end{bmatrix}. \tag{8.14}$$

Define $\varsigma(t) = (I_{N-1} \otimes I)\psi_F(t) - (\mathbf{1}_{N-1} \otimes I)\xi_1(t)$ and $\zeta(t) = \left[\xi_1^T(t), \varsigma^T(t)\right]^T$. One can get that $\zeta(t) = \left(T^{-1} \otimes I\right)\psi(t)$ and

$$\psi(t) = (T \otimes I)\zeta(t). \tag{8.15}$$

Substitute (8.15) into (8.12) and premultiply $T^{-1} \otimes I_2$ for the both sides of (8.12). One gets

$$\begin{aligned} \dot{\zeta}(t)=&\begin{bmatrix} B_1 B_2^T + B_2\alpha & 0 \\ 0 & I_{N-1} \otimes \left(B_1 B_2^T + B_2\alpha\right) \end{bmatrix} \zeta(t) \\ &+\left(\begin{bmatrix} 0 & 0 \\ 0 & L^{ff}_{\sigma(t)} \end{bmatrix} \otimes B_2 K\right) \zeta(t) \\ &+\begin{bmatrix} 0 \\ I_{N-1} \otimes B_1 B_2^T \end{bmatrix} h_F(t)-\begin{bmatrix} 0 \\ I_{N-1} \otimes B_1 B_1^T \end{bmatrix} \dot{h}_F(t). \end{aligned} \tag{8.16}$$

Recalling that $\zeta(t) = \left[\xi_1^T(t), \varsigma^T(t)\right]^T$, one can obtain from (8.16) that

$$\dot{\xi}_1(t) = (B_1 B_2^T + B_2\alpha)\xi_1(t), \tag{8.17}$$

and

$$\begin{aligned} \dot{\varsigma}(t)=&(I_{N-1} \otimes (B_1 B_2^T + B_2\alpha) + L^{ff}_{\sigma(t)} \otimes B_2 K)\varsigma(t) \\ &+(I_{N-1} \otimes B_1 B_2^T)h_F(t) - (I_{N-1} \otimes B_1 B_1^T)\dot{h}_F(t). \end{aligned} \tag{8.18}$$

A procedure to design the time-varying formation tracking protocol (8.3) with switching and directed graphs is proposed in Algorithm 8.1.

Algorithm 8.1. *The time-varying formation tracking protocol (8.3) can be designed by the following four steps.*

1) For a given formation specified by $h_F(t)$ for the follower UAVs, test the following time-varying formation tracking feasibility condition:

$$\lim_{t\to\infty} \left(h_{iv}(t) - \dot{h}_{ix}(t)\right) = 0 \; (i \in F). \tag{8.19}$$

If condition (8.19) is satisfied, continue; else stop the algorithm and the desired time-varying formation is infeasible for UAV swarm system (8.5).

2) For a given $\varepsilon > 0$, compute a positive definite matrix P from the inequality

$$(B_1 B_2^T + B_2 \alpha)P + P(B_1 B_2^T + B_2 \alpha)^T - B_2 B_2^T + \varepsilon P < 0. \tag{8.20}$$

3) Choose a sufficiently large constant δ satisfying $\delta > 1/(2\bar{\gamma})$, in which $\bar{\gamma} = \min\{\gamma_{\sigma(t)}, \sigma(t) \in \{1, 2, \ldots, p\}\}$.
4) Let $K = -\delta B_2^T P^{-1}$.

The following theorem shows that the desired formation tracking can be achieved by UAV swarm systems under the proposed controller.

Theorem 8.1. *Suppose that Assumption 8.1 holds. UAV swarm system (8.1) with switching and directed graphs achieves the time-varying formation tracking under the protocol (8.3) designed by Algorithm 8.1 if the time-varying formation tracking feasibility condition (8.19) is satisfied and the dwell time $\tau_0 > \ln \mu / \varepsilon$, where $\mu = \max\{\lambda_{\max}(\Upsilon_{\bar{i}}^{-1} \Upsilon_{\bar{j}}), \bar{i}, \bar{j} \in \{1, 2, \ldots, p\}\}$ with $\lambda_{\max}(\Upsilon_{\bar{i}}^{-1} \Upsilon_{\bar{j}})$ representing the maximum eigenvalue of $\Upsilon_{\bar{i}}^{-1} \Upsilon_{\bar{j}}$.*

Proof. Let

$$\psi_T(t) = (T \otimes I)\left[\xi_1^T(t), 0\right]^T, \tag{8.21}$$

$$\psi_{\bar{T}}(t) = (T \otimes I)\left[0, \varsigma^T(t)\right]^T. \tag{8.22}$$

Note that $\left[\xi_1^T(t), 0\right]^T = e_1 \otimes \xi_1(t)$, where $e_{\bar{i}} \in \mathbb{R}^N$ with 1 as its first entry and 0 elsewhere. From (8.21), one has that

$$\psi_T(t) = T e_1 \otimes \xi_1(t) = \mathbf{1}_N \otimes \xi_1(t). \tag{8.23}$$

It holds from (8.15), (8.21) and (8.22) that

$$\psi(t) = \psi_T(t) + \psi_{\bar{T}}(t), \tag{8.24}$$

in which $\psi_T(t)$ and $\psi_{\bar{T}}(t)$ are linearly independent. From (8.23) and (8.24), we get that

$$\psi_{\bar{T}}(t) = \xi(t) - \begin{bmatrix} 0 \\ I \end{bmatrix} h_F(t) - \mathbf{1}_N \otimes \xi_1(t), \tag{8.25}$$

that is,

$$\psi_{\bar{T}}(t) = \begin{bmatrix} 0 \\ \xi_F(t) - h_F(t) - \mathbf{1}_{N-1} \otimes \xi_1(t) \end{bmatrix}. \tag{8.26}$$

From (8.26), it follows that $\lim_{t \to \infty} \psi_{\bar{T}}(t) = 0$ is equivalent to $\lim_{t \to \infty}(\xi_F(t) - h_F(t) - \mathbf{1}_{N-1} \otimes \xi_1(t)) = 0$, which means that UAV swarm system (8.5) with switching and directed graphs realizes the time-varying formation tracking if and only if

$$\lim_{t\to\infty} \psi_{\bar{T}}(t) = 0. \tag{8.27}$$

Since $T \otimes I$ is nonsingular, it follows from (8.22) that (8.27) is equivalent to

$$\lim_{t\to\infty} \varsigma(t) = 0. \tag{8.28}$$

Therefore, $\varsigma(t)$ depicts the time-varying formation tracking error.

Note that the time-varying formation tracking feasibility condition (8.19) is satisfied. It holds from (8.19) that

$$\lim_{t\to\infty} \left(B_1 B_2^T h_i(t) - B_1 B_1^T \dot{h}_i(t) \right) = 0 \ (i \in F). \tag{8.29}$$

The augmented form of (8.29) can be derived as

$$\lim_{t\to\infty} \left((I_{N-1} \otimes B_1 B_2^T) h_F(t) - (I_{N-1} \otimes B_1 B_1^T) \dot{h}_F(t) \right) = 0. \tag{8.30}$$

Analyze the stability of the following linear switched system

$$\dot{\phi}(t) = \left(I_{N-1} \otimes \left(B_1 B_2^T + B_2 \alpha \right) + \left(L_{\sigma(t)}^{ff} \otimes B_2 K \right) \right) \phi(t). \tag{8.31}$$

Construct a piecewise Lyapunov functional candidate as follows

$$V(t) = \phi^T(t) \left(\Upsilon_{\sigma(t)} \otimes P^{-1} \right) \phi(t), \tag{8.32}$$

where $\Upsilon_{\sigma(t)}$ and P are given as (8.6) and (8.20), respectively. Notice that each graph $G_{\sigma(t)}$ is fixed in $t \in [t_k, t_{k+1})$ ($k \in \{0, 1, \dots\}$). Differentiating $V(t)$ along the trajectory of the linear switched system (8.31), one gets that for any $t \in [t_k, t_{k+1})$

$$\dot{V}(t) = \phi^T(t)(\Upsilon_{\sigma(t)} \otimes \Xi_B + \Xi_L)\phi(t), \tag{8.33}$$

in which $\Xi_B = (B_1 B_2^T + B_2 \alpha)^T P^{-1} + P^{-1}(B_1 B_2^T + B_2 \alpha)$ and $\Xi_L = (L_{\sigma(t)}^{ff})^T \Upsilon_{\sigma(t)} \otimes (B_2 K)^T P^{-1} + \Upsilon_{\sigma(t)} L_{\sigma(t)}^{ff} \otimes P^{-1} B_2 K$. Substituting $K = -\delta B_2^T P^{-1}$ into (8.33), one has

$$\dot{V}(t) = \phi^T(t)(\Upsilon_{\sigma(t)} \otimes \Xi_B + \bar{\Xi}_L)\phi(t), \tag{8.34}$$

where $\bar{\Xi}_L = -\delta((L_{\sigma(t)}^{ff})^T \Upsilon_{\sigma(t)} + \Upsilon_{\sigma(t)} L_{\sigma(t)}^{ff}) \otimes P^{-1} B_2 B_2^T P^{-1}$. Let $\phi(t) = (I_{N-1} \otimes P)\tilde{\phi}(t)$. It follows from (8.34) that

$$\dot{V}(t) = \tilde{\phi}^T(t)(\Upsilon_{\sigma(t)} \otimes \tilde{\Xi}_B + \tilde{\Xi}_L)\tilde{\phi}(t), \tag{8.35}$$

where $\tilde{\Xi}_B = P(B_1 B_2^T + B_2 \alpha)^T + (B_1 B_2^T + B_2 \alpha)P$ and $\tilde{\Xi}_L = -\delta((L_{\sigma(t)}^{ff})^T \Upsilon_{\sigma(t)} + \Upsilon_{\sigma(t)} L_{\sigma(t)}^{ff}) \otimes B_2 B_2^T$. It holds from (8.6), (8.20), and (8.35) that

$$\dot{V}(t) \leq \tilde{\phi}^T(t)(\Upsilon_{\sigma(t)} \otimes (B_2 B_2^T - \varepsilon P) - 2\delta\gamma_{\sigma(t)} \Upsilon_{\sigma(t)} \otimes B_2 B_2^T)\tilde{\phi}(t). \quad (8.36)$$

Since $\delta > 1/(2\bar{\gamma})$ and $\bar{\gamma} = \min\{\gamma_{o(t)}, \sigma(t) \in \{1, 2, \ldots, p\}\}$, one has that

$$\dot{V}(t) < \tilde{\phi}^T(t)\left(-\varepsilon\Upsilon_{\sigma(t)} \otimes P\right)\tilde{\phi}(t). \quad (8.37)$$

It can be further obtained from $\tilde{\phi}(t) = (I_{N-1} \otimes P^{-1})\phi(t)$ that for any $t \in [t_k, t_{k+1})$

$$\dot{V}(t) < -\varepsilon\phi^T(t)\left(\Upsilon_{\sigma(t)} \otimes P^{-1}\right)\phi(t)$$
$$= -\varepsilon V(t). \quad (8.38)$$

Because the graph is fixed in $t \in [t_k, t_{k+1})$, one gets that

$$V(t) < e^{-\varepsilon(t-t_k)}V(t_k). \quad (8.39)$$

Notice that the continuity of $\phi(t)$ implies that $\phi(t_k) = \phi(t_k^-)$. It can be obtained from (8.32) that

$$V(t_k) \leq \mu V(t_k^-). \quad (8.40)$$

From (8.39) and (8.40), one has

$$V(t) < \mu e^{-\varepsilon(t-t_k)}V(t_k^-). \quad (8.41)$$

Since $V(t_k^-) < e^{-\varepsilon(t_k - t_{k-1})}V(t_{k-1})$, it holds from (8.41) that

$$V(t) < \mu e^{-\varepsilon(t-t_{k-1})}V(t_{k-1}). \quad (8.42)$$

Using the recursion method, one has from (8.42) that

$$V(t) < \mu^k e^{-\varepsilon(t-t_0)}V(t_0). \quad (8.43)$$

It can be proved that

$$t - t_0 = t - t_k + (t_k - t_{k-1}) + \cdots + (t_1 - t_0) \geq k\tau_0. \quad (8.44)$$

Since $\mu = \max\{\lambda_{\max}(\Upsilon_{\tilde{i}}^{-1}\Upsilon_{\tilde{j}}), \tilde{i}, \tilde{j} \in \{1, 2, \ldots, p\}\}$, we get that $\mu \geq 1$. Then it can be obtained from (8.43) and (8.44) that

$$V(t) < e^{\left(\frac{\ln\mu}{\tau_0} - \varepsilon\right)t}V(0). \quad (8.45)$$

If $\tau_0 > \ln\mu/\varepsilon$, then $\ln\mu/\tau_0 - \varepsilon < 0$. One obtains that $\lim_{t\to\infty} V(t) = 0$, which leads to the asymptotical stability of the linear switched system (8.31).

From (8.18), (8.30) and (8.31), one gets that $\lim_{t\to\infty}\varsigma(t) = 0$; namely, the time-varying formation tracking error converges to zero as $t \to \infty$. Therefore, the time-varying formation tracking for UAV swarm system (8.1) with switching and directed graphs is realized by the given protocol (8.3). This completes the proof for Theorem 8.1. $\qquad\square$

Remark 8.2. *Constraint (19) reveals that the expected time-varying formation tracking should be consistent with the dynamics of every UAV. Unlike the switching undirected case, there exists a minimal threshold for the dwell time or the switching frequency in switching directed cases. Notice that the time-varying formation specified by $h_F(t)$ will bring both the formation information and its derivative $\dot{h}_F(t)$ to the design and analysis of the time-varying formation tracking protocol. For each UAV, the proposed time-varying formation tracking protocol (8.3) depends on the neighbouring relative information, which means that the protocol (8.3) can be implemented in a distributed form. Thus, the proposed algorithm has less computation cost than the centralized coordination approaches.*

8.2.3 Simulation and Experimental Results

In this subsection, an experimental platform composed of quadrotor UAVs and UWB positioning system is introduced firstly. Then, the proposed time-varying formation tracking method is utilized to deal with the target enclosing problem on this platform.

FIGURE 8.1: Formation tracking experimental platform based on quadrotor UAVs.

The main components of the experimental platform are shown in Fig. 8.1, where there are four quadrotor UAVs that have flight control system (FCS), a UWB positioning system, and a ground control station (GCS). Each quadrotor has 65 cm tip-to-tip wingspan. The ground weight and maximum take-off weight of each quadrotor UAV are about 1600 g and 2300 g, respectively. The hang time is between 10 to 18 min with a typical value of 12 min. The processor for the FCS is a user-programmable DSP belonging to the TMS320F28335 series. The acceleration and the attitude of each quadrotor UAV are measured using a three-axis magnetometer, a one-axis gyroscopes, and a three-axis accelerometer. The UWB positioning system provides the position and velocity of each UAV in the horizontal plane (i.e., the XY plane). The altitude of each quadrotor UAV is obtained by a barometer and a laser range module. The key flight parameters are recorded in onboard micro SD card. The interactions among quadrotor UAVs and the GCS are realized by wireless communication network based on Zigbee modules. The remote control (RC) radio transmitter and receiver are kept in each FCS for emergency cases. The architecture for the quadrotor UAV hardware system is given in Fig. 8.2.

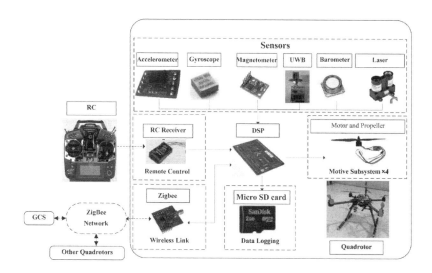

FIGURE 8.2: Hardware architecture of the quadrotor UAV.

The UWB positioning system is composed of four base stations with antennas, four user devices, a switcher and a GCS. It is developed based on the DecaWave company's DW1000 wireless transceiver chips using the time of arrival (TOA) measurement of wireless signals. The very narrow pulse width of UWB offers a high resolution of timing that can reaches hundreds or tens of picoseconds. Each quadrotor UAV takes a user device which broadcast ultra-wide band signal (position request) periodically. At the same time, the four base stations receive the signal and measure its TOA respectively. The base stations are connected to the GCS via a switcher. These TOA values are aggregated to the GCS to be processed to generate the user position result using positioning algorithms. The velocity is further estimated by integrated the differencing position results and the acceleration provided by the inertial navigation system using Kalman filtering algorithm. Note that the UWB positioning system is applicable to both outdoor and indoor environment, while the GPS can only be used in the outdoor unsheltered scenes and the Vicon

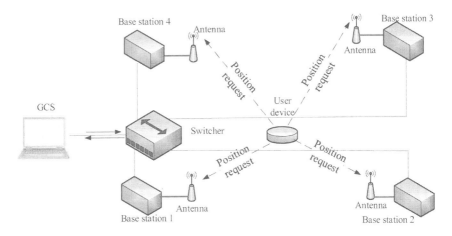

FIGURE 8.3: Hardware architecture of the UWB positioning system.

system only works indoors. Fig. 8.3 gives the hardware structure of the UWB positioning system.

The GCS has two roles. The first role is to communicate with the quadrotor UAVs via Zigbee network. Control commands can be transferred to a specified UAV or broadcasted to all the quadrotor UAVs. To monitor the flight state of all the quadrotors, the state of each UAV is required to send to the GCS once per second. The second role is to collect data from UWB base stations and calculate the position results based on the data. The main components applied in the UAV experimental platform are provided in TABLE 8.1.

Quadrotor UAVs 2, 3 and 4 are assigned to be follower UAVs and the first one is chosen as the leader UAV (target). These follower UAVs are demanded to enclose the target by utilizing the proposed time-varying formation tracking method. For simplicity, the target enclosing is carried out in the horizontal XY plane; namely, $n = 2$. The dynamics of the follower UAVs are depicted as (8.1b), where one has that $\xi_i(t) = [x_{iX}(t), v_{iX}(t), x_{iY}(t), v_{iY}(t)]^T$ and $u_i(t) = [u_{iX}(t), u_{iY}(t)]^T$ $(i = 2, 3, 4)$. The dynamics of the leader UAV is depicted as (8.1a) with $\xi_1(t) = [x_{1X}(t), v_{1X}(t), x_{1Y}(t), v_{1Y}(t)]^T$. The yaw angle and the altitude of each UAV are controlled to be appropriate constants using separate PID controllers. The trajectory dynamics along x-axis and y-axis in the outer loops are governed by the time-varying formation tracking controller described by (8.3) with a frequency of 10 Hz. Using the PD controllers in [160] with a frequency of 500 Hz, the attitudes in the inner loops can track the control commands generated by the outer loops.

TABLE 8.1: Main components applied in the quadrotor UAV experimental platform.

Components	Part numbers	Key features
UWB	DW1000	Position resolution: ± 3 cm; Position conversion rate: 200Hz
Laser rangefinder	LIDAR Lite V3	Range: 40m; Accuracy: ± 2.5cm
Magnetometer	HMC5983	Range: 8Gs; Accuracy:± 2mGs
Gyroscope	ADXRS610	Range: $\pm 300°$/s; Accuracy: ± 6mV/$°$/s
Accelerometer	HQ7001	Range: ± 6g; Accuracy: 16-bit AD
Zigbee	DTK1605H	Maximum transmission: 115200bps; Communication range: 1600m
ESC	Flycker-30A	Rated current: 30A; Maximum control rate: 500Hz
Motor	Flycker KV750	Bearing length: 10mm; Outer diameter: 28mm
Propeller	APC 1147	Propeller pitch: 4.7-inch; Diameter: 11-inch
Battery	3S-5AH-30C	Voltage: 11.1V; Capacity: 5000mAh

The time-varying formation for the three follower UAVs is specified by

$$h_i(t) = \begin{bmatrix} 5\cos(0.2t + 2\pi(i-2)/3) \\ -\sin(0.2t + 2\pi(i-2)/3) \\ 5\sin(0.2t + 2\pi(i-2)/3) \\ \cos(0.2t + 2\pi(i-2)/3) \end{bmatrix} (i = 2,3,4).$$

Considering the limitation of the flying space, set the damping constants $\alpha_x = -3$ and $\alpha_v = -3$. It can be obtained that the eigenvalues of $B_1 B_2^T + B_2 \alpha$ are $-1.5+0.87j$ and $-1.5-0.87j$ ($j^2 = -1$), which means that the motion mode of the target is stable. Assume that there are three potential directed topologies G_1, G_2 and G_3 shown in Fig. 8.4. During $t \in [0s, 100s]$, the graph among the four quadrotor UAVs is selected from G_1, G_2 and G_3 with a frequency of τ_0.

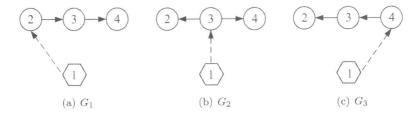

(a) G_1 (b) G_2 (c) G_3

FIGURE 8.4: Switching directed interaction topologies.

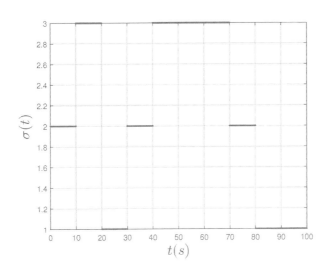

FIGURE 8.5: Switching signal of the graphs.

We can obtain that the time-varying formation tracking feasibility condition (8.19) is satisfied. Using Algorithm 8.1, the gain matrix K can be

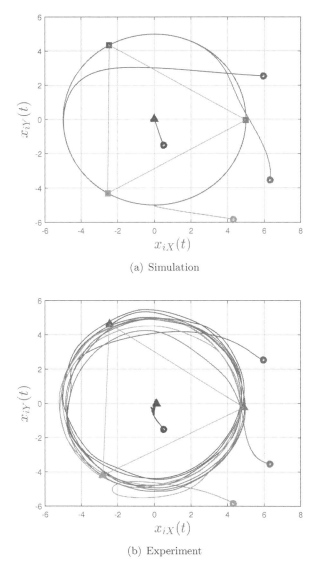

(a) Simulation

(b) Experiment

FIGURE 8.6: Position trajectories of the four quadrotor UAVs within $t = 100$s.

obtained as $K = I_2 \otimes [\ -1.8263 \quad -2.2778\ ]$ and the dwell time can be obtained as $\tau_0 = 10$s. Select the initial states of the four quadrotor UAVs as $x_1(t) = [0.49, 0.01, -1.5, -0.15]^T$, $x_2(t) = [6.31, 0, -3.53, -0.02]^T$,

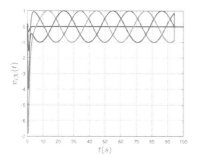
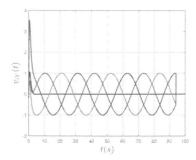

(a) Velocity trajectories along X-axis   (b) Velocity trajectories along Y-axis

FIGURE 8.7: Velocity trajectories of the four quadrotor UAVs within t=100s in the simulation.

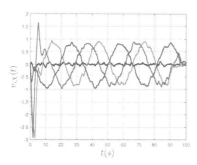
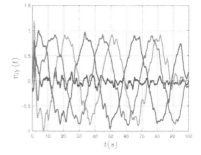

(a) Velocity trajectories along X-axis   (b) Velocity trajectories along Y-axis

FIGURE 8.8: Velocity trajectories of the four quadrotor UAVs within t=100s in the experiment.

$x_3(t) = [5.94, 0.01, 2.55, -0.06]^T$ and $x_4(t) = [4.32, -0.05, -5.83, 0.07]^T$. Fig. 8.5 depicts the switching signal within $t = 100s$.

Figs. 8.6-8.8 present the position and velocity trajectories of the four quadrotor UAVs in the simulation and experiment respectively, where the initial positions of the four quadrotor UAVs are described by circles, and the final positions of the followers are described by squares, and the one for the target is denoted by black triangle. Fig. 8.9 shows the curves of the time-varying formation tracking error within $t = 100s$. Fig. 8.10 depicts a captured image of four UAVs in the experiment. It follows from Figs. 8.5-8.10 that three follower quadrotors form the desired circular formation while enclosing the target subject to the switching directed graphs in both simulation and experiment, which means that the predefined time-varying formation tracking is realized by the practical quadrotot UAV swarm

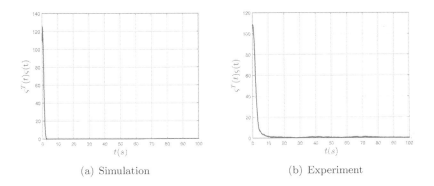

(a) Simulation   (b) Experiment

FIGURE 8.9: Curves of the formation tracking error within t=100s.

FIGURE 8.10: A captured image of the four quadrotor UAVs in target enclosing.

system using the UWB positioning technology. The experiment video is shown at http://v.youku.com/v_show/id_XMzA2ODYwMTI3Mg==.html or https://youtu.be/4xUuUrbPsoM.

8.3 Formation-containment Control for UAV Swarm Systems

Formation-containment control problems for quadrotor UAV systems with directed topologies are studied in this section, where the states of leaders form desired formation and the states of followers converge to the convex hull spanned by those of the leaders. Firstly, formation-containment control protocols are constructed based on the neighbouring information of UAVs. Then, sufficient conditions for UAV swarm systems to achieve formation-containment are presented. An explicit expression to describe the relationship among the states of followers, the time-varying formation for the leaders, and the formation reference is derived. Moreover, an approach to determine the gain matrices of the formation-containment protocol is proposed by solving an algebraic Riccati equation. Finally, a formation-containment platform with five quadrotor UAVs is introduced, and both simulation and experimental results are presented to demonstrate the effectiveness of the obtained results.

8.3.1 Problem Description

For a UAV swarm system with N quadrotor UAVs, the interaction topology of the UAV swarm system can be described by a directed graph G with node i representing the UAV i and edge ε_{ij} denoting the interaction between UAV i and j. The interaction strength from UAV i to UAV j can be described by the adjacency element w_{ji}. On the formation-containment control level, the dynamics of the UAV i, $i \in \{1, 2, \ldots, N\}$ can be modeled by the following double-integrator model:

$$\begin{cases} \dot{x}_i(t) = v_i(t), \\ \dot{v}_i(t) = u_i(t), \end{cases} \tag{8.46}$$

where $x_i(t) \in \mathbb{R}^n$ is the position vector, $v_i(t) \in \mathbb{R}^n$ is the velocity vector, and $u_i(t) \in \mathbb{R}^n$ is the control input vector. For simplicity of description, n is assumed to be 1, if not otherwise specified. However, all the results hereafter are still valid for higher dimensional space by using the Kronecker product.

Remark 8.3. *As discussed in Section 2.6, the dynamics of a quadrotor UAV with six degrees of freedom can be classified into trajectory dynamics and attitude dynamics. Due to the fact that the trajectory dynamics have much larger time constants than the attitude dynamics, the control of a quadrotor UAV in the horizontal plane (i.e., the XY plane) can be implemented with an inner-loop/outer-loop structure; that is, the outer-loop drives the UAV towards the desired position while the inner-loop tracks the attitude. In this section, the formation-containment problem for quadrotor UAV swarm systems only concerns the positions and velocities. Therefore, the dynamics of the leader and follower UAVs in the outer-loop can be approximately modeled by (8.46).*

Let $\phi_i(t) = [x_i(t), v_i(t)]^T$, $B_1 = [1, 0]^T$, and $B_2 = [0, 1]^T$. Then, UAV swarm system (8.46) can be rewritten as

$$\dot{\phi}_i(t) = B_1 B_2^T \phi_i(t) + B_2 u_i(t). \tag{8.47}$$

Definition 8.2. *UAVs in the system (8.47) are classified into leaders and followers. A UAV is called a leader if its neighbours are only leaders and it coordinates with its neighbours to achieve a desired formation. A UAV is called a follower if it has at least one neighbour in the UAV swarm system and it coordinates with its neighbours to achieve the containment.*

Assume that in UAV swarm system (8.47) there are M ($M < N$) followers with states $\phi_i(t)$ ($i = 1, 2, \ldots, M$) and $N - M$ leaders with states $\phi_j(t)$ ($j = M + 1, M + 2, \ldots, N$). Denote by $F = \{1, 2, \ldots, M\}$ and $E = \{M + 1, M + 2, \ldots, N\}$ the follower subscript set and leader subscript set, respectively. A time-varying formation for leaders is specified by a vector $h_E(t) = [h_{M+1}^T(t), h_{M+2}^T(t), \ldots, h_N^T(t)]^T \in \mathbb{R}^{2(N-M)}$ with $h_i(t) = [h_{ix}(t), h_{iv}(t)]^T$ ($i \in E$) piecewise continuously differentiable.

Definition 8.3. *Leaders in UAV swarm system (8.47) are said to achieve time-varying formation $h_E(t)$ if there exists a vector-valued function $r(t) \in \mathbb{R}^2$ such that*

$$\lim_{t \to \infty} (\phi_i(t) - h_i(t) - r(t)) = 0 \ (i \in E), \tag{8.48}$$

where $r(t)$ is called a formation reference function.

In Definition 8.3, $h_i(t)$ ($i \in E$) is used to characterize the desired time-varying formation configuration and $r(t)$ is a representation of the macroscopic movement of the whole formation. To explain the roles of $h_i(t)$ and $r(t)$ more clearly, consider the following illustration example.

Illustrative example 8.1. *Consider a UAV swarm system with four quadrotors in the XY plane. For simplicity, the states of the four quadrotors are required to achieve a time-invariant square formation with edge $\sqrt{2}l$. To specify the given formation shape, the vector $h_E = [h_1^T, h_2^T, h_3^T, h_4^T]^T$ can be chosen as $h_1 = [-l, 0]^T$, $h_2 = [0, l]^T$, $h_3 = [l, 0]^T$ and $h_4 = [0, -l]^T$. If Definition 8.3 is satisfied, it follows from (8.48) that $\lim_{t \to \infty}((\phi_i(t) - \phi_j(t)) - (h_i - h_j)) = 0$ ($i, j = 1, 2, 3, 4$), which means that the two squares formed by h_i and $\phi_i(t)$ are congruent with each other. Therefore, the given square formation is achieved by the UAV swarm system. Fig. 8.11 depicts the geometric relationships of $\phi_i(t)$, h_i ($i = 1, 2, 3, 4$) and $r(t)$. From Fig. 8.11, one gets that the formation reference $r(t)$ can be used to describe the macroscopic movement of the whole formation, and h_i denotes the relative offset vector of $\phi_i(t)$ relative to $r(t)$.*

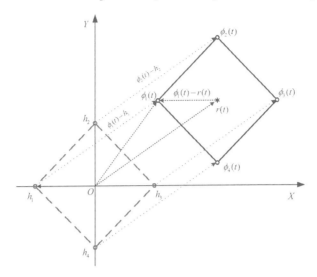

FIGURE 8.11: Illustration example for square formation with four UAVs.

Definition 8.4. *UAV swarm system (8.47) is said to achieve containment if for any $j \in F$, there exist non-negative constants α_{jk} ($k \in E$) satisfying $\sum_{k=M+1}^{N} \alpha_{jk} = 1$ such that*

$$\lim_{t \to \infty} \left(\phi_j(t) - \sum_{k=M+1}^{N} \alpha_{jk} \phi_k(t) \right) = 0. \qquad (8.49)$$

Definition 8.5. *UAV swarm system (8.47) is said to achieve formation-containment if there exist a vector-valued function $r(t) \in \mathbb{R}^2$ and non-negative constants α_{jk} ($k \in E$) satisfying $\sum_{k=M+1}^{N} \alpha_{jk} = 1$ such that for any $i \in E$ and $j \in F$, (8.48) and (8.49) hold simultaneously.*

Remark 8.4. *From Definitions 8.3, 8.4, and 8.5, one sees that for UAV swarm system (8.47), if $M = 0$, formation-containment problems become the formation problems. If $h(t) \equiv 0$ and leaders have no neighbours (that is, for all $i, j \in E$, $w_{ij} = 0$), the formation-containment problem becomes the containment problem. Therefore, for UAV swarm system (8.47), formation problems and containment problems can be regarded as special cases of the formation-containment problems discussed in this section.*

Consider the following formation-containment protocols for UAV swarm system (8.47)

$$u_i(t) = K_1 \phi_i(t) + K_2 \sum_{j \in N_i} w_{ij} \left(\phi_i(t) - \phi_j(t) \right), \ i \in F, \qquad (8.50)$$

$$\begin{aligned} u_i(t) = K_1 \phi_i(t) + K_3 \sum_{j \in N_i} w_{ij} \left((\phi_i(t) - h_i(t)) \right. \\ \left. -(\phi_j(t) - h_j(t)) \right), \ i \in E, \end{aligned} \qquad (8.51)$$

where $K_l = [k_{l1}, k_{l2}]$ $(l = 1, 2, 3)$ are constant gain matrices to be determined later.

Remark 8.5. *In protocols (8.50) and (8.51), the gain matrix K_1 can be chosen to expand the feasible formation set or specify the motion modes of the formation reference. K_2 and K_3 can be used to drive the states of followers to converge to the convex hull formed by those of leaders and propel the states of leaders to achieve the desired time-varying formation, respectively. It should be pointed out that by choosing $K_1 = 0$, protocols (8.50) and (8.51) become the ones using only relative information of neighbours.*

In this section, the following three problems for UAV swarm system (8.47) with protocols (8.51) and (8.52) are mainly studied: (i) under what conditions formation-containment can be achieved; (ii) how to design protocols (8.51) and (8.52) to achieve formation-containment; and (iii) how to demonstrate the theoretical results on practical quardrotor experiment platform.

8.3.2 Formation-containment Analysis and Protocol Design

In this subsection, firstly, sufficient conditions for UAV swarm system (8.47) under protocols (8.50) and (8.51) to achieve formation-containment are proposed. Then, an explicit expression to describe the relationship among the states of followers, the time-varying formation of the leaders, and the formation reference is derived. Finally, an approach to design the protocols (8.50) and (8.51) is presented.

Let G_E denote the interaction topology among leaders. Under Definition 8.2, the Laplacian matrix corresponding to G has the following form

$$L = \begin{bmatrix} L_1 & L_2 \\ 0 & L_3 \end{bmatrix},$$

where $L_1 \in \mathbb{R}^{M \times M}$, $L_2 \in \mathbb{R}^{M \times (N-M)}$, and $L_3 \in \mathbb{R}^{(N-M) \times (N-M)}$ which is the Laplacian matrix corresponding to G_E.

Assumption 8.2. *The interaction topology among leaders has a spanning tree.*

Assumption 8.3. *For each follower, there exists at least one leader that has a directed path to it.*

Under Assumption 8.3, the following lemma can be obtained.

Lemma 8.2 ([134]). *If the directed interaction topology G satisfies Assumption 8.3, then all the eigenvalues of L_1 have positive real parts, each entry of $-L_1^{-1}L_2$ is nonnegative, and each row of $-L_1^{-1}L_2$ has a sum equal to one.*

Let $\phi_F(t) = [\phi_1^T(t), \phi_2^T(t), \ldots, \phi_M^T(t)]^T$ and $\phi_E(t) = [\phi_{M+1}^T(t), \phi_{M+2}^T(t),$

$\ldots, \phi_N^T(t)]^T$. Under the protocols (8.50) and (8.51), UAV swarm system (8.47) can be written in a compact form as

$$
\begin{aligned}
\dot{\phi}_F(t) = & \left(I_M \otimes (B_1 B_2^T + B_2 K_1) + L_1 \otimes B_2 K_2\right) \phi_F(t) \\
& + (L_2 \otimes B_2 K_2) \phi_E(t),
\end{aligned}
\tag{8.52}
$$

$$
\begin{aligned}
\dot{\phi}_E(t) = & \left(I_{N-M} \otimes (B_1 B_2^T + B_2 K_1) + L_3 \otimes B_2 K_3\right) \phi_E(t) \\
& - (L_3 \otimes B_2 K_3) h_E(t).
\end{aligned}
\tag{8.53}
$$

Let λ_i $(i \in E)$ be the eigenvalue of the matrix L_3, where $\lambda_{M+1} = 0$ with the associated eigenvector $\bar{u}_{M+1} = \mathbf{1}$ and $0 < \mathrm{Re}(\lambda_{M+2}) \leq \cdots \leq \mathrm{Re}(\lambda_N)$. There exists a non-singular matrix $U_E = [\bar{u}_{M+1}, \bar{u}_{M+2}, \ldots, \bar{u}_N]$ with $U_E^{-1} = [\tilde{u}_{M+1}^H, \tilde{u}_{M+2}^H, \ldots, \tilde{u}_N^H]^H$ such that $U_E^{-1} L_3 U_E = J_E$, where J_E is the Jordan canonical form of L_3. By the structure of U_E, J_E has the form of $J_E = \mathrm{diag}\{0, \bar{J}_E\}$, where $\bar{J}_E$ consists of Jordan blocks corresponding to λ_i $(i = M+2, M+3, \ldots, N)$. Let $\tilde{U}_E = [\tilde{u}_{M+2}^H, \tilde{u}_{M+3}^H, \ldots, \tilde{u}_N^H]^H$. Define $U_F \in \mathbb{C}^{M \times M}$ as a non-singular matrix such that $U_F^{-1} L_1 U_F = \Lambda_F$, where Λ_F is an upper-triangular matrix with diagonal entries λ_i $(i = 1, 2, \ldots, M)$ and $\mathrm{Re}(\lambda_1) \leq \mathrm{Re}(\lambda_2) \leq \cdots \leq \mathrm{Re}(\lambda_M)$.

Lemma 8.3 ([125]). *The system $\dot{X} = MX$, where M is a 2×2 complex matrix with characteristic polynomial $f(s) = s^2 + a_1 s + a_2$, is asymptotically stable if and only if $\mathrm{Re}(a_1) > 0$ and $\mathrm{Re}(a_1)\mathrm{Re}(a_1 \bar{a}_2) - \mathrm{Im}(a_2)^2 > 0$.*

The following theorem gives sufficient conditions for UAV swarm system (8.47) under the proposed protocols to achieve formation-containment.

Theorem 8.2. *UAV swarm system (8.47) under protocols (8.50) and (8.51) achieves formation-containment if the following conditions hold simultaneously*
(i) For any $i \in E$ and $j \in N_i$, the desired formation satisfies the following formation feasibility constraint

$$
\lim_{t \to \infty} \left((B_1 B_2^T + B_2 K_1)(h_i(t) - h_j(t)) - (\dot{h}_i(t) - \dot{h}_j(t)) \right) = 0;
\tag{8.54}
$$

(ii) For any $i \in \{M+2, M+3, \ldots, N\}$, it holds that

$$
\begin{cases}
k_{12} + \mathrm{Re}(\lambda_i) k_{32} < 0, \\
\left(k_{12} + \mathrm{Re}(\lambda_i) k_{32}\right) \Psi_i + \mathrm{Im}(\lambda_i)^2 k_{31}^2 < 0,
\end{cases}
\tag{8.55}
$$

where $\Psi_i = k_{12} k_{11} - \mathrm{Re}(\lambda_i) \left(k_{12} k_{31} + k_{11} k_{32}\right) + \left(\mathrm{Re}(\lambda_i)^2 + \mathrm{Im}(\lambda_i)^2\right) k_{31} k_{32}$;
(iii) For any $i \in F$, one can obtain that

$$
\begin{cases}
k_{12} + \mathrm{Re}(\lambda_i) k_{22} < 0, \\
\left(k_{12} + \mathrm{Re}(\lambda_i) k_{22}\right) \Psi_i + \mathrm{Im}(\lambda_i)^2 k_{21}^2 < 0,
\end{cases}
\tag{8.56}
$$

where $\Psi_i = k_{12} k_{11} - \mathrm{Re}(\lambda_i) \left(k_{12} k_{21} + k_{11} k_{22}\right) + \left(\mathrm{Re}(\lambda_i)^2 + \mathrm{Im}(\lambda_i)^2\right) k_{21} k_{22}$.

Proof. Let $\bar{\phi}_i(t) = \phi_i(t) - h_i(t)$ $(i \in E)$ and $\bar{\phi}_E(t) = [\bar{\phi}_{M+1}^T(t), \bar{\phi}_{M+2}^T(t), \ldots,$ $\bar{\phi}_N^T(t)]^T$. Then, the system (8.53) can be rewritten as

$$\dot{\bar{\phi}}_E(t) = \left(I_{N-M} \otimes (B_1 B_2^{\ T} + B_2 K_1) + L_3 \otimes B_2 K_3\right) \bar{\phi}_E(t)$$
$$+ \left(I_{N-M} \otimes (B_1 B_2^{\ T} + B_2 K_1)\right) h_E(t) \qquad (8.57)$$
$$- (I_{N-M} \otimes I_2)\dot{h}_E(t).$$

Let $\theta_E(t) = (\tilde{u}_{M+1} \otimes I_2)\bar{\phi}_E(t)$ and $\zeta_E(t) = (\tilde{U}_E \otimes I_2)\bar{\phi}_E(t)$, then the system (8.57) can be transformed into

$$\dot{\theta}_E(t) = \left(B_1 B_2^{\ T} + B_2 K_1\right) \theta_E(t)$$
$$+ \left(\tilde{u}_{M+1} \otimes (B_1 B_2^{\ T} + B_2 K_1)\right) h_E(t) \qquad (8.58)$$
$$- (\tilde{u}_{M+1} \otimes I_2)\dot{h}_E(t),$$

$$\dot{\zeta}_E(t) = \left(I_{N-M-1} \otimes (B_1 B_2^{\ T} + B_2 K_1) + \bar{J}_E \otimes B_2 K_3\right) \zeta_E(t)$$
$$+ (\tilde{U}_E \otimes (B_1 B_2^{\ T} + B_2 K_1))h_E(t) - (\tilde{U}_E \otimes I_2)\dot{h}_E(t). \qquad (8.59)$$

If condition (i) holds, there is

$$\lim_{t \to \infty} \left((L_3 \otimes (B_1 B_2^{\ T} + B_2 K_1))h_E(t) - (L_3 \otimes I)\,\dot{h}_E(t)\right) = 0. \qquad (8.60)$$

Substituting $L_3 = U_E J_E U_E^{-1}$ into (8.60) and then pre-multiplying the both sides of (8.60) by $U_E^{-1} \otimes I_2$ yields

$$\lim_{t \to \infty} \left(\left(\bar{J}_E \tilde{U}_E \otimes (B_1 B_2^{\ T} + B_2 K_1)\right) h_E(t)\right.$$
$$\left. - \left(\bar{J}_E \tilde{U}_E \otimes I_2\right) \dot{h}_E(t)\right) = 0. \qquad (8.61)$$

Because G_E has a spanning tree, by Lemma 2.2 and the structure of J_E, $\bar{J}_E$ is nonsingular. Pre-multiplying the both sides of (8.61) by $\bar{J}_E^{-1} \otimes I_2$, one has

$$\lim_{t \to \infty} \left(\left(\tilde{U}_E \otimes (B_1 B_2^{\ T} + B_2 K_1)\right) h_E(t) - \left(\tilde{U}_E \otimes I_2\right) \dot{h}_E(t)\right) = 0. \qquad (8.62)$$

Consider the following $N - M - 1$ subsystems

$$\dot{\bar{\zeta}}_i(t) = \left(B_1 B_2^{\ T} + B_2 K_1 + \lambda_i B_2 K_3\right) \bar{\zeta}_i(t), \qquad (8.63)$$

where $i = M + 2, M + 3, \ldots, N$. The characteristic polynomial of subsystems (8.63) are $f_i(s) = s^2 - (k_{12} + \lambda_i k_{32})s - (k_{11} + \lambda_i k_{31})$. If condition (ii) holds, by Lemma 8.3, it can be obtained that the $N - M - 1$ subsystems described by (8.63) are asymptotically stable. Based on (8.63) and the structure of $\bar{J}_E$, it holds that the system described by

$$\dot{\tilde{\zeta}}_E(t) = \left(I_{N-M-1} \otimes (B_1 B_2^{\ T} + B_2 K_1) + \bar{J}_E \otimes B_2 K_3\right) \tilde{\zeta}_E(t), \qquad (8.64)$$

is asymptotically stable. From (8.59), (8.62) and (8.64), one gets that

$$\lim_{t \to \infty} \zeta_E(t) = 0. \tag{8.65}$$

Define

$$\bar{\phi}_{EC}(t) = \mathbf{1} \otimes \theta_E(t), \tag{8.66}$$

$$\bar{\phi}_{E\bar{C}}(t) = \bar{\phi}_E(t) - \bar{\phi}_{EC}(t). \tag{8.67}$$

Let $e_1 \in \mathbb{R}^{N-M}$ be a vector with 1 as its first component and 0 elsewhere. Because $[\theta_E^H(t), 0]^H = e_1 \otimes \theta_E(t)$ and $U_E e_1 = \mathbf{1}$, one has

$$\bar{\phi}_{EC}(t) = (U_E \otimes I)[\theta_E^H(t), 0]^H. \tag{8.68}$$

Note that $\bar{\phi}_E(t) = (U_E \otimes I)[\theta_E^H(t), \zeta_E^H(t)]^H$. From (8.66) and (8.67), it holds that

$$\bar{\phi}_{E\bar{C}}(t) = (U_E \otimes I)[0, \zeta_E^H(t)]^H. \tag{8.69}$$

Since $U_E \otimes I$ is nonsingular, from (8.68) and (8.69), one gets that $\bar{\phi}_{EC}(t)$ and $\bar{\phi}_{E\bar{C}}(t)$ are linearly independent. Therefore, it follows from (8.65), (8.67) and (8.69) that

$$\lim_{t \to \infty} \left(\bar{\phi}_E(t) - \mathbf{1} \otimes \theta_E(t) \right) = 0, \tag{8.70}$$

that is

$$\lim_{t \to \infty} \left(\phi_i(t) - h_i(t) - \theta_E(t) \right) = 0 \ (i \in E). \tag{8.71}$$

From (8.71), one gets that UAV swarm system (8.53) achieves the predefined time-varying formation specified by $h_E(t)$.

Let

$$\varphi_i(t) = \sum_{j \in N_i} w_{ij} \left(\phi_i(t) - \phi_j(t) \right) \ (i \in F),$$

and $\varphi_F(t) = [\varphi_1^T(t), \varphi_2^T(t), \ldots, \varphi_M^T(t)]^T$, then one gets

$$\varphi_F(t) = (L_1 \otimes I_2) \phi_F(t) + (L_2 \otimes I_2) \phi_E(t). \tag{8.72}$$

From (8.52), (8.53) and (8.72), it can be obtained that

$$\begin{aligned} \dot{\varphi}_F(t) = & \left(I_M \otimes (B_1 B_2^T + B_2 K_1) + (L_1 \otimes B_2 K_2) \right) \varphi_F(t) \\ & + \left((L_2 L_3) \otimes (B_2 K_3) \right) \left(\phi_E(t) - h_E(t) \right). \end{aligned} \tag{8.73}$$

Let $\bar{\varphi}_F(t) = (U_F^{-1} \otimes I) \varphi_F(t) = [\bar{\varphi}_1^H(t), \bar{\varphi}_2^H(t), \ldots, \bar{\varphi}_M^H(t)]^H$. Then system (8.73) can be transformed into

$$\begin{aligned} \dot{\bar{\varphi}}_F(t) = & \left(I_M \otimes (B_1 B_2^T + B_2 K_1) + (\Lambda_F \otimes B_2 K_2) \right) \bar{\varphi}_F(t) \\ & + \left((U_F^{-1} L_2 L_3) \otimes (B_2 K_3) \right) \left(\phi_E(t) - h_E(t) \right). \end{aligned} \tag{8.74}$$

When the leaders achieve time-varying formation $h_E(t)$, there is

$$\lim_{t \to \infty} (L_3 \otimes (B_2 K_3)) (\phi_E(t) - h_E(t))$$
$$= \lim_{t \to \infty} (L_3 \otimes (B_2 K_3)) (\mathbf{1} \otimes r(t)) \tag{8.75}$$

Since $L_3 \mathbf{1} = 0$, one has

$$\lim_{t \to \infty} \left((U_F^{-1} L_2 L_3) \otimes (B_2 K_3) \right) (\phi_E(t) - h_E(t)) = 0. \tag{8.76}$$

With a similar analysis as for system (8.64), it can be verified that if condition (iii) holds, then $I_M \otimes (B_1 B_2{}^T + B_2 K_1) + (\Lambda_F \otimes B_2 K_2)$ is Hurwitz. Therefore, $\lim_{t \to \infty} \bar{\varphi}_F(t) = 0$. Since U_F is nonsingular, one has

$$\lim_{t \to \infty} \varphi_F(t) = 0. \tag{8.77}$$

It follows from (8.72) and (8.77) that

$$\lim_{t \to \infty} \left(\phi_F(t) - \left(-L_1^{-1} L_2 \otimes I_2 \right) \phi_E(t) \right) = 0. \tag{8.78}$$

Based on Lemma 8.2, from (8.71) and (8.78), one gets that UAV swarm system (8.47) under protocols (8.50) and (8.51) achieves formation-containment. The proof of Theorem 8.2 is completed. □

Remark 8.6. *In the theoretical analysis, for simplicity of description, the dimension n is assumed to be 1. It should be pointed out that all the results in Theorem 8.2 can be extended to the higher dimensional space directly by using the Kronecker product. Theorem 8.2 presents sufficient conditions for UAV swarm system (8.47) with directed topologies under protocols (8.50) and (8.51) to achieve formation-containment. From (8.54), one gets that not all the formation can be achieved by UAV swarm system (8.47), and the desired formation should satisfy the formation feasibility constraint (8.54) determined by the dynamics of each UAV. Moreover, the application of K_1 can expand the feasible formation set.*

Remark 8.7. *It should be pointed out that the formation reference function $r(t)$ is a representation of the macroscopic movement of the whole formation-containment. From (8.48) and (8.71), one gets that $\lim_{t \to \infty}(\theta_E(t) - r(t)) = 0$, which means that the formation reference is determined by $\theta_E(t)$. By solving the differential equation (8.58), an explicit expression of the formation reference $r(t)$ can be obtained. Moreover, from (8.58), one gets that the motion modes of the formation reference can be specified by assigning the eigenvalues of $B_1 B_2{}^T + B_2 K_1$ at the desired locations in the complex plane using K_1.*

It is proved that the states of followers can converge to the convex hull formed by those of leaders in Theorem 8.2. Furthermore, the following theorem shows the explicit relationship among the states of followers, the time-varying formation of the leaders, and the formation reference.

Theorem 8.3. *If conditions (i), (ii) and (iii) in Theorem 8.2 are satis-fied, UAV swarm system (8.47) under protocols (8.50) and (8.51) achieves formation-containment, and the states of followers satisfy*

$$\lim_{t \to \infty} \left(\phi_i(t) - \sum_{j=M+1}^{N} l_{ij} h_j(t) - \theta_E(t) \right) = 0, \tag{8.79}$$

where $i \in F$ and l_{ij} is the entries of $-L_1^{-1}L_2$.

Proof. If conditions (i), (ii) and (iii) in Theorem 8.2 are satisfied, it holds that UAV swarm system (8.47) under protocols (8.50) and (8.51) achieves formation-containment. From the proof of Theorem 8.2, one gets that (8.70) and (8.78) hold. From (8.70) and (8.78), one has

$$\lim_{t \to \infty} \left(\phi_F(t) - \left(-L_1^{-1}L_2 \otimes I_2 \right) \left(h_E(t) + \mathbf{1} \otimes \theta_E(t) \right) \right) = 0. \tag{8.80}$$

It follows from Lemma 8.2 that

$$-L_1^{-1}L_2 \mathbf{1} = \mathbf{1}. \tag{8.81}$$

From (8.80) and (8.81), it can be obtained that

$$\lim_{t \to \infty} \left(\phi_F(t) - \left(-L_1^{-1}L_2 \otimes I_2 \right) h_E(t) - \mathbf{1} \otimes \theta_E(t) \right) = 0, \tag{8.82}$$

which means that (8.79) holds. This completes the proof of Theorem 8.3. □

   Theorem 8.3 shows that the states of followers are jointly determined by the topology, the time-varying formation $h_E(t)$ and the formation reference. From (8.79), one gets that the states of followers keep a time-varying formation specified by the convex combination of the formation $h_E(t)$ for the leaders with respect to $r(t)$.

   From Theorems 8.2 and 8.3, one gets that the state $\theta_E(t)$ of subsystem (8.58) determines the formation reference function $r(t)$ which represents the macroscopic movement of the whole formation-containment. Gain matrix K_1 can be chosen to satisfy the formation feasibility constraint (8.54) or specify the motion modes of the formation reference $r(t)$ by assigning the eigenvalues of $B_1 B_2^T + B_2 K_1$ at the desired locations in the complex plane. After K_1 is chosen, the following theorem provides an approach to determine the gain matrices K_2 and K_3.

Theorem 8.4. *If condition (i) in Theorem 8.2 holds, then UAV swarm system (8.47) achieves formation-containment by protocols (8.50) and (8.51) with $K_2 = -\alpha[\text{Re}(\lambda_1)]^{-1}R^{-1}B_2^T P$ and $K_3 = -\beta[\text{Re}(\lambda_{M+2})]^{-1}R^{-1}B_2^T P$, where $\alpha > 0.5$ and $\beta > 0.5$ are given constants, and P is the positive definite solution to the algebraic Riccati equation*

$$P(B_1 B_2^T + B_2 K_1) + (B_1 B_2^T + B_2 K_1)^T P - P B_2 R^{-1} B_2^T P + I = 0, \tag{8.83}$$

with $R^T = R > 0$.

Proof. Since $(B_1 B_2^T, B_2)$ is stabilizable and $(I, B_1 B_2^T)$ is observable, for any given $R^T = R > 0$, algebraic Riccati equation (8.83) has a unique solution $P^T = P > 0$. Firstly, consider the stabilities of the $N - M - 1$ subsystems described by (8.63). Construct the following Lyapunov function candidates

$$V_i(t) = \bar{\zeta}_i^H(t) P \bar{\zeta}_i(t) \ (i = M + 2, M + 3, \dots, N). \tag{8.84}$$

Taking the derivative of $V_i(t)$ along the trajectories of subsystems (8.63) gives

$$\begin{aligned} \dot{V}_i(t) = & \bar{\zeta}_i^H(t) \Big(\big(B_1 B_2{}^T + B_2 K_1\big)^T P + P \big(B_1 B_2{}^T + B_2 K_1\big) \Big) \bar{\zeta}_i(t) \\ & + \bar{\zeta}_i^H(t) \left(\lambda_i^H (B_2 K_3)^T P + \lambda_i P B_2 K_3 \right) \bar{\zeta}_i(t). \end{aligned} \tag{8.85}$$

Substituting $K_3 = -\beta [\text{Re}(\lambda_{M+2})]^{-1} R^{-1} B_2^T P$ and

$$P(B_1 B_2{}^T + B_2 K_1) + (B_1 B_2{}^T + B_2 K_1)^T P = P B_2 R^{-1} B_2^T P - I$$

into (8.85) one has

$$\dot{V}_i(t) = -\bar{\zeta}_i^H(t)\bar{\zeta}_i(t) + (1 - 2\beta[\text{Re}(\lambda_{M+2})]^{-1}\text{Re}(\lambda_i))\bar{\zeta}_i^H(t)(P B_2 R^{-1} B_2^T P)\bar{\zeta}_i(t). \tag{8.86}$$

Since $1 - 2\beta[\text{Re}(\lambda_{M+2})]^{-1}\text{Re}(\lambda_i) < 0$, it follows from (8.86) that $\lim_{t\to\infty} \bar{\zeta}_i(t) = 0$, which means that the $N - M - 1$ subsystems described by (8.63) are asymptotically stable. Since condition (i) is satisfied, one gets $\lim_{t\to\infty} \zeta_E(t) = 0$. With a similar analysis as for subsystems (18), it can be verified that $K_2 = -\alpha[\text{Re}(\lambda_1)]^{-1} R^{-1} B_2^T P$ can guarantee that $B_1 B_2{}^T + B_2 K_1 + \lambda_i B_2 K_2$ $(i \in F)$ is Hurwitz, which means that $I_M \otimes (B_1 B_2{}^T + B_2 K_1) + (\Lambda_F \otimes B_2 K_2)$ is Hurwitz. From the proof of Theorem 8.2, one can conclude that UAV swarm system (8.47) achieves formation-containment by protocols (8.50) and (8.51). This completes the proof of Theorem 8.4. □

From (8.83), one sees that to determine the gain matrices K_2 and K_3, only an algebraic Riccati equation needs to be solved. Moreover, the existence of K_2 and K_3 can be guaranteed. Therefore, the approach to design the formation-containment protocols (8.50) and (8.51) in this section is quite practical.

8.3.3 Simulation and Experimental Results

In this subsection, firstly a quadrotor formation-containment platform with five quadrotors is introduced. Then, both a numerical simulation and a practical experiment are carried out on the quadrotor formation-containment platform with five quadrotors in Example 8.1. Moreover, to demonstrate the scalability of the obtained results, a simulation result with six leaders and eight followers in the three dimensional space is given as Example 8.2. Due to the quantity limitation of the quadrotor UAVs in the lab, experimental results are not presented in Example 8.2.

FIGURE 8.12: Quadrotor formation-containment platform.

The quadrotor formation-containment platform consists of 5 quadrotors equipped with onboard flight control systems (FCS) and a ground control station (GCS), as shown in Fig. 8.12. The hardware of the quadrotor is developed based on the frame of the commercial X650V from Xaircraft with a tip-to-tip wingspan of 65cm, a battery life of 12 minutes and a maximum take-off weight of 1800g. The FCS is developed based on a TMS320F28335 DSP with a clock frequency of 135MHz. The attitude and acceleration of the quadrotor are estimated by three one-axis gyroscopes, a three-axis accelerometer and a three-axis magnetometer. A GPS module with an accuracy of 1.2m CEP is employed to obtain the position and velocity of the quadrotor with an output rate of 10Hz. The height of the quadrotor is measured by an ultrasonic range finder for the near the ground case or a barometer for the far away from the ground case. A 2GB micro SD card is mounted onboard for data logging. Zigbee modules are employed in each quadrotor and the GCS for data transmission. A RC receiver is kept in each quadrotor to deal with the emergency situation. The GCS is developed based on Labview 8.2.1 in Windows environment by which the flight parameters of all quadrotors can be monitored and control commands can be sent to quadrotors. Fig. 8.13 describes the hardware structure of the quadrotor formation-containment platform.

Example 8.1. Simulation and experiment with five quadrotors

The formation-containment control is implemented in the horizontal plane ($n = 2$); that is, movements of followers and leaders along X and Y axes are controlled by the formation-containment protocols (8.50) and (8.51) respectively with a rate of 5Hz. The height of each quadrotor is specified to be

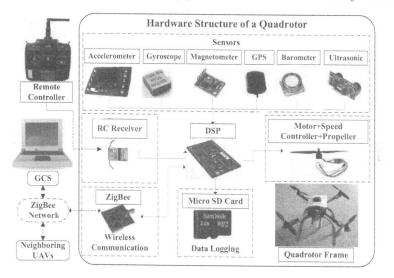

FIGURE 8.13: Hardware structure of the quadrotor formation-containment platform.

constant. The pitch, roll and yaw angles of each quadrotor are controlled by three decoupled PD controllers shown in [160] with a rate of 500Hz respectively as the inner loop. Using the Kronecker product, the dynamics of the quadrotor swarm system in two-dimensional space can be described by (8.47) with $\phi_i(t) = [x_{iX}(t), v_{iX}(t), x_{iY}(t), v_{iY}(t)]^T$, $u_i(t) = [u_{iX}(t), u_{iY}(t)]^T$ ($i = 1, 2, \ldots, N$), $B_1 = I_2 \otimes [1, 0]^T$, $B_2 = I_2 \otimes [0, 1]^T$ and $h_i(t) = [h_{ixX}(t), h_{ivX}(t), h_{ixY}(t), h_{ivY}(t)]^T$ ($i \in E$).

Quadrotors 1 and 2 are assigned as followers while quadrotors 3, 4 and 5 are specified to be leaders; that is, $M = 2$ and $N = 5$. The directed interaction topology of the quadrotor swarm system is shown in Fig. 8.14. For simplicity, the interaction topology is 0-1 weighted.

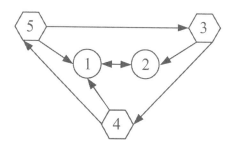

FIGURE 8.14: Directed interaction topology G_1.

The three leaders are required to achieve a formation specified by

$$h_3(t) = \begin{bmatrix} 10 \\ 0 \\ 0 \\ 0 \end{bmatrix}, \quad h_4(t) = \begin{bmatrix} 10\cos(2\pi/3) \\ 0 \\ 10\sin(2\pi/3) \\ 0 \end{bmatrix},$$

$$h_5(t) = \begin{bmatrix} 10\cos(4\pi/3) \\ 0 \\ 10\sin(4\pi/3) \\ 0 \end{bmatrix}.$$

If $h_i(t)$ $(i = 3, 4, 5)$ are achieved, then the positions of three leaders keep a regular triangle with edge $10\sqrt{3}$m while the velocities converge to a common value. Moreover, the positions of the two followers are required to converge to the triangle formation formed by those of the leaders, and the velocities of the followers reach an agreement with those of the leaders.

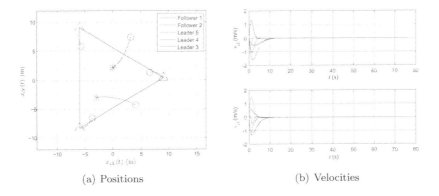

(a) Positions   (b) Velocities

FIGURE 8.15: Position and velocity trajectories for five quadrotors in simulation.

Due to the limitation of flight space and the requirement of performing the experiment within a visual range, the motion modes of the formation reference $r(t)$ are designed to be stable by choosing $K_1 = I_2 \otimes [0, -0.6]$ to specify the eigenvalues of $B_1 B_2^T + B_2 K_1$ at 0, 0, -0.6 and -0.6. In this configuration, when the desired formation-containment is achieved, the five quadrotors will be stationary. Let $K_2 = I_2 \otimes [-0.5, -0.8]$ and $K_3 = I_2 \otimes [-0.8, -0.8]$. It can be verified that the conditions in Theorem 8.2 are satisfied. Choose the initial states for all the quadrotors as $\phi_1(0) = [3.23, -0.04, 7.38, -0.01]^T$, $\phi_2(0) = [4.12, -0.01, -4.34, -0.19]^T$, $\phi_3(0) = [6.62, -0.26, 1.32, 0.13]^T$, $\phi_4(0) = [-5.66, 0.20, 5.86, 0.14]^T$ and $\phi_5(0) = [-3.66, 0.04, -6.57, 0.08]^T$.

Figs. 8.15 and 8.16 show the position and velocity trajectories of the five quadrotors in the simulation and experiment within 75s, where the initial states of the quadrotors are marked by "∘" and the final states of leaders and

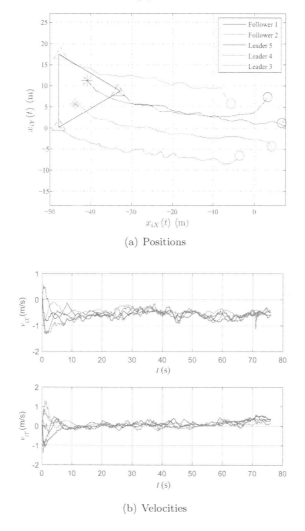

(a) Positions

(b) Velocities

FIGURE 8.16: Position and velocity trajectories for five quadrotors in experiment.

followers are denoted by "$\triangle$" and "$*$", respectively. Moreover, the convex hull formed by leaders is marked by solid lines. Fig. 8.17 shows an image of the formation-containment.

Let the formation error $e_E(t) = \phi_E(t) - h_E(t) - \mathbf{1} \otimes \theta_E(t)$ and the containment error $e_C(t) = \phi_F(t) - (-L_1^{-1}L_2 \otimes I_6)\phi_E(t)$. The curves of the formation error $e_E(t)$ for the leader quadrotors and the containment error $e_C(t)$ for the follower quadrotors in simulation and experiment are depicted in Figs. 8.18 and 8.19, respectively. From Figs. 8.15(a) and 8.16(a), one sees that the positions of leaders keep the desired regular triangle formation and the

FIGURE 8.17: Formation-containment image for five quadrotors in experiment.

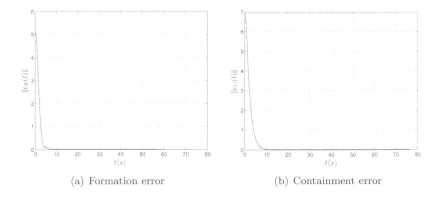

(a) Formation error

(b) Containment error

FIGURE 8.18: Curves of the formation error for the leader quadrotors and the containment error for the follower quadrotors in simulation.

positions of followers stay in the convex hull formed by those of the leaders in both the simulation and the experiment. Figs. 8.15(b) and 8.16(b) show that all the velocities of the quadrotors reach an agreement. Therefore, the predefined formation-containment for the five quadrotors are achieved in both the simulation and experiment. It should be pointed out that due to the existence of wind in the experiment, there is a drift in the position of the five quadrotors. However, the formation-containment is still realized under the disturbance of the wind. The video of the experiment can be found at http://v.youku.com/v_show/id_XNzA5ODc3NjA4.html.html. or https://www.youtube.com/watch?v=-amro6wFOxY.

Example 8.2. Simulation with fourteen quadrotors

Assume that there are eight followers and six leaders in the quadrotor

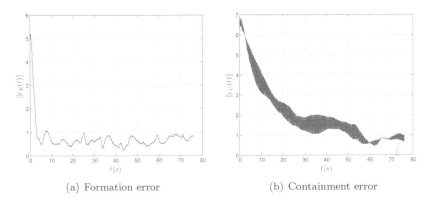

(a) Formation error   (b) Containment error

FIGURE 8.19: Curves of the formation error for the leader quadrotors and the containment error for the follower quadrotors in experiment.

swarm system with directed interaction topology G_2 as shown in Fig. 8.20. These agents move in the three dimensional space (i.e., the XYZ space). Using the Kronecker product, the dynamics of the quadrotor swarm system in three dimensional space can be described by (8.47) with $\phi_i(t) = [x_{iX}(t), v_{iX}(t), x_{iY}(t), v_{iY}(t), x_{iZ}(t), v_{iZ}(t)]^T$, $u_i(t) = [u_{iX}(t), u_{iY}(t), u_{iZ}(t)]^T$ $(i = 1, 2, \ldots, N)$, $B_1 = I_3 \otimes [1, 0]^T$, $B_2 = I_3 \otimes [0, 1]^T$ and $h_i(t) = [h_{ixX}(t), h_{ivX}(t), h_{ixY}(t), h_{ivY}(t), h_{ixZ}(t), h_{ivZ}(t)]^T$ $(i \in E)$. The leaders are required to achieve the following time-varying formation:

$$
h_i(t) = \begin{bmatrix}
50\cos\left(0.5t + (i-9)\pi/3\right) \\
-25\sin\left(0.5t + (i-9)\pi/3\right) \\
25\sqrt{2}\sin\left(0.5t + (i-9)\pi/3\right) \\
12.5\sqrt{2}\cos\left(0.5t + (i-9)\pi/3\right) \\
-25\sqrt{2}\sin\left(0.5t + (i-9)\pi/3\right) \\
-12.5\sqrt{2}\cos\left(0.5t + (i-9)\pi/3\right)
\end{bmatrix} \quad (i = 9, 10, \ldots, 14).
$$

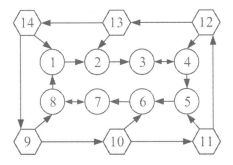

FIGURE 8.20: Directed interaction topology G_2.

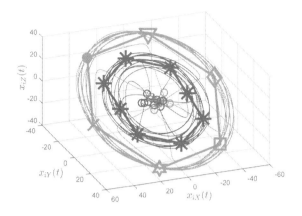

(a) Positions

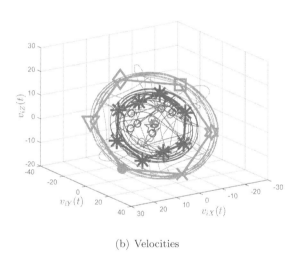

(b) Velocities

FIGURE 8.21: Position and velocity trajectories for fourteen quadrotors within $t = 80$s.

From $h_E(t)$, one gets that if the predefined formation is realized, both the positions and velocities of the six leaders will keep a regular hexagon formation while keep rotating around the formation reference with a velocity of 0.5rad/s. The motion modes of the formation reference can be designed to be oscillating by choosing $K_1 = I_3 \otimes [-0.25, 0]$ to specify the eigenvalues of $B_1 B_2{}^T + B_2 K_1$ at $0.5j$, $0.5j$, $0.5j$, $-0.5j$, $-0.5j$ and $-0.5j$ ($j^2 = -1$). In this case, the whole formation-containment will move periodically. It can be verified that condition (i) in Theorem 8.2 is satisfied. Choose $\alpha = 0.6$ and $\beta = 0.55$. Using the approach in Theorem 8.4, K_2 and K_3 can be obtained as

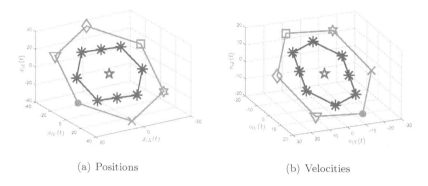

(a) Positions   (b) Velocities

FIGURE 8.22: Snapshots of position and velocity trajectories at $t = 70$s.

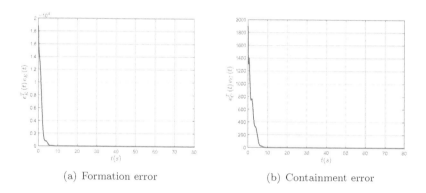

(a) Formation error   (b) Containment error

FIGURE 8.23: Curves of the formation error for the leaders and the containment error for the followers.

$K_2 = I_3 \otimes [-0.6435, -1.3191]$ and $K_3 = I_3 \otimes [-0.8589, -1.7605]$. For simplicity, the initial position components and velocity components of each UAV are given by $30(\Theta - 0.5)$, where Θ is a pseudorandom value distributed in (0,1).

Fig. 8.21 shows the numerical simulation results of the fourteen quadrotors within $t = 80$s, where the initial states are marked by "○", the final states of followers are represented by "*", the final states of the leaders are denoted by "●", "△", "◇", "□", "x", and hexagram respectively, and the final states of the formation reference are marked by the pentagram. Fig. 8.22 shows the snapshots of position and velocity trajectories at $t = 70$s. Fig. 8.23 gives the curves of the formation error $e_E(t)$ and the containment error $e_C(t)$. From Figs. 8.21–8.23, the following phenomena can be found: 1) both the position and velocity components of leaders form the regular hexagon formation; 2) the regular hexagon formation keep rotating around the oscillating formation reference; and 3) the states of followers not only converge to the convex hull spanned by those of the leaders but also form hexagon formation. Therefore, the desired formation-containment is achieved by the UAV swarm systems in the three dimensional space.

8.4 Time-varying Formation Tracking for UAV-UGV Heterogeneous Swarm Systems

In this section, time-varying formation tracking problems for UAV-UGV heterogeneous swarm systems are studied. Based on the distributed observer for the virtual leader, a formation tracking controller is proposed and an algorithm to design the control parameters is given. Then, the proposed control approach is applied to air-ground cooperative reconnaissance scene, where both simulation and experiment results are presented to verify the effectiveness of the theoretical results.

8.4.1 Problem Description

Consider a UAV-UGV heterogeneous swarm system composed of N vehicles, where F_A and F_G denote the UAV set and the UGV set, respectively. As discussed in Section 2.6, the dynamics models of UAVs and UGVs can be described by

$$\begin{cases} \dot{x}_i = A_i x_i + B_i u_i, \\ y_i = C_i x_i, \end{cases} \tag{8.87}$$

where $x_i \in \mathbb{R}^{n_i}$, $u_i \in \mathbb{R}^{m_i}$, and $y_i \in \mathbb{R}^p$ $(i = 1, 2, \cdots, N)$ denote the state, control input, and output vectors, respectively. Since the height channel of a quadrotor UAV can be controlled separately, formation tracking problems for UAVs and UGVs in the horizontal plane (i.e., the XY plane) are considered in this section. Specifically, for the i-th UAV $(i \in F_A)$, it can be verified that $x_i = [p_{Xi}, v_{Xi}, p_{Yi}, v_{Yi}]^T$, $u_i = [u_{Xi}, u_{Yi}]^T$, $y_i = [p_{Xi}, p_{Yi}]^T$, $A_i = I_2 \otimes \begin{bmatrix} 0 & 1 \\ 0 & 0 \end{bmatrix}$, $B_i = I_2 \otimes \begin{bmatrix} 0 \\ 1 \end{bmatrix}$, and $C_i = I_2 \otimes \begin{bmatrix} 1 & 0 \end{bmatrix}$. For the i-th UGV $(i \in F_G)$, we can obtain that $x_i = [p_{Xi}, p_{Yi}]^T$, $u_i = [u_{Xi}, u_{Yi}]^T$, $y_i = [p_{Xi}, p_{Yi}]^T$, $A_i = I_2 \otimes [0]$, $B_i = I_2$, and $C_i = I_2$.

A virtual leader labelled by 0 is introduced to generate the macroscopic reference trajectory of the heterogeneous swarm system (8.87). The dynamics model of the virtual leader is described by

$$\begin{cases} \dot{q}_0 = A_0 q_0, \\ y_0 = C_0 q_0, \end{cases} \tag{8.88}$$

where $q_0 \in \mathbb{R}^4$ and $y_0 \in \mathbb{R}^2$ represent the state and output of the virtual leader, respectively. In detail, for UAV and UGV swarm system in this section, $y_0 = [p_{X0}, p_{Y0}]^T$ denotes the position of the virtual leader in the XY plane.

Let $h = [h_1^T, h_2^T, \ldots, h_N^T]^T$, where $h_i \in \mathbb{R}^4$ denotes the desired offset vector relative to q_0 for follower i ($i \in \{1, 2, \ldots, N\}$). Define $h_{yi} = C_0 h_i$ as the desired time-varying output formation vector, and it is required that h_{yi} is piecewise continuously differentiable.

Definition 8.6. *For the heterogeneous swarm system (8.87) with the virtual leader (8.88), for any bounded initial system, if*

$$\lim_{t \to \infty} (y_i - h_{yi} - y_0) = 0, \tag{8.89}$$

then heterogeneous UAV and UGV swarm system is said to achieve the desired time-varying output formation tracking.

Let the graph G denote the interaction topology among the heterogeneous swarm system (8.87) and the virtual leader (8.88). The virtual leader only sends out information, and the UAVs and UGVs are regarded as followers.

Assumption 8.4. *The graph G has a spanning tree rooted by the virtual leader.*

Under Assumption 8.4, the Laplacian matrix L can be divided as $L = \begin{bmatrix} 0 & 0 \\ L_2 & L_1 \end{bmatrix}$, where $L_2 \in \mathbb{R}^{N \times 1}$ and $L_1 \in \mathbb{R}^{N \times N}$. It follows from Lemma 2.2 that all eigenvalues of L_1 have positive real parts.

Assumption 8.5. *The regulator equations*

$$\begin{cases} X_i A_0 = A_i X_i + B_i U_i \\ 0 = C_i X_i - C_0 \end{cases} \tag{8.90}$$

have solution pairs (X_i, U_i), $i = 1, 2, \ldots, N$.

8.4.2 Formation Tracking Controller Design

For the i-th follower UAV or UGV, consider the following time-varying formation tracking controller:

$$\begin{cases} \dot{\hat{q}}_i = A_0 \hat{q}_i - \eta \left[w_{i0} (\hat{q}_i - q_0) + \sum_{j=1}^{M+N} w_{ij} (\hat{q}_i - \hat{q}_j) \right], \\ u_i = K_{1i} x_i + K_{2i} (\hat{q}_i + h_i) + r_i, \end{cases} \tag{8.91}$$

where η is a positive constant to be determined, K_{1i} and K_{2i} are constant gain matrices to be designed, $\hat{q}_i$ is the distributed estimation state on q_0 of the i-th follower, and r_i is the time-varying formation tracking compensation input to expand the feasible formation set.

The following algorithm is given to design the formation controller (8.91).

Algorithm 8.2. *The time-varying formation tracking controller (8.91) can be determined by the following four steps.*

Step 1. *Choose X_i and U_i ($i = 1, 2, \ldots, N$) such that the regulator equations (8.90) hold.*

Step 2. *For each UAV or UGV, check the following formation tracking feasibility condition:*

$$X_i \left(A_0 h_i - \dot{h}_i \right) + B_i r_i = 0. \tag{8.92}$$

If there exist r_i ($i = 1, 2, \ldots, N$) such that the feasible condition (8.92) holds for each follower, then continue. Otherwise the given formation h_i is not feasible under the controller (8.91) and the algorithm stops.

Step 3. *Select K_{1i} to make $A_i + B_i K_{1i}$ Hurwitz and let $K_{2i} = U_i - K_{1i} X_i$.*

Step 4. *Choose sufficiently large η such that $\eta > \frac{R_{\max}(A_0)}{R_{\min}(L_1)}$, where $R_{\max}(A_0)$ denotes the maximum of real part of eigenvalues of A_0 and $R_{\min}(L_1)$ represents the minimum of real part of eigenvalues of L_1.*

Similar to Theorems 5.1 and 5.2, we can prove that the desired formation tracking can be achieved by heterogeneous swarm system under the proposed controller (8.91). Thus, the detailed proof of the following theorem is omitted.

Theorem 8.5. *Suppose that Assumptions 8.4 and 8.5 hold. If the formation tracking feasibility condition (8.92) is satisfied, then heterogeneous swarm system (8.87) and (8.88) can realize the expected time-varying output formation tracking under the distributed controller (8.91) determined by Algorithm 8.2.*

8.4.3 Simulation and Experimental Results

In this subsection, a heterogeneous swarm system composed of one UGV and two UAVs is considered. The UAVs are labelled by 1 and 2, and the UGV is denoted by 3, i.e., $F_A = \{1, 2\}$ and $F_G = \{3\}$. The directed graph G is shown in Fig. 8.24. For UAVs 1 and 2, the system matrices in (8.87) are set as $A_i = I_2 \otimes \begin{bmatrix} 0 & 1 \\ 0 & 0 \end{bmatrix}$, $B_i = I_2 \otimes \begin{bmatrix} 0 \\ 1 \end{bmatrix}$, and $C_i = I_2 \otimes \begin{bmatrix} 1 & 0 \end{bmatrix}$ ($i \in F_A$). For UGV 3, we can get $A_i = I_2 \otimes [0]$, $B_i = I_2$, and $C_i = I_2$ ($i \in F_G$). Moreover, the model (8.88) of the virtual leader is chosen as $A_0 = I_2 \otimes \begin{bmatrix} 0 & 1 \\ 0 & 0 \end{bmatrix}$ and $C_0 = I_2 \otimes \begin{bmatrix} 1 & 0 \end{bmatrix}$ in this example.

Firstly, simulation results on formation tracking control for the UAV-UGV swarm system are given. Air-ground cooperative reconnaissance scene is considered, where the UGV is required to track the virtual leader precisely and the UAVs need to rotate around the virtual leader. To specify this relative

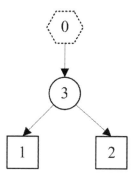

FIGURE 8.24: Directed graph.

movement relationship, the desired offset vectors h_i can be chosen as

$$h_1 = [r_1 \cos(\omega t), -r_1\omega \sin(\omega t), r_1 \sin(\omega t), -r_1\omega \cos(\omega t)]^T,$$
$$h_2 = [r_2 \cos(\omega t + \pi), -r_2\omega \sin(\omega t + \pi), r_2 \sin(\omega t + \pi), -r_2\omega \cos(\omega t + \pi)]^T,$$
$$h_3 = [0, 0, 0, 0]^T,$$

where r_i ($i \in F_A$) denotes the radius of rotation and ω represents the rotation speed. In this simulation, choose $r_1 = 1.5$m, $r_2 = 3$m, and $\omega = 0.5$rad/s. Besides, since the height channel of a quadrotor can be controlled separately, UAV-1 and UAV-2 are required to fly at a constant height 5m.

Based on Algorithm 8.2, design the control parameters in the formation tracking controller (8.91). Choose the following matrices X_i and U_i such that the regulator equations (8.90) hold:

$$X_1 = I_4, \ U_1 = 0_{2\times 4},$$
$$X_2 = I_4, \ U_2 = 0_{2\times 4},$$
$$X_3 = I_2 \otimes [\ 1 \quad 0\ ], \ U_3 = I_2 \otimes [\ 0 \quad 1\ ].$$

For the given vectors h_i, we can verify that the formation feasibility condition (8.92) holds for all followers, and the formation tracking compensation inputs can be designed as

$$v_1(t) = [-r_1\omega^2 \cos(\omega t), -r_1\omega^2 \sin(\omega t)]^T,$$
$$v_2(t) = [-r_2\omega^2 \cos(\omega t + \pi), -r_2\omega^2 \sin(\omega t + \pi)]^T,$$
$$v_3(t) = 0.$$

Moreover, the gain matrices K_{1i} and K_{2i} are set as $K_{11} = I_2 \otimes [\ -2 \quad -2\ ]$, $K_{21} = I_2 \otimes [\ 2 \quad 2\ ]$, $K_{12} = I_2 \otimes [\ -2 \quad -2\ ]$, $K_{22} = I_2 \otimes [\ 2 \quad 2\ ]$, $K_{13} = -0.5I_2$, and $K_{23} = I_2 \otimes [\ 0.5 \quad 1\ ]$. Let $\eta = 1$. The initial position and velocity of UAVs and UGV are generated by random numbers. The initial position and

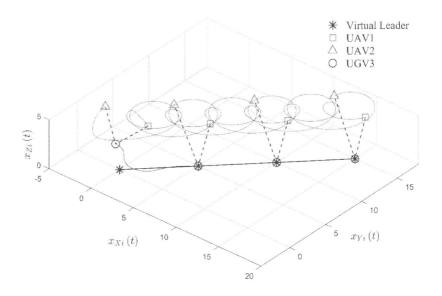

FIGURE 8.25: Position trajectories of heterogeneous swarm system in the XYZ space.

velocity of the virtual leader are chosen as $(0m, 0m)$ and $(0.2m/s, 0.2m/s)$, respectively.

Fig. 8.25 shows the position trajectories of heterogeneous swarm system in the XYZ space, where the virtual leader is denoted by asterisk, and UAV-1, UAV-2, and UGV-3 are represented by square, triangle, and circle, respectively. Fig. 8.26 gives the projection trajectories of the swarm system in the XY plane. Define the time-varying output formation tracking error as $e_i = y_i - h_{yi} - y_0$ $(i \in \{1, 2, 3\})$. Fig. 8.27 depicts the curves of $\|e_i\|$. From Figs. 8.25-8.27, we can see that the heterogeneous UAV and UGV swarm system can achieve the desired time-varying formation tracking under the proposed controller.

In the following, experiment results on formation tracking control for UAV-UGV heterogeneous swarm system will be given. As shown in Fig. 8.28, the experiment platform is composed of two UAVs, one UGV, a ground control station (GCS), an indoor UWB positioning system, and a WIFI router. The GCS is only applied to monitor running status and collect experimental data of vehicles and does not directly participate in formation operation. Four UWB base stations are installed around the flight arena and one UWB label is

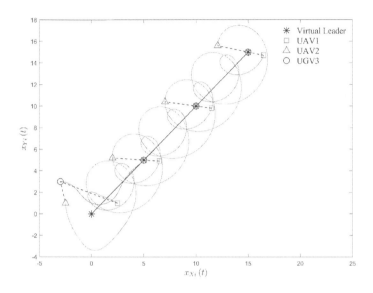

FIGURE 8.26: Projection position trajectories of heterogeneous swarm system in the XY plane.

attached on each UAV or UGV. Using this system, the position information in horizontal plane can be obtained with an accuracy of 10cm at a frequency of 50Hz. Each UAV and UGV can run the proposed formation tracking controller (5) by using the onboard processor and obtain the distributed estimator information of neighbouring robots through WIFI communication. The system structures of UAV and UGV are shown in Figs. 8.29 and 8.30, respectively.

A collaborative reconnaissance task using one UGV and two UAVs is carried out in the experiment by using the proposed formation tracking approach. When the reconnaissance task starts, two UAVs will take off from the UGV and fly at constant heights. Then, UAV and UGV heterogeneous system is required to achieve the desired time-varying formation tracking, where two UAVs will rotate around the UGV with different radii. Meanwhile, the camera in each UAV will look down to search the target.

The desired time-varying formation in the experiment has the same structure as the simulation example, i.e., the UGV is required to track the virtual leader precisely and the UAVs need to rotate around the virtual leader. For UAV-1 and UAV-2, choose $r_1 = 1.7$m, $r_2 = 2.1$m, and $\omega = 0.1$ rad/s. Using Algorithm 8.2 to design the formation controller (8.91), choose

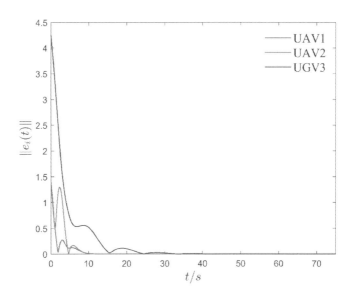

FIGURE 8.27: Time-varying formation tracking errors of heterogeneous swarm system.

$K_{11} = I_2 \otimes \begin{bmatrix} -1 & -1 \end{bmatrix}$, $K_{21} = I_2 \otimes \begin{bmatrix} 1 & 1 \end{bmatrix}$, $K_{12} = I_2 \otimes \begin{bmatrix} -1 & -1 \end{bmatrix}$, $K_{22} = I_2 \otimes \begin{bmatrix} 1 & 1 \end{bmatrix}$, $K_{13} = -0.1I_2$, and $K_{23} = I_2 \otimes \begin{bmatrix} 0.1 & 1 \end{bmatrix}$. The initial velocities of UGV and UAVs are set to zero. The virtual leader is required to do uniform linear motion with the initial position $(5.5m, 4.0m)$ and the initial velocity $(0m/s, 0.05m/s)$ in the XY plane. The other parameters are selected to be the same as the simulation example.

Fig. 8.31 shows the position trajectories within 150s in the XY plane of UGV and UAVs in the experiment, where the UAV-1 and UAV-2 are denoted by square and triangle respectively, and the UGV-3 is represented by circle. The time-varying formation tracking errors $\|e_i\|$ $(i = 1, 2, 3)$ of the UAVs and UGV are given in Fig. 8.32. Moreover, Fig. 8.33 gives a captured image of the UAV-UGV swarm system during the experiment. From Figs. 8.31-8.33, we can conclude that the desired time-varying formation tracking is realized by the heterogeneous UAV-UGV swarm system under the proposed controller in the experiment. Due to the positioning error of UWB system, the control error of vehicles, and so on, the small formation tracking errors in Fig. 8.32 are reasonable and acceptable for practical applications. The experiment video is shown at youtube.com/watch?v=zrMvz8sgHOI or v.youku.com/v_show/id_XNTgwMTU1NDQyNA.

FIGURE 8.28: UAV and UGV formation experiment platform.

8.5 Conclusions

The formation tracking approaches proposed in the previous chapters were applied to the practical cooperative experiment platforms composed of UAVs and UGVs in this chapter. How to modify the general formation controllers to meet the characteristics of UAV and UGV swarm system was given, and the formation controller design and stability analysis were provided. The system composition, hardware structure, and software framework of the experimental platform were introduced. Several formation tracking experiments were carried out for UAV and UGV swarm system to further verify the effectiveness of the theoretical results. The contents in this chapter are mainly based on [50] and [64].

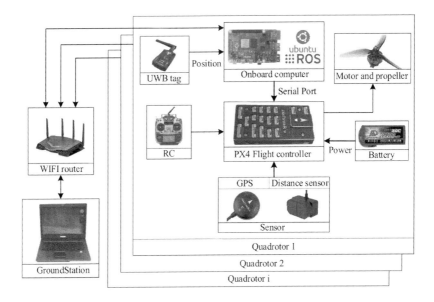

FIGURE 8.29: System structure of the quadrotor UAV.

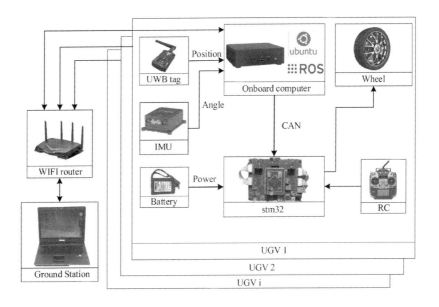

FIGURE 8.30: System structure of the UGV.

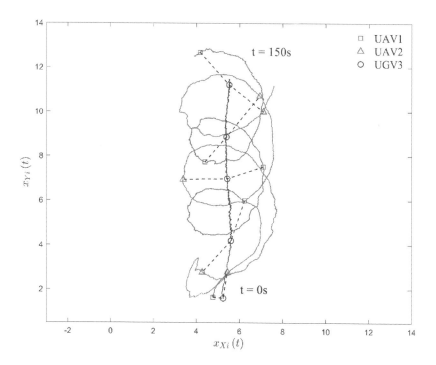

FIGURE 8.31: Position trajectories in the XY plane of UGV and UAVs in the experiment.

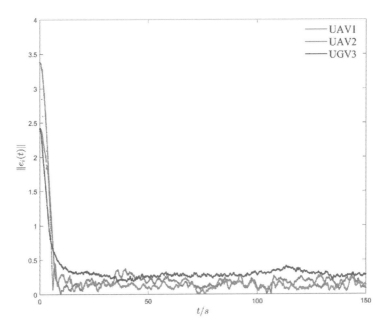

FIGURE 8.32: Time-varying formation tracking errors $\|e_i\|$ in the experiment.

FIGURE 8.33: A captured image of the UAV-UGV swarm system in the experiment.

Chapter 9

Conclusions and Future Prospects

9.1 Conclusions

In cross-domain collaborative applications, including air-ground coordination and air-sea coordination, how to achieve time-varying formation tracking for heterogeneous swarm systems is a crucial technical problem, which has important theoretical value and practical significance. This book investigates the time-varying formation tracking control problems for high-order heterogeneous swarm systems, where several specific cooperative forms, including leaderless formation control, formation tracking control with a single leader, formation tracking control with multiple leaders, and formation-containment tracking control, are considered. The main results of this book are summarized as follow.

(1) This book starts from time-varying formation tracking for homogeneous swarm systems, where basic concepts, formation analysis, and controller design are given. Time-varying formation problems for homogeneous swarm systems with switching directed interaction topologies are investigated firstly. Necessary and sufficient conditions to achieve time-varying formations are proposed, where a description of the feasible time-varying formation set and approaches to expand the feasible formation set are given. An algorithm to design the formation control protocol for homogeneous swarm systems is presented. Then, time-varying formation tracking problems for homogeneous linear swarm systems with multiple leaders are studied based on the well-informed follower assumption. A formation tracking protocol is constructed using only neighbouring relative information. Necessary and sufficient conditions for swarm systems with multiple leaders to achieve time-varying formation tracking are proposed by utilizing the properties of the Laplacian matrix. An approach to design the formation tracking protocol is presented by solving an algebraic Riccati equation. The proposed approach can be applied to solve the formation tracking problems with a single leader directly.

(2) Time-varying formation tracking control problems for weak heterogeneous swarm systems with matched/mismatched disturbances are studied respectively. For the case with matched disturbances, a robust adaptive time-varying formation tracking protocol and an algorithm to design the parameters in a distributed manner are proposed. Then, formation tracking feasible

DOI: 10.1201/9781003263470-9

conditions, an approach to expand the feasible formation set, and sufficient conditions for swarm systems to achieve the desired formation tracking are given. For the case with mismatched disturbances, based on the finite-time disturbance observer, the integral sliding mode control, and the super-twisting algorithm, a continuous time-varying formation tracking protocol using the neighbouring interaction is presented, and the finite-time convergence of the output formation tracking errors of high-order swarm systems is proved.

(3) For high-order heterogeneous swarm systems with a non-autonomous leader of unknown input, time-varying formation tracking control problems are discussed. Based on the output regulation control and the sliding mode control, a hierarchical formation tracking control strategy composed of the distributed observer and the local tracking controller is provided. Using the neighbouring interaction, a distributed time-varying output formation tracking protocol with the adaptive compensation capability for the unknown input of the leader is proposed. Considering the features of heterogeneous dynamics, the time-varying formation tracking feasible constraints are provided, and a compensation input is applied to expand the feasible formation set. It is proved that the outputs of the heterogeneous swarm systems can achieve the desired formation tracking. As a result, the limitations of the existing results which require that the swarm system is homogeneous and the leader has no control input are overcome.

(4) For high-order heterogeneous swarm systems with multiple leaders, time-varying formation tracking control problems with directed switching topologies and multiple leaders' incomplete information are investigated respectively. For the case with directed switching topologies, based on the well-informed follower assumption, a distributed time-varying output formation tracking protocol is designed. Sufficient conditions to achieve formation tracking with multiple leaders are given by using the piecewise Lyapunov stability theory. Furthermore, the well-informed follower assumption is removed, and the formation tracking problems with incomplete information of multiple leaders are discussed. A distributed observer is designed for each follower to estimate the dynamical matrices and the states of multiple leaders, and an adaptive algorithm is proposed to solve the regulator equations in finite time. Then, a fully distributed time-varying output formation tracking protocol and a design algorithm are proposed. It is proved the desired formation tracking with multiple leaders can be achieved by heterogeneous swarm systems without requiring the well-informed follower assumption.

(5) For high-order heterogeneous swarm systems with different intra-layer cooperative control objectives and inter-layer coordination couplings, the definition and the framework of formation-containment tracking control are presented. A tracking-leader with time-varying input is applied to generate the macroscopic reference trajectory for the whole swarm systems. So the shortcoming of the existing formation-containment control approaches which cannot control effectively the macroscopic movement of the entire system is overcome. Considering the influences of switching topologies, based on the robust

adaptive estimation approach and the predefined containment control strategy, a distributed formation-containment tracking protocol and a multi-step design algorithm are proposed. With inter-layer coordination couplings, sufficient conditions for heterogeneous swarm systems on switching graphs to achieve formation-containment tracking are given.

(6) To overcome the barriers between complex cooperative theories and actual physical systems, time-varying formation tracking control approaches proposed in the previous chapters are applied to practical cooperative experiment platforms composed of UAVs and UGVs. The proposed general formation controllers are modified to meet the characteristics of UAV and UGV swarm system, and the formation controller design algorithms and stability analysis of closed-loop swarm systems are provided. Then, the system composition, hardware structure, and software framework of the UAV-UGV experimental platforms are introduced. Several formation tracking experiments are carried out for UAV and UGV swarm systems to further verify the effectiveness of the theoretical results.

9.2 Future Prospects

Formation tracking control for heterogeneous swarm systems is a hot topic full of continuous development, renewal, and vitality. Besides the problems discussed in this book, there exist several remaining open problems which require further attention. Some future prospects are given as follow.

(1) With the continuous breakthrough of machine learning algorithm and the rapid improvement of computing power, it is possible to deeply combine the formation control approaches with artificial intelligence technology. It is interesting to explore intelligent formation tracking control strategies for heterogeneous swarm systems based on reinforcement learning or deep reinforcement learning.

(2) Formation control should not be only limited to the bottom control level, and the integration of decision-making and control layers will enhance the intelligence of heterogeneous swarm systems. Combined with the distributed optimization approaches, it is worthy of further investigation on optimal formation control, where the online optimization design of formation configuration and macroscopic motion trajectory of the swarm system can be carried out by using neighbouring information interaction.

(3) There exist both intra-cluster cooperation and inter-cluster competition in swarm attack-defense confrontation scenes, where cluster time-varying formation tracking control under cooperation and competition mechanism has potential application values and is also a challenging problem to be solved. It is significant to further study the cluster formation tracking control problems for heterogeneous swarm systems based on N-cluster non-cooperative game

theory. Besides, how to use differential game strategy to design distributed formation controllers for swarm systems in the confrontation scenes is an interesting research topic.

Bibliography

[1] E. Bonabeau, M. Dorigo, and G. Theraulaz. *Swarm Intelligence: From Natural to Artificial Systems.* Oxford University Press, Oxford, 1999.

[2] US Department of Defense. Unmanned systems integrated roadmap 2017–2042. 2018.

[3] X.W. Dong. *Formation and Containment Control for High-order Linear Swarm Systems.* Springer, Berlin, 2015.

[4] K.K. Oh, M.C. Park, and H.S. Ahn. A survey of multi-agent formation control. *Automatica*, 53:424–440, 2015.

[5] L. Briñón-Arranz, L. Schenato, and A. Seuret. Distributed source seeking via a circular formation of agents under communication constraints. *IEEE Transactions on Control of Network Systems*, 3(2):104–115, 2015.

[6] R.H. Zheng, Y.H. Liu, and D. Sun. Enclosing a target by nonholonomic mobile robots with bearing-only measurements. *Automatica*, 53:400–407, 2015.

[7] N. Nigam, S. Bieniawski, I. Kroo, and J. Vian. Control of multiple UAVs for persistent surveillance: Algorithm and flight test results. *IEEE Transactions on Control Systems Technology*, 20(5):1236–1251, 2011.

[8] X.W. Dong, Y. Zhou, Z. Ren, and Y.S. Zhong. Time-varying formation tracking for second-order multi-agent systems subjected to switching topologies with application to quadrotor formation flying. *IEEE Transactions on Industrial Electronics*, 64(6):5014–5024, 2017.

[9] X.W. Dong and G.Q. Hu. Time-varying formation tracking for linear multiagent systems with multiple leaders. *IEEE Transactions on Automatic Control*, 62(7):3658–3664, 2017.

[10] Y.Z. Hua, X.W. Dong, L. Han, Q.D. Li, and Z. Ren. Formation-containment tracking for general linear multi-agent systems with a tracking-leader of unknown control input. *Systems & Control Letters*, 122:67–76, 2018.

[11] J. Chen, X. Zhang, B. Xin, and H. Fang. Coordination between unmanned aerial and ground vehicles: A taxonomy and optimization perspective. *IEEE Transactions on Cybernetics*, 46(4):959–972, 2015.

[12] M. Lindemuth, R. Murphy, E. Steimle, and et al. Sea robot-assisted inspection. *IEEE Robotics & Automation Magazine*, 18(2):96–107, 2011.

[13] T. Balch and R.C. Arkin. Behavior-based formation control for multirobot teams. *IEEE Transactions on Robotics and Automation*, 14(6):926–939, 1998.

[14] M.A. Lewis and K.H. Tan. High precision formation control of mobile robots using virtual structures. *Autonomous Robots*, 4(4):387–403, 1997.

[15] P.K.C. Wang. Navigation strategies for multiple autonomous mobile robots moving in formation. *Journal of Robotic Systems*, 8(2):177–195, 1991.

[16] R.W. Beard, J. Lawton, and F.Y. Hadaegh. A coordination architecture for spacecraft formation control. *IEEE Transactions on Control Systems Technology*, 9(6):777–790, 2001.

[17] J. Baillieul and P.J. Antsaklis. Control and communication challenges in networked real-time systems. *Proceedings of the IEEE*, 95(1):9–28, 2007.

[18] R. Olfati-Saber and R.M. Murray. Consensus problems in networks of agents with switching topology and time-delays. *IEEE Transactions on Automatic Control*, 49(9):1520–1533, 2004.

[19] W. Ren and R.W. Beard. Consensus seeking in multiagent systems under dynamically changing interaction topologies. *IEEE Transactions on Automatic Control*, 50(5):655–661, 2005.

[20] Z.K. Li, Z.S. Duan, G.R. Chen, and L. Huang. Consensus of multiagent systems and synchronization of complex networks: A unified viewpoint. *IEEE Transactions on Circuits and Systems I: Regular Papers*, 57(1):213–224, 2010.

[21] W. Ren. Consensus strategies for cooperative control of vehicle formations. *IET Control Theory & Applications*, 1(2):505–512, 2007.

[22] F. Xiao, L. Wang, J. Chen, and Y.P. Gao. Finite-time formation control for multi-agent systems. *Automatica*, 45(11):2605–2611, 2009.

[23] C.L. Liu and Y.P. Tian. Formation control of multi-agent systems with heterogeneous communication delays. *International Journal of Systems Science*, 40(6):627–636, 2009.

[24] A. Abdessameud and A. Tayebi. Formation control of vtol unmanned aerial vehicles with communication delays. *Automatica*, 47(11):2383–2394, 2011.

[25] H.B. Du, S.H. Li, and X.Z. Lin. Finite-time formation control of multiagent systems via dynamic output feedback. *International Journal of Robust and Nonlinear Control*, 23(14):1609–1628, 2013.

[26] D.Y. Meng, Y.M. Jia, J.P. Du, and J. Zhang. On iterative learning algorithms for the formation control of nonlinear multi-agent systems. *Automatica*, 50(1):291–295, 2014.

[27] G. Lafferriere, A. Williams, J. Caughman, and J.J.P. Veerman. Decentralized control of vehicle formations. *Systems & Control Letters*, 54(9):899–910, 2005.

[28] J.A. Fax and R.M. Murray. Information flow and cooperative control of vehicle formations. *IEEE Transactions on Automatic Control*, 49(9):1465–1476, 2004.

[29] C.Q. Ma and J.F. Zhang. On formability of linear continuous-time multiagent systems. *Journal of Systems Science and Complexity*, 25(1):13–29, 2012.

[30] X.H. Ge and Q.L. Han. Distributed formation control of networked multi-agent systems using a dynamic event-triggered communication mechanism. *IEEE Transactions on Industrial Electronics*, 64(10):8118–8127, 2017.

[31] L. Brinón-Arranz, A. Seuret, and C. Canudas-de Wit. Cooperative control design for time-varying formations of multi-agent systems. *IEEE Transactions on Automatic Control*, 59(8):2283–2288, 2014.

[32] X.W. Dong, J.X. Xi, G. Lu, and Y.S. Zhong. Formation control for high-order linear time-invariant multiagent systems with time delays. *IEEE Transactions on Control of Network Systems*, 1(3):232–240, 2014.

[33] X.W. Dong and G.Q. Hu. Time-varying formation control for general linear multi-agent systems with switching directed topologies. *Automatica*, 73:47–55, 2016.

[34] X.W. Dong, Q.D. Li, Q.L. Zhao, and Z. Ren. Time-varying group formation analysis and design for general linear multi-agent systems with directed topologies. *International Journal of Robust and Nonlinear Control*, 27(9):1640–1652, 2017.

[35] X.W. Dong, Z.Y. Shi, G. Lu, and Y.S. Zhong. Time-varying output formation control for high-order linear time-invariant swarm systems. *Information Sciences*, 298:36–52, 2015.

[36] X.W. Dong and G.Q. Hu. Time-varying output formation for linear multiagent systems via dynamic output feedback control. *IEEE Transactions on Control of Network Systems*, 4(2):236–245, 2017.

[37] R. Wang, X.W. Dong, Q.D. Li, and Z. Ren. Distributed time-varying formation control for linear swarm systems with switching topologies using an adaptive output-feedback approach. *IEEE Transactions on Systems, Man, and Cybernetics: Systems*, 49(12):2664–2675, 2019.

[38] R. Wang, X.W. Dong, Q.D. Li, and Z. Ren. Distributed time-varying output formation control for general linear multiagent systems with directed topology. *IEEE Transactions on Control of Network Systems*, 6(2):609–620, 2018.

[39] Y.Z. Hua, X.W. Dong, Q.D. Li, and Z. Ren. Distributed fault-tolerant time-varying formation control for high-order linear multi-agent systems with actuator failures. *ISA Transactions*, 71:40–50, 2017.

[40] Y.C. Cao, W.W. Yu, W. Ren, and G.R. Chen. An overview of recent progress in the study of distributed multi-agent coordination. *IEEE Transactions on Industrial Informatics*, 9(1):427–438, 2013.

[41] T.F. Liu and Z.P. Jiang. Distributed formation control of nonholonomic mobile robots without global position measurements. *Automatica*, 49(2):592–600, 2013.

[42] P. Wang and B.C. Ding. Distributed RHC for tracking and formation of nonholonomic multi-vehicle systems. *IEEE Transactions on Automatic Control*, 59(6):1439–1453, 2014.

[43] X. Yu and L. Liu. Distributed formation control of nonholonomic vehicles subject to velocity constraints. *IEEE Transactions on Industrial Electronics*, 63(2):1289–1298, 2016.

[44] M. Porfiri, D.G. Roberson, and D.J. Stilwell. Tracking and formation control of multiple autonomous agents: A two-level consensus approach. *Automatica*, 43(8):1318–1328, 2007.

[45] W. Ren and N. Sorensen. Distributed coordination architecture for multi-robot formation control. *Robotics and Autonomous Systems*, 56(4):324–333, 2008.

[46] C. Wang and G.M. Xie. Limit-cycle-based decoupled design of circle formation control with collision avoidance for anonymous agents in a plane. *IEEE Transactions on Automatic Control*, 62(12):6560–6567, 2017.

[47] X. Yu, X. Xu, L. Liu, and G. Feng. Circular formation of networked dynamic unicycles by a distributed dynamic control law. *Automatica*, 89:1–7, 2018.

[48] X. Yu and L. Liu. Cooperative control for moving-target circular formation of nonholonomic vehicles. *IEEE Transactions on Automatic Control*, 62(7):3448–3454, 2017.

[49] L. Han, X.W. Dong, Q.D. Li, and Z. Ren. Formation tracking control for time-delayed multi-agent systems with second-order dynamics. *Chinese Journal of Aeronautics*, 30(1):348–357, 2017.

[50] X.W. Dong, Y.F. Li, C. Lu, G.Q. Hu, Q.D. Li, and Z. Ren. Time-varying formation tracking for UAV swarm systems with switching directed topologies. *IEEE Transactions on Neural Networks and Learning Systems*, 30(12):3674–3685, 2019.

[51] Y.Z. Hua, X.W. Dong, Q.D. Li, and Z. Ren. Distributed time-varying formation robust tracking for general linear multiagent systems with parameter uncertainties and external disturbances. *IEEE Transactions on Cybernetics*, 47(8):1959–1969, 2017.

[52] Y.Z. Hua, X.W. Dong, L. Han, Q.D. Li, and Z. Ren. Finite-time time-varying formation tracking for high-order multiagent systems with mismatched disturbances. *IEEE Transactions on Systems, Man, and Cybernetics: Systems*, 50(10):3795–3803, 2020.

[53] J.L. Yu, X.W. Dong, Q.D. Li, and Z. Ren. Practical time-varying formation tracking for second-order nonlinear multiagent systems with multiple leaders using adaptive neural networks. *IEEE Transactions on Neural Networks and Learning Systems*, 29(12):6015–6025, 2018.

[54] H.S. Su, J.X. Zhang, and X. Chen. A stochastic sampling mechanism for time-varying formation of multiagent systems with multiple leaders and communication delays. *IEEE Transactions on Neural Networks and Learning Systems*, 30(12):3699–3707, 2019.

[55] Z.Y. Meng, W. Ren, and Z. You. Distributed finite-time attitude containment control for multiple rigid bodies. *Automatica*, 46(12):2092–2099, 2010.

[56] H.Y. Liu, G.M. Xie, and L. Wang. Necessary and sufficient conditions for containment control of networked multi-agent systems. *Automatica*, 48(7):1415–1422, 2012.

[57] Z.K. Li, W. Ren, X.D. Liu, and M.Y. Fu. Distributed containment control of multi-agent systems with general linear dynamics in the presence of multiple leaders. *International Journal of Robust and Nonlinear Control*, 23(5):534–547, 2013.

[58] X.W. Dong, F.L. Meng, Z.Y. Shi, G. Lu, and Y.S. Zhong. Output containment control for swarm systems with general linear dynamics: A dynamic output feedback approach. *Systems & Control Letters*, 71:31–37, 2014.

[59] H.Y. Liu, L. Cheng, M. Tan, Z.G. Hou, Z.Q. Cao, and M. Wang. Containment control with multiple interacting leaders under switching topologies. In *Proceedings of the 32nd Chinese Control Conference*, pages 7093–7098. IEEE, 2013.

[60] S. Liu, L.H. Xie, and H.S. Zhang. Containment control of multi-agent systems by exploiting the control inputs of neighbours. *International Journal of Robust and Nonlinear Control*, 24(17):2803–2818, 2014.

[61] L. Han, X.W. Dong, Q.D. Li, and Z. Ren. Formation-containment control for second-order multi-agent systems with time-varying delays. *Neurocomputing*, 218:439–447, 2016.

[62] B.J. Zheng and X.W. Mu. Formation-containment control of second-order multi-agent systems with only sampled position data. *International Journal of Systems Science*, 47(15):3609–3618, 2016.

[63] Y.J. Wang, Y.D. Song, and M. Krstic. Collectively rotating formation and containment deployment of multiagent systems: A polar coordinate-based finite time approach. *IEEE Transactions on Cybernetics*, 47(8):2161–2172, 2017.

[64] X.W. Dong, Y.Z. Hua, Y. Zhou, Z. Ren, and Y.S. Zhong. Theory and experiment on formation-containment control of multiple multirotor unmanned aerial vehicle systems. *IEEE Transactions on Automation Science and Engineering*, 16(1):229–240, 2019.

[65] D.Y. Li, W. Zhang, W. He, C.J. Li, and S.S. Ge. Two-layer distributed formation-containment control of multiple Euler-Lagrange systems by output feedback. *IEEE Transactions on Cybernetics*, 49(2):675–687, 2019.

[66] C.J. Li, L.M. Chen, Y.N. Guo, and G.F. Ma. Formation-containment control for networked Euler-Lagrange systems with input saturation. *Nonlinear Dynamics*, 91(2):1307–1320, 2018.

[67] X.W. Dong, Z.Y. Shi, G. Lu, and Y.S. Zhong. Formation-containment analysis and design for high-order linear time-invariant swarm systems. *International Journal of Robust and Nonlinear Control*, 25(17):3439–3456, 2015.

[68] X.W. Dong, Q.D. Li, Z. Ren, and Y.S. Zhong. Formation-containment control for high-order linear time-invariant multi-agent systems with time delays. *Journal of the Franklin Institute*, 352(9):3564–3584, 2015.

[69] X.W. Dong, Q.D. Li, Z. Ren, and Y.S. Zhong. Output formation-containment analysis and design for general linear time-invariant multi-agent systems. *Journal of the Franklin Institute*, 353(2):322–344, 2016.

[70] Y. Zheng, Y.J. Zhu, and L.F. Wang. Consensus of heterogeneous multi-agent systems. *IET Control Theory & Applications*, 5(16):1881–1888, 2011.

[71] C.L. Liu and F. Liu. Stationary consensus of heterogeneous multi-agent systems with bounded communication delays. *Automatica*, 47(9):2130–2133, 2011.

[72] Y.S. Zheng and L. Wang. Consensus of heterogeneous multi-agent systems without velocity measurements. *International Journal of Control*, 85(7):906–914, 2012.

[73] Y.S. Zheng and L. Wang. Finite-time consensus of heterogeneous multi-agent systems with and without velocity measurements. *Systems & Control Letters*, 61(8):871–878, 2012.

[74] J.M. Kim, J.B. Park, and Y.H. Choi. Leaderless and leader-following consensus for heterogeneous multi-agent systems with random link failures. *IET Control Theory & Applications*, 8(1):51–60, 2014.

[75] Y.Z. Feng, S.Y. Xu, F.L. Lewis, and B.Y. Zhang. Consensus of heterogeneous first-and second-order multi-agent systems with directed communication topologies. *International Journal of Robust and Nonlinear Control*, 25(3):362–375, 2015.

[76] D.A.B. Lombana and M. Di Bernardo. Multiplex PI control for consensus in networks of heterogeneous linear agents. *Automatica*, 67:310–320, 2016.

[77] Y.Z. Lv, Z.K. Li, and Z.S. Duan. Distributed PI control for consensus of heterogeneous multiagent systems over directed graphs. *IEEE Transactions on Systems, Man, and Cybernetics: Systems*, 50(4):1602–1609, 2020.

[78] Y.Z. Lv, Z.K. Li, and Z.S. Duan. Fully distributed adaptive PI controllers for heterogeneous linear networks. *IEEE Transactions on Circuits and Systems II: Express Briefs*, 65(9):1209–1213, 2018.

[79] P. Wieland, R. Sepulchre, and F. Allgöwer. An internal model principle is necessary and sufficient for linear output synchronization. *Automatica*, 47(5):1068–1074, 2011.

[80] W.F. Hu, L. Liu, and G. Feng. Output consensus of heterogeneous linear multi-agent systems by distributed event-triggered/self-triggered strategy. *IEEE Transactions on Cybernetics*, 47(8):1914–1924, 2017.

[81] X. Chen and Z.Y. Chen. Robust perturbed output regulation and synchronization of nonlinear heterogeneous multiagents. *IEEE Transactions on Cybernetics*, 46(12):3111–3122, 2016.

[82] Y.Y. Qian, L. Liu, and G. Feng. Output consensus of heterogeneous linear multi-agent systems with adaptive event-triggered control. *IEEE Transactions on Automatic Control*, 64(6):2606–2613, 2019.

[83] Y.F. Su and J. Huang. Cooperative output regulation of linear multi-agent systems. *IEEE Transactions on Automatic Control*, 57(4):1062–1066, 2012.

[84] G.S. Seyboth, W. Ren, and F. Allgöwer. Cooperative control of linear multi-agent systems via distributed output regulation and transient synchronization. *Automatica*, 68:132–139, 2016.

[85] Z.K. Li, M.Z.Q. Chen, and Z.T. Ding. Distributed adaptive controllers for cooperative output regulation of heterogeneous agents over directed graphs. *Automatica*, 68:179–183, 2016.

[86] M.B. Lu and L. Liu. Distributed feedforward approach to cooperative output regulation subject to communication delays and switching networks. *IEEE Transactions on Automatic Control*, 62(4):1999–2005, 2017.

[87] H. Cai, F.L. Lewis, G.Q. Hu, and J. Huang. The adaptive distributed observer approach to the cooperative output regulation of linear multi-agent systems. *Automatica*, 75:299–305, 2017.

[88] W.F. Hu, L. Liu, and G. Feng. Event-triggered cooperative output regulation of linear multi-agent systems under jointly connected topologies. *IEEE Transactions on Automatic Control*, 64(3):1317–1322, 2019.

[89] X.L. Wang, Y.G. Hong, J. Huang, and Z.P. Jiang. A distributed control approach to a robust output regulation problem for multi-agent linear systems. *IEEE Transactions on Automatic control*, 55(12):2891–2895, 2010.

[90] Y.F. Su, Y.G. Hong, and J. Huang. A general result on the robust cooperative output regulation for linear uncertain multi-agent systems. *IEEE Transactions on Automatic Control*, 58(5):1275–1279, 2013.

[91] C. Huang and X.D. Ye. Cooperative output regulation of heterogeneous multi-agent aystems: An H_∞ criterion. *IEEE Transactions on Automatic Control*, 59(1):267–273, 2014.

[92] F.A. Yaghmaie, F.L. Lewis, and R. Su. Output regulation of linear heterogeneous multi-agent systems via output and state feedback. *Automatica*, 67:157–164, 2016.

[93] F.A. Yaghmaie, R. Su, F.L. Lewis, and S. Olaru. Bipartite and cooperative output synchronizations of linear heterogeneous agents: A unified framework. *Automatica*, 80:172–176, 2017.

[94] X.W. Li, Y.C. Soh, L.H. Xie, and F.L. Lewis. Cooperative output regulation of heterogeneous linear multi-agent networks via H_∞ performance allocation. *IEEE Transactions on Automatic Control*, 64(2):683–696, 2019.

[95] H.B. Duan, Y.P. Zhang, and S.Q. Liu. Multiple UAVs/UGVs heterogeneous coordinated technique based on Receding Horizon Control (RHC) and velocity vector control. *Science China Technological Sciences*, 54(4):869–876, 2011.

[96] R. Rahimi, F. Abdollahi, and K. Naqshi. Time-varying formation control of a collaborative heterogeneous multi agent system. *Robotics and Autonomous Systems*, 62(12):1799–1805, 2014.

[97] A. Aghaeeyan, F. Abdollahi, and H.A. Talebi. UAV-UGVs cooperation: With a moving centre based trajectory. *Robotics and Autonomous Systems*, 63:1–9, 2015.

[98] B.X. Mu, K.W. Zhang, and Y. Shi. Integral sliding mode flight controller design for a quadrotor and the application in a heterogeneous multi-agent system. *IEEE Transactions on Industrial Electronics*, 64(12):9389–9398, 2017.

[99] Y. Xu, D.L. Luo, Y.C. You, and H.B. Duan. Distributed adaptive affine formation control for heterogeneous linear networked systems. *IEEE Access*, 7:23354–23364, 2019.

[100] Y. Xu, D.L. Luo, D.Y. Li, Y.C. You, and H.B. Duan. Affine formation control for heterogeneous multi-agent systems with directed interaction networks. *Neurocomputing*, 330:104–115, 2019.

[101] X.L. Wang. Distributed formation output regulation of switching heterogeneous multi-agent systems. *International Journal of Systems Science*, 44(11):2004–2014, 2013.

[102] X.K. Liu, Y.W. Wang, J.W. Xiao, and W. Yang. Distributed hierarchical control design of coupled heterogeneous linear systems under switching networks. *International Journal of Robust and Nonlinear Control*, 27(8):1242–1259, 2017.

[103] S.B. Li, J. Zhang, X.L. Li, F. Wang, X.Y. Luo, and X.P. Guan. Formation control of heterogeneous discrete-time nonlinear multi-agent systems with uncertainties. *IEEE Transactions on Industrial Electronics*, 64(6):4730–4740, 2017.

[104] F.A. Yaghmaie, R. Su, F.L. Lewis, and L.H. Xie. Multiparty consensus of linear heterogeneous multiagent systems. *IEEE Transactions on Automatic Control*, 62(11):5578–5589, 2017.

[105] S. Zuo, Y.D. Song, F.L. Lewis, and A. Davoudi. Adaptive output formation-tracking of heterogeneous multi-agent systems using time-varying L_2-gain design. *IEEE Control Systems Letters*, 2(2):236–241, 2018.

[106] Y.Z. Hua, X.W. Dong, G.Q. Hu, Q.D. Li, and Z. Ren. Distributed time-varying output formation tracking for heterogeneous linear multi-agent systems with a nonautonomous leader of unknown input. *IEEE Transactions on Automatic Control*, 64(10):4292–4299, 2019.

[107] S. Zuo, Y.D. Song, F.L. Lewis, and A. Davoudi. Time-varying output formation containment of general linear homogeneous and heterogeneous multiagent systems. *IEEE Transactions on Control of Network Systems*, 6(2):537–548, 2019.

[108] Y.Z. Hua, X.W. Dong, J.B. Wang, Q.D. Li, and Z. Ren. Time-varying output formation tracking of heterogeneous linear multi-agent systems with multiple leaders and switching topologies. *Journal of the Franklin Institute*, 356(1):539–560, 2019.

[109] Y.Z. Hua, X.W. Dong, Q.D. Li, and Z. Ren. Distributed adaptive formation tracking for heterogeneous multiagent systems with multiple non-identical leaders and without well-informed follower. *International Journal of Robust and Nonlinear Control*, 30(6):2131–2151, 2020.

[110] H. Haghshenas, M.A. Badamchizadeh, and M. Baradarannia. Containment control of heterogeneous linear multi-agent systems. *Automatica*, 54:210–216, 2015.

[111] H.J. Chu, L.X. Gao, and W.D. Zhang. Distributed adaptive containment control of heterogeneous linear multi-agent systems: An output regulation approach. *IET Control Theory & Applications*, 10(1):95–102, 2016.

[112] S. Zuo, Y.D. Song, F.L. Lewis, and A. Davoudi. Adaptive output containment control of heterogeneous multi-agent systems with unknown leaders. *Automatica*, 92:235–239, 2018.

[113] H.J. Liang, Y. Zhou, H. Ma, and Q. Zhou. Adaptive distributed observer approach for cooperative containment control of nonidentical networks. *IEEE Transactions on Systems, Man, and Cybernetics: Systems*, 49(2):299–307, 2019.

[114] S. Zuo, Y.D. Song, F.L. Lewis, and A. Davoudi. Output containment control of linear heterogeneous multi-agent systems using internal model principle. *IEEE Transactions on Cybernetics*, 47(8):2099–2109, 2017.

[115] Y.R. Cong, Z.G. Feng, H.W. Song, and S.M. Wang. Containment control of singular heterogeneous multi-agent systems. *Journal of the Franklin Institute*, 355(11):4629–4643, 2018.

[116] Y.L. Yang, H. Modares, D.C. Wunsch, and Y.X. Yin. Optimal containment control of unknown heterogeneous systems with active leaders. *IEEE Transactions on Control Systems Technology*, 27(3):1228–1236, 2019.

[117] J.H. Qin, Q.C. Ma, X.H. Yu, and Y. Kang. Output containment control for heterogeneous linear multiagent systems with fixed and switching topologies. *IEEE Transactions on Cybernetics*, 49(12):4117–4128, 2019.

[118] Y.W. Wang, X.K. Liu, J.W. Xiao, and X.N. Lin. Output formation-containment of coupled heterogeneous linear systems under intermittent communication. *Journal of the Franklin Institute*, 354(1):392–414, 2017.

[119] Y.W. Wang, X.K. Liu, J.W. Xiao, and Y.J. Shen. Output formation-containment of interacted heterogeneous linear systems by distributed hybrid active control. *Automatica*, 93:26–32, 2018.

[120] W. Jiang, G.G. Wen, Z.X. Peng, T.W. Huang, and A. Rahmani. Fully distributed formation-containment control of heterogeneous linear multiagent systems. *IEEE Transactions on Automatic Control*, 64(9):3889–3896, 2019.

[121] R.A. Horn and C.R. Johnson. *Topics in Matrix Analysis*. Cambridge University Press, Cambridge, 1994.

[122] G. Hardy, J. Littlewood, and G. Polya. *Inequalities*. Cambridge University Press, Cambridge, 1952.

[123] C.J. Qian and W. Lin. A continuous feedback approach to global strong stabilization of nonlinear systems. *IEEE Transactions on Automatic Control*, 46(7):1061–1079, 2001.

[124] S. Boyd, L.E. Ghaoui, E. Feron, and V. Balakrishnan. *Linear Matrix Inequalities in System and Control Theory*. SIAM, Philadelphia, 1994.

[125] R.L. Williams and D.A. Lawrence. *Linear State-space Control Systems*. John Wiley & Sons, Hoboken, 2007.

[126] L. Zadeh and C. Desoer. *Linear System Theory: the State Space Approach*. Courier Dover Publications, New York, 2008.

[127] H.K. Khalil. *Nonlinear Systems*. Prentice Hall, Upper Saddle River, 2002.

[128] S.P. Bhat and D.S. Bernstein. Finite-time stability of continuous autonomous systems. *SIAM Journal on Control and Optimization*, 38(3):751–766, 2000.

[129] I. Bayezit and B. Fidan. Distributed cohesive motion control of flight vehicle formations. *IEEE Transactions on Industrial Electronics*, 60(12):5763–5772, 2013.

[130] A. Karimoddini, H. Lin, B.M. Chen, and T.H. Lee. Hybrid threed-imensional formation control for unmanned helicopters. *Automatica*, 49(2):424–433, 2013.

[131] L.I. Allerhand and U. Shaked. Robust stability and stabilization of linear switched systems with dwell time. *IEEE Transactions on Automatic Control*, 56(2):381–386, 2011.

[132] I. Saboori and K. Khorasani. H_∞ consensus achievement of multi-agent systems with directed and switching topology networks. *IEEE Transactions on Automatic Control*, 59(11):3104–3109, 2014.

[133] L. Huang. *Linear Algebra in System and Control Theory*. Science Press, Beijing, 1984.

[134] W. Ren and Y.C. Cao. *Distributed Coordination of Multi-agent Networks*. Springer-Verlag, London, 2011.

[135] A.P. Aguiar and J.P. Hespanha. Trajectory-tracking and path-following of underactuated autonomous vehicles with parametric modeling uncertainty. *IEEE Transactions on Automatic Control*, 52(8):1362–1379, 2007.

[136] J. Guo, G.F. Yan, and Z.Y. Lin. Local control strategy for moving-target-enclosing under dynamically changing network topology. *Systems & Control Letters*, 59(10):654–661, 2010.

[137] Y.J. Wang, Y.D. Song, M. Krstic, and C.Y. Wen. Fault-tolerant finite time consensus for multiple uncertain nonlinear mechanical systems under single-way directed communication interactions and actuation failures. *Automatica*, 63:374–383, 2016.

[138] Z.K. Li, Z.S. Duan, and F.L. Lewis. Distributed robust consensus control of multi-agent systems with heterogeneous matching uncertainties. *Automatica*, 50(3):883–889, 2014.

[139] G. Wheeler, C.Y. Su, and Y. Stepanenko. A sliding mode controller with improved adaptation laws for the upper bounds on the norm of uncertainties. *Automatica*, 34(12):1657–1661, 1998.

[140] H.S. Wu. Adaptive robust tracking and model following of uncertain dynamical systems with multiple time delays. *IEEE Transactions on Automatic Control*, 49(4):611–616, 2004.

[141] J.J.E. Slotine and W.P. Li. *Applied Nonlinear Control*. Prentice Hall, Englewood Cliffs, 1991.

[142] S.H. Yu, X.H. Yu, B. Shirinzadeh, and Z.H. Man. Continuous finite-time control for robotic manipulators with terminal sliding mode. *Automatica*, 41(11):1957–1964, 2005.

[143] Y.G. Hong, J.K. Wang, and D.Z. Cheng. Adaptive finite-time control of nonlinear systems with parametric uncertainty. *IEEE Transactions on Automatic control*, 51(5):858–862, 2006.

[144] Y.B. Shtessel, I.A. Shkolnikov, and A. Levant. Smooth second-order sliding modes: Missile guidance application. *Automatica*, 43(8):1470–1476, 2007.

[145] J.A. Moreno and M. Osorio. Strict Lyapunov functions for the super-twisting algorithm. *IEEE Transactions on Automatic Control*, 57(4):1035–1040, 2012.

[146] S.P. Bhat and D.S. Bernstein. Geometric homogeneity with applications to finite-time stability. *Mathematics of Control, Signals and Systems*, 17(2):101–127, 2005.

[147] S. Mondal, R. Su, and L.H. Xie. Heterogeneous consensus of higher-order multi-agent systems with mismatched uncertainties using sliding mode control. *International Journal of Robust and Nonlinear Control*, 27(13):2303–2320, 2017.

[148] X.Y. Wang, S.H. Li, X.H. Yu, and J. Yang. Distributed active anti-disturbance consensus for leader-follower higher-order multi-agent systems with mismatched disturbances. *IEEE Transactions on Automatic Control*, 62(11):5795–5801, 2017.

[149] J. Huang. *Nonlinear Output Regulation: Theory and Applications*. SIAM, Philadelphia, 2004.

[150] Z. Qu. *Cooperative Control of Dynamical Systems: Applications to Autonomous Vehicles*. Springer-Verlag, London, 2009.

[151] Y.Z. Lv, Z.K. Li, Z.S. Duan, and Jie Chen. Distributed adaptive output feedback consensus protocols for linear systems on directed graphs with a leader of bounded input. *Automatica*, 74:308–314, 2016.

[152] Z.K. Li, X.D. Liu, W. Ren, and L.H. Xie. Distributed tracking control for linear multiagent systems with a leader of bounded unknown input. *IEEE Transactions on Automatic Control*, 58(2):518–523, 2013.

[153] Y.T. Tang, Y.G. Hong, and X.H. Wang. Distributed output regulation for a class of nonlinear multi-agent systems with unknown-input leaders. *Automatica*, 62:154–160, 2015.

[154] Z.K. Li, G.H. Wen, Z.S. Duan, and W. Ren. Designing fully distributed consensus protocols for linear multi-agent systems with directed graphs. *IEEE Transactions on Automatic Control*, 60(4):1152–1157, 2015.

[155] Z. Feng, G.Q. Hu, and G.H. Wen. Distributed consensus tracking for multi-agent systems under two types of attacks. *International Journal of Robust and Nonlinear Control*, 26(5):896–918, 2016.

[156] Y.F. Su and J. Huang. Cooperative output regulation with application to multi-agent consensus under switching network. *IEEE Transactions on Systems, Man, and Cybernetics-Part B: Cybernetics*, 42(3):864–875, 2012.

[157] H. Cai and J. Huang. The leader-following consensus for multiple uncertain euler-lagrange systems with an adaptive distributed observer. *IEEE Transactions on Automatic Control*, 61(10):3152–3157, 2016.

[158] Z.K. Li, W. Ren, X.D. Liu, and M.Y. Fu. Consensus of multi-agent systems with general linear and lipschitz nonlinear dynamics using distributed adaptive protocols. *IEEE Transactions on Automatic Control*, 58(7):1786–1791, 2013.

[159] Y. Dong, J. Chen, and J. Huang. A self-tuning adaptive distributed observer approach to the cooperative output regulation problem for networked multi-agent systems. *International Journal of Control*, 92(8):1796–1804, 2019.

[160] A. Tayebi and S. McGilvray. Attitude stabilization of a VTOL quadrotor aircraft. *IEEE Transactions on Control Systems Technology*, 14(3):562–571, 2006.

Index

Printed and bound by CPI Group (UK) Ltd, Croydon, CR0 4YY

24/10/2024

01778622-0001